W9-DBA-273

PEOPLE AND THE ENVIRONMENT

Approaches for Linking Household and Community Surveys to Remote Sensing and GIS

PEOPLE AND THE ENVIRONMENT

Approaches for Linking Household and Community Surveys to Remote Sensing and GIS

edited by

Jefferson Fox
East-West Center - Honolulu

Ronald R. Rindfuss
University of North Carolina - Chapel Hill

Stephen J. Walsh
University of North Carolina - Chapel Hill

Vinod Mishra
East-West Center - Honolulu

KLUWER ACADEMIC PUBLISHERS
Boston / Dordrecht / London

Distributors for North, Central and South America:
Kluwer Academic Publishers
101 Philip Drive
Assinippi Park
Norwell, Massachusetts 02061 USA
Telephone (781) 871-6600
Fax (781) 681-9045
E-Mail: kluwer@wkap.com

Distributors for all other countries:
Kluwer Academic Publishers Group
Post Office Box 322
3300 AH Dordrecht, THE NETHERLANDS
Telephone 31 786 576 000
Fax 31 786 576 474
E-Mail: services@wkap.nl

Electronic Services <http://www.wkap.nl>

Library of Congress Cataloging-in-Publication Data

People and the environment: approaches for linking household and community surveys to remote sensing and GIS/ edited by Jefferson Fox....[et al.].
p.cm.
Includes bibliographical references and index.
ISBN 1-40207-322-4 (alk. Paper)
1. Land use—Data processing. 2. Land use—Remote sensing. 3. Land use-Environmental aspects—Data processing. 4. Land use—Environmental aspects—Remote sensing. 5. Information storage and retrieval systems—Land use. 6. Household surveys—Data processing. 7. Social Surveys—Data processing. 8. Economic Surveys—Data processing 9. Land use—Maps. 10. Digital mapping. I. Fox, Jefferson, 1951-

HD108.15 .P46 2003
333.73'13'0285-dc21

2002035709

Printed in the United States of America.

CONTENTS

LIST OF FIGURES

Chapter 1

Chapter 2

Chapter 3

Chapter 4

Chapter 5

Chapter 6

Chapter 7

Chapter 9

Chapter 10

LIST OF TABLES

Chapter 2

Chapter 3

Chapter 4

Chapter 6

LIST OF CONTRIBUTORS

Li An, Department of Fisheries and Wildlife, Michigan State University, East Lansing, Michigan

William G. Axinn, Department of Sociology and Institute for Social Research, University of Michigan, Ann Arbor, Michigan

Francis Baquero, Ecociencia, Quito, Ecuador

Jennifer S. Barber, Institute for Social Research, University of Michigan, Ann Arbor, Michigan

Sandra S. Batie, Department of Agricultural Economics, Michigan State University, East Lansing, Michigan

Richard E. Bilsborrow, Department of Biostatistics and Carolina Population Center, University of North Carolina, Chapel Hill, North Carolina

Randall B. Boone, Natural Resource Ecology Laboratory, Colorado State University, Fort Collins, Colorado

Eduardo Brondizio, Anthropological Center for Training and Research on Global Environmental Change, Indiana University at Bloomington, Indiana

Shauna B. BurnSilver, Natural Resource Ecology Laboratory, Colorado State University, Fort Collins, Colorado

Kelley A. Crews-Meyer, Department of Geography, University of Texas, Austin, Texas

Dao Minh Truong, Center for Resource and Environmental Studies, National University of Vietnam, Hanoi, Vietnam

Glenn P. Dolcemascolo, Asian Disaster Preparedness Center, Bangkok, Thailand

Deanna Donovan, Environmental Studies, East-West Center, Honolulu, Hawaii

Barbara Entwisle, Department of Sociology and Carolina Population Center, University of North Carolina, Chapel Hill, North Carolina

Jefferson Fox, Environmental Studies, East-West Center, Honolulu, Hawaii

Brian G. Frizzelle, Carolina Population Center, University of North Carolina, Chapel Hill, North Carolina

Kathleen A. Galvin, Natural Resource Ecology Laboratory and Department of Anthropology, Colorado State University, Fort Collins, Colorado

Jacqueline Geoghegan, Department of Economics and George Perkins Marsh Institute, Clark University, Worcester, Massachusetts

Thomas Giambelluca, Geography Department, University of Hawaii at Manoa, Honolulu, Hawaii

Richard E. Groop, Department of Geography, Michigan State University, East Lansing, Michigan

Eric F. Lambin, Department of Geography, University of Louvain, Louvain-la-Neuve, Belgium

Le Trong Cuc, Center for Resource and Environmental Studies, National University of Vietnam, Hanoi, Vietnam

Stephen Leisz, Institute of Geography, University of Copenhagen, Denmark and the Center for Agricultural Research and Ecological Studies, Hanoi, Vietnam

Zai Liang, Department of Sociology, State University of New York at Albany, New York

Marc A. Linderman, Department of Fisheries and Wildlife, Michigan State University, East Lansing, Michigan

Jianguo Liu, Department of Fisheries and Wildlife, Michigan State University, East Lansing, Michigan

George P. Malanson, Department of Geography, University of Iowa, Iowa City, Iowa

Stephen J. McGregor, Carolina Population Center and Department of Geography, University of North Carolina, Chapel Hill, North Carolina

Angela G. Mertig, Department of Fisheries and Wildlife and Department of Sociology, Michigan State University, East Lansing, Michigan

Joseph P. Messina, Department of Geography, Michigan State University, East Lansing, Michigan

Vinod Mishra, Research Program, East-West Center, Honolulu, Hawaii

Emilio F. Moran, Anthropological Center for Training and Research on Global Environmental Change, Indiana University at Bloomington, Indiana

Zhiyun Ouyang, Department of Systems Ecology, Research Center for Eco-Environmental Sciences, Chinese Academy of Sciences, Beijing, China

William K. T. Pan, Department of Biostatistics, University of North Carolina, Chapel Hill, North Carolina

Donald Plondke, Environmental Studies, East-West Center, Honolulu, Hawaii

Pramote Prasartkul, Institute for Population and Social Research, Mahidol University, Bangkok, Thailand

Jiaguo Qi, Department of Geography, Michigan State University, East Lansing, Michigan

Terry Rambo, Center for Southeast Asia Studies, Kyoto University, Japan

Ronald R. Rindfuss, Department of Sociology and Carolina Population Center, University of North Carolina, Chapel Hill, North Carolina

Yothin Sawangdee, Institute for Population and Social Research, Mahidol University, Bangkok, Thailand

Andréa Siqueira, Anthropological Center for Training and Research on Global Environmental Change, Indiana University at Bloomington, Indiana

Gregory N. Taff, Department of Geography and Carolina Population Center, University of North Carolina, Chapel Hill, North Carolina

Tran Duc Vien, Center for Agricultural Research and Ecological Studies, Hanoi Agricultural University, Hanoi, Vietnam

Billie L. Turner II, Graduate School of Geography and George Perkins Marsh Institute, Clark University, Worcester, Massachusetts

John B. Vogler, Research Program, East-West Center, Honolulu, Hawaii

Stephen J. Walsh, Department of Geography and Carolina Population Center, University of North Carolina, Chapel Hill, North Carolina

Alan Ziegler, Environmental Engineering and Water Resources Program, Princeton University, Princeton, New Jersey

FOREWORD

In recent years a number of projects have attempted to link social science data at household and community levels to remotely sensed data on land-use and land-cover change. These projects have been funded by a variety of sources, which within the United States include the National Science Foundation, National Institutes of Health, National Aeronautics and Space Administration, and the MacArthur Foundation. These projects have been located primarily in the developing world, including Asia, Latin America, and Africa and have involved scientists from a variety of disciplines, including sociology, anthropology, demography, economics, public policy, geography, ecology, and remote sensing, which itself is broadly interdisciplinary. These projects have been largely independent of one another and even though there has been some *ad hoc* consulting among subgroups of investigators, there has not been any careful review of the methods used to link people and pixels (the land parcels they use) as well as the associated challenges and opportunities.

The book is a collection of papers presented at a workshop on "Human Actions and Land-Use/Land-Cover Change," held at the East-West Center in Honolulu, January 3-8, 2002. The workshop brought together members of international research teams that have conducted successful projects linking social science data at household and community levels with remotely sensed data. These teams represented research projects in China, Ecuador, Vietnam, Thailand, Mexico, Brazil, Kenya, and Cameroon. Each team prepared a paper detailing the methodological and practical issues involved in conducting their project. Before the workshop, the four coeditors prepared a background paper (included as the introductory chapter in this volume) and a list of key questions to help guide each team in preparing their papers.

This book is an attempt to carefully review how each research team linked household and community level social data with remotely sensed and other spatial data. Which approaches worked and which did not? To what extent are approaches developed in one region transferable to another region? What are the starting points: is land sampled and then households interviewed or are households sampled and subsequently linked to the land? What levels of precision are needed in the linkage to do meaningful change analysis? What types of modeling have been used and how successfully? What uncertainty exists in these data and what validation efforts are required? How might remote sensing inform social surveys and social surveys inform remote sensing? What are the spatial and temporal scale issues in studying the effects of human activity on land-use and land-cover change? What surprises were encountered and what were the lessons learned?

The book addresses a need for a comprehensive and rigorous treatment of linking across thematic domains (e.g., social, biophysical, and geographical) and across space and time scales for research and study within the context of human-environment interactions. The human dimensions research community, land-use and land-cover change programs, and human and landscape ecology communities collectively view landscapes within spatially-explicit perspectives, where people are viewed as agents of landscape change that shape and are shaped by the landscape, and where landscape form and functions are assessed within a space-time context. Current researchers and those following this early group of integrative scientists face challenges in conducting this type of research, but the potential rewards for insight are substantial. This volume is an effort in that direction.

The book is comprised of 11 chapters. Chapter 1 is based on a background paper that the coeditors wrote to identify some of the key issues in linking household and community data with remotely sensed data. Chapters 2-9 are based on selected case studies from Mexico, Brazil, Ecuador, Thailand, Kenya, Vietnam, Cameroon and Kenya, and China. Chapters 10-11 review the case studies presented in Chapters 2-9 from different disciplinary perspectives. Chapter 10 is written from an ecological perspective and Chapter 11 from a sociological perspective. The case studies presented in this volume represent a cross-section of approaches and methods for linking household and community level social data with remotely sensed spatial data.

SHORT BIOSKETCHES OF EDITORS

Jefferson Fox is a Senior Fellow and Coordinator of Environmental Studies at the East-West Center in Honolulu. Research interests include understanding land-use and land-cover change in Asia and the social context of spatial information technology especially when it is used to help local communities map their land claims as well as their land-use practices. He has been instrumental in establishing GIS/remote sensing laboratories in numerous universities and organizations across the Asia/Pacific region.

Ronald R. Rindfuss is a Professor in the Department of Sociology and a Fellow at the Carolina Population Center, University of North Carolina, Chapel Hill. His research interests include fertility, aspects of the life course, and population and the environment. He is currently working on projects in Thailand, Norway, Japan and the United States.

Stephen J. Walsh is a Professor in the Department of Geography and a Fellow at the Carolina Population Center, University of North Carolina, Chapel Hill. Research interests include GIS, remote sensing, spatial analysis, physical geography, and population-environment interactions. Ongoing studies include research in Ecuador, Thailand, and North Carolina and Montana, USA.

Vinod Mishra is a Fellow in Population and Health Studies at the East-West Center, Honolulu. Dr. Mishra is broadly interested in population and environment interactions and his current research focuses on the human impacts on land-use and land-cover and on the effects of air pollution on health. Dr. Mishra has also worked on many reproductive and child health issues.

ACKNOWLEDGEMENTS

The workshop on which this book is based was made possible by a grant from the National Science Foundation (Grant No. BCS-0083474, Biocomplexity Incubation Activity). We thank Thomas Baerwald, our NSF Program Officer, for his assistance and cooperation. We are also thankful to the International Human Dimensions Programme on Global Environmental Change (IHDP) in Bonn, Germany for funding the travel of two participants from Europe and to the support provided by their Program Officer, Debra Meyer Wefering. At the East-West Center, we are grateful for the administrative and fiscal assistance provided by Arlene Hamasaki, Penny Higa, Margaret McGowen, and Karen Yamamoto. We are also indebted to Lee Motteler for his skillful and timely copyediting; Josh Taylor (www.geocities.com/joshtheillustrator) for his cover design; Philip Page for producing the CD-ROM; and John Vogler for his efforts on the workshop web page. We also benefited from comments and editorial suggestions provided by Glenn Dolcemascolo, Deanna Donovan, and Krisna Suryanata. We also wish to thank Rebecca Clark, National Institute for Child Health and Development of the US National Institutes of Health and to Garik Gutman, Land-Cover and Land-Use Change Program of the US National Aeronautics and Space Administration for their comments and discussions made during the workshop. Finally, we thank the East-West Center for hosting the workshop and the Carolina Population Center, University of North Carolina - Chapel Hill for general support, assistance, and cooperation on behalf of the workshop and the book project.

Chapter 1

LINKING HOUSEHOLD AND REMOTELY SENSED DATA

Methodological and Practical Problems

Ronald R. Rindfuss
Department of Sociology and Carolina Population Center, University of North Carolina
ron_rindfuss@unc.edu
Stephen J. Walsh
Department of Geography and Carolina Population Center, University of North Carolina
Vinod Mishra
East-West Center
Jefferson Fox
East-West Center
Glenn P. Dolcemascolo
Asian Disaster Preparedness Center

1. INTRODUCTION

Changes in global land cover (biophysical attributes typically observed remotely) and land use (human purpose applied to these attributes) are occurring at a rate, magnitude, and spatial extent unprecedented in human history (Lambin et al. 2001). When aggregated globally these changes impact biodiversity (Sala et al. 2000), contribute to local and regional climate change (Chase et al. 1999) as well as to global warming (Houghton et al. 1999), and affect the ability of biological systems to support human needs (Vitousek et al. 1997). Such changes also determine in part the vulnerability of places and people to climatic, economic, ecologic, or sociopolitical perturbations (Kasperson et al. 1995).

Land-use and land-cover change has become part of the global science agenda on environmental change. Research activities fall under the auspices of the International Geosphere-Biosphere Program (IGBP), the International Human Dimensions Program (IHDP), and the Intergovernmental Panel on Climate Change (IPCC)—especially the Land Use/Cover Change program jointly sponsored by IGBP and IHDP and various parts of the IGBP's Global

Change and the Terrestrial Ecosystem (GCTE) program (Turner 2001). That human activity has a profound effect on patterns of land cover is now undisputed (e.g., Lambin et al. 2001; Giest and Lambin 2001). But it is critical to move beyond this accepted fact to understanding when and under what circumstances human behavior and land cover are interrelated.

Studying the effects of human activities on land-use/cover change typically involves joining social science data with remotely sensed and other spatial data (e.g., Walsh et al. 1999; Liverman et al. 1998; Turner and Meyer 1991; Fox et al. 1995; Skole et al. 1994; Moran et al. 1994; Guyer and Lambin 1993). These are quite different types of data that are typically collected by scientists with very diverse orientations, and linking social science, natural science, and spatial science data has proven a major challenge. This chapter and this volume address the issue of linkage across these scientific domains at a micro scale.

A common solution to linking social, natural, and spatial data has been to use census data gathered at the household level, aggregate them to some administrative boundary, and link them to remotely sensed and GIS data for the same administrative unit. One can then relate changes in land cover to changes in demographic and socioeconomic indicators. While much can be learned from such linkages at the administrative level, there are also some drawbacks. One is that remotely sensed data provide information on land *cover* but not necessarily on land *use*. Another is that numerous land-use decisions are made at the household level—and others at the community level—and aggregating up to administrative units renders household and community decision making invisible.

Further, there is no reason why an association at the administrative level would necessarily be the same as that found at the household level. This is a well-known problem that goes by different names in different disciplines. To fix terms, we will use the definition of "ecological fallacy" given by Gibson et al. (2000): ecological fallacies are those that impute the cause of lower-level (or micro) patterns to be the same as those operating at a higher (or macro) level. Using aggregated census data to impute household-level relationships is perhaps the earliest identified form of ecological fallacy (Robinson 1950).

A fundamental premise of this volume is that households—and household decision making—are critical to understanding the changes in land use and land cover that are occurring throughout the world and contributing to such problems as global warming, loss of biodiversity, and increased vulnerability. This volume brings together research teams that have linked household-level social science data and remotely sensed (and other spatially explicit) data, sometimes at the household level and sometimes at the community level. These researchers come from diverse backgrounds ranging from sociology to ecology and from geography to agronomy. They have diverse theories and substantive interests, yet they all recognize a need to link specific remotely sensed pixels and spatial coordinates to local decision makers. Why? It is because many important land-use/cover decisions are made at the micro level.

This book examines methodological and practical issues that face researchers who design studies linking microhuman behavior and remotely sensed data. In laying out these issues, we hope to begin a process that leads to a better understanding of the theoretical and substantive implications of the methodological and design decisions being made by those in the emerging land-use/land-cover change field. We start the chapter with a brief theoretical discussion to place the methodological focus of the volume in broader perspective. The remainder of the chapter examines methodological and practical issues that affect how we choose to link social, natural, and spatial science data at the micro level.

2. THEORETICAL FRAMEWORKS AND SUBSTANTIVE QUESTIONS

The guiding theoretical frameworks used by the authors in this volume come from various disciplines and, because each project represents an interdisciplinary team, numerous and sometimes overlapping theories are often used in a single project. For example, both Lambin (this volume) and Turner and Geoghegan (this volume) use economic approaches, while Walsh and his collaborators (this volume), Liu and his collaborators (this volume), and BurnSilver and her collaborators (this volume) work within landscape ecology. One common theme that runs through all the frameworks employed by the authors is an emphasis on multiplicity: multiple responses to social change, multiple levels of analysis, multiple aspects of the life course of individuals, households, and land parcels, multiple connections in social and geographical space, and multiple ties between people and land in rural agricultural areas.

To set the orientation of the present volume in the broader theoretical and substantive space within which land-use change research is set, consider Figure 1. This figure is not meant to represent any single theory or framework but rather is a stylized version of many. It is here for heuristic purposes. Figure 1, as is the practice with numerous authors (e.g., Meyer and Turner 1992; Ojima et al. 1994; and Lambin et al. 1999), has both proximate causes and more distal driving forces of land-use and land-cover change. Proximate causes include human activities (land uses) that directly affect the environment and thus constitute proximate sources of change. They can also include biophysical or other factors. Proximate causes are usually seen to operate at the local level. Underlying driving forces (or social and biophysical processes) are fundamental forces that underpin the more obvious or proximate causes of land-use and land-cover change. They consist of a complex set of social, political, economic, ecologic, technological, cultural, and other variables that constitute initial conditions in the human-environmental relations that are structural (or systematic) in nature. In terms of spatial scale, underlying drivers may operate directly at the local level or indirectly from the national or even global level.

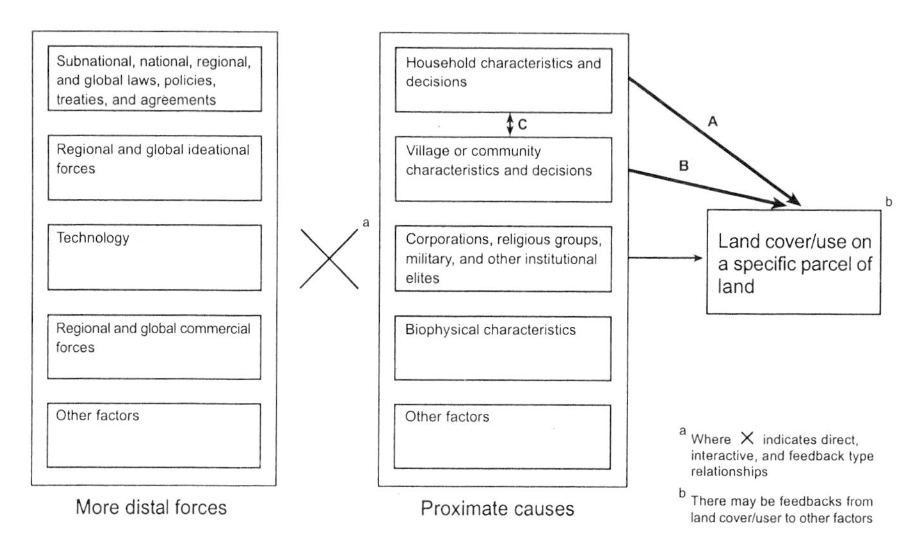

Figure 1. A conceptual model of factors influencing land-cover and land-use change

The authors of the chapters in this volume and their research programs are concerned with the complexity represented in Figure 1. For example, Fox and his collaborators (this volume) place their work in the political ecology perspective, where relationships are locally contextualized but also subject to the influences of actors outside the community, including national policies and changes in global economies. All the boxes shown in Figure 1 are important and, indeed, many of them are mentioned throughout this volume. But our emphasis—our niche—is the three boldface lines in the upper right-hand corner of Figure 1: the methodological and practical issues involved in linking household-level data to information on local land use. Such links are absolutely essential to convincingly test the complex issues represented in Figure 1 and the theoretical perspectives from which Figure 1 was distilled.

Some of the studies in this volume link household-level data to land-use data indirectly by going through the village or community—lines B and C in Figure 1. These include Fox et al., Lambin, and Liu et al. (this volume).

Some of the studies in this volume directly link household-level data to remotely sensed data—the line labeled "A." These include Walsh et al., Turner and Geoghegan, Moran et al., and Rindfuss et al. (this volume). These four studies also have the ability to link indirectly through the community level—lines B and C—but that is not emphasized in the discussions presented here. BurnSilver et al. (this volume) link households to herding patterns practiced by pastoralists. They associate landscape conditions mapped through remote sensing techniques and locate the dynamic pattern of the herds through GPS technology.

We now turn to the methodological and practical issues involved in collecting data that provide the links represented by lines A, B, and C in Figure 1.

3. METHODOLOGICAL ISSUES

As previously stated, individually and together, the guiding theoretical frameworks for documenting land-use and land-cover change emphasize multiplicity on both the land and human sides. As such, there is a push for data collection strategies that incorporate the myriad aspects of individuals and land parcels, including their biographies, the households within which individuals live, the neighborhoods within which parcels of land are located, and the broader context within which households and neighborhoods are located. These contexts include time as well as space parameters of the social, biophysical, and geographical domains. Put differently, the guiding theoretical issues addressed by these authors are data hungry and demanding of innovative data collection strategies that go beyond traditional surveys and the application of digital spatial technologies with emphasis on remote sensing. Creating such data sets "from scratch" is admittedly difficult, and the studies represented in this volume are complex. The authors emphasize the household (community) land-cover/land-use linkages rather than all the complexity in their entire design.

Another theme that is clear from the entire collection of essays is that there is no single linking design that would fit all situations. To see this, consider the different demands that *settled agriculture* would require in terms of research design compared to *pastoralism.* The BurnSilver et al. (this volume) approach of following the herds as they go from grazing spot to grazing spot is appropriate for a setting where pastoralists use common lands, but it would be inappropriate in settings where there are settled agriculture and secure land titles. Or to take a different example, the pattern of land ownership forced quite different strategies in Amazonia (Moran et al. this volume and Walsh et al. this volume) compared to the pattern in

Thailand (Rindfuss et al. this volume). The general point is that it is critical for research teams to know as much as possible about their study sites prior to designing the linkage research plan.

3.1 Remotely Sensed Data and Resolutions

Fundamental to linking people and land is landscape characterization and spatial representation using elements of *geographic information science* (GISc), a term used to describe a set of spatial digital technologies. One of the technologies at the forefront is remote sensing. Here we focus on how these technologies are being used to link people to the land and vice versa, as well as to represent their relationships within a spatially explicit context. Just to fix terms, *remote sensing,* simply defined, is the gathering of data through a sensor mounted on an aircraft or a satellite.

In remote sensing, the spatial, spectral, temporal, and radiometric resolutions are central factors in landscape characterization because they define the sensor systems and reconnaissance platforms of a particular mapping mission. *Spatial resolution* refers to the ground area simultaneously sensed (Instantaneous Field of View) through either active or passive sensors; *spectral resolution* is the wavelengths of the electromagnetic spectrum within which sensors operate; *temporal resolution* is the periodicity or return interval of a satellite within its prescribed orbit or an aircraft within its assigned flight path; and *radiometric resolution* refers to the range of intensity levels used to quantify spectral responses assessed by the respective sensor systems.

Broadly stated, spectral resolution differentiates the optical systems (e.g., Landsat Thematic Mapper) from the nonoptical systems (e.g., Synthetic Aperture Radar). Most efforts at characterizing land-use/land-cover change have tended to focus on optical systems that generally extend from the visible to the near- or middle-infrared spectral regions. Nearly all the essays in this volume use optical systems data—most notably the Landsat Thematic Mapper. Increasingly, radar systems are being used for landscape characterizations, particularly in environments likely to have heavy cloud coverage. LIDAR systems are also being used to construct horizontal and vertical profiles of vegetation for land-use/land-cover change characterizations, and hyperspectral data are also being used to sample across a large number of spectral channels (e.g., the 210 channel of AVIRIS) for feature characterization.

The historical depth of sensor systems is also important so that short- and long-term observations can be accommodated using the same (or similar) sensor package. This temporal depth allows the retrospective formulation of image time series to correspond with the temporal depth in social surveys.

Spatial resolution sets the areal dimension of the pixel or picture element. In mapping terms, the pixel determines the minimum mapping unit—the size of the smallest landscape feature that can be distinguished and mapped. The remotely sensed data itself, as determined by the various remote sensing resolutions, come to the user as a series of row-column matrices. An initial and crucial link involves taking this raw data and linking it to the land it covers (typically referred to as *georeferencing*). The data are georeferenced through the use of a series of distributed control points located on the satellite image, base maps, and/or referenced in the field through GPS techniques. What results is the generation of a spatially explicit set of data matrices that are matched to a Cartesian coordinate system to serve as the mapping base for subsequent image displays and analyses. In a number of environments, however, the georeferencing process is severely constrained by a lack of quality ground control points (i.e., stable landscape features such as road intersections), and this can increase the spatial error. Any spatial operation, such as overlay analysis or distance/orientation computations, would be affected by this added degree of spatial uncertainty.

Traditionally, the natural sciences have relied upon remote sensing systems and ground survey techniques to place measurements and/or observations within some spatially explicit context. Social scientists are now equipping survey teams with GPS technologies for geographically referencing respondents to spatial coordinate systems as well as important landscape, community, or household features. In many instances, the georeferenced satellite data set may become the base map for the social survey, or at least the image may serve as an important planning and orientation document used in navigation and framing the local to regional geography of the study area (Walsh et al. this volume). They can also be used during the actual interview to orient respondents to spatially explicit questions (e.g., Moran et al. this volume). Even though social data can be georeferenced, however, the fundamental difference compared to remotely sensed data is that people can and do move across the landscape; they can affect multiple land parcels. Further, households can affect the landscape directly (planting a crop in previously forested land) or indirectly (buying a type of forest product that leads to further timbering in some distant forest).

In addition to remotely sensed data, the projects represented in this volume use other types of spatially explicit data, such as road maps, soil maps, and cadastral maps. Today, much of this type of data that previously had to be digitized, collected in the field, or generated directly are available for purchase from commercial vendors or government organizations. For instance, shape files (e.g., outlines of political borders), base layers (e.g., digital elevation models, transportation networks, and hydrography), and some special purpose data layers (e.g., meteorological stations and stream gauges) can be acquired. Since these coverages are already in Cartesian space—normally the Universal Transverse Mercator (UTM) earth coordinate system—the link to the remotely sensed data is relatively straightforward, and the further link to the social data falls out as a result of the linkage previously established.

Finally, while it may be obvious, it is worth noting that once the remotely sensed data are georeferenced they can be linked straightforwardly to other georeferenced data layers. The reason is that except for highly unusual, typically cataclysmic events, land does not move. Hence, it is possible to coregister various georeferenced data layers, and the Cartesian coordinate system functions as if it were an ID system because each pixel is referenced within a spatially explicit context.

3.2 Social Science Data and Spatial Context

The field of social science encompasses a broad arena of social science disciplines, many with their own data collection methodologies. These include both quantitative and qualitative techniques, some of which have their own literatures and quite specialized practitioners. For example, numerous techniques have been developed for following individuals over time if they are part of a panel (longitudinal) study (e.g., Clarridge et al. 1978; Call et al. 1982; Booth and Johnson 1985; Ribisl et al. 1996). Qualitative methodologies include participant observation, key informants, focus groups, and semistructured interviews. Other social scientists use participatory assessment techniques (Chambers et al. 1989) such as sketch mapping, crop calendars, and various participatory ranking techniques to develop a set of hypotheses about the causes and consequences of events that shape peoples' lives and livelihoods (e.g., Colfer et al. 1999).

Many efforts to combine social science and remotely sensed data for understanding land-use and land-cover change have used quantitative social science data collection methods (see Moran et al. and Turner et al. in this volume). Others incorporate qualitative data collection methods (see Fox et al. in this volume). And some have used both (Rindfuss et al. this volume). One challenge that faces the land-use and land-cover change research community is working out methods for integrating quantitative and qualitative data, within a spatial database, for accurately capturing and portraying the causes of change.

Spatial resolution and *context* (in its geographic sense) are not terms most social scientists use frequently, yet all social science research implicitly has a resolution and a context (or scale): the resolution is the smallest unit of analyses (individual, household, community) and the context is the areal dimension for which it is relevant (village, region, country). In ecology, these concepts are termed *grain* and *extent.* Demographic data are a common form of social science data available for most countries and, frequently, for regions within a country.

Censuses attempt to enumerate all individuals within a country, and typically they are household based—that is, households are enumerated, and then information is collected on all individuals living in each household. There are three problems with censuses from the perspective of linking households to the land for which they are major decision makers. First, censuses are conducted infrequently, with once a decade being the most common periodicity. Second, in virtually all countries, household-level census data are considered confidential and not released to land-use/land-cover change researchers. Rather, the data are aggregated to a high enough level to protect confidentiality and then released—negating the possibility of examining individual households. Third, with the exception of some agricultural censuses, the typical census does not have links to the land the household owns or uses.

Social scientists studying land-use and land-cover change at the household level will consequently want to develop their own sample surveys—in essence going to households and asking questions. This is the case for all the studies reported in this volume. Researchers designing sample surveys need to be particularly conscious of how the temporal periodicity and spatial extent of the survey matches with the time period and spatial extent available from the remotely sensed data sources.

3.3 Linking through Spatial Approaches

Much depends on the direction that one is seeking to link—from the land to people or people to land. Commonly, a remote sensing analyst will want to start from the landscape, having land-cover types or vegetation greenness levels serving as dependent variables. An example might be modeling variation in the Normalized Difference Vegetation Index (NDVI) as a consequence of resource endowments, technology available at the household level, and the geographic access of households to their fields, markets, or other communities. Remote sensing measures can just as easily serve as independent variables and demographic or socioeconomic characteristics as the dependent variables. Regardless of the direction analyses might take, establishing links between the two types of data is confounded by the mismatch in their characteristics. For instance, households might be spatially described as a set of discrete point locations. A collection of households forming a nuclear village might also be spatially described in a similar manner, whereas NDVI of land use/land cover is normally characterized as a continuous coverage of data values represented at some pixel resolution unit and spatially articulated through geographic coordinates linked to the image array.

There has been relatively little published work that links specific household units to satellite-based measures of land cover on plots used by these households. Moran et al. (this volume) and Walsh et al. (this volume) link household-level social data with land-cover data. In both studies, farmers live on the land that they farm. Walsh et al. (this volume) further describe the added challenge of tracking land parcels over time that have become associated with multiple households through land subdivision and the necessity of then linking "many" households to "one" land parcel. Relating land-use/land-cover change to a single household at a specified time period and then associating land-use/land-cover change for the same land parcel to multiple households at a subsequent time period further complicates the timing and nature of household links to landscape patterns and dynamics.

It is frequently the case that households are organized in a nuclear settlement structure with farmland arrayed outwardly surrounding the settlement core. In such settings, social surveys may indicate the number of parcels farmed by each household, but seldom does the survey indicate the location of the parcels, whether the parcels are rented or owned, and land-use and land-cover change history. Rindfuss et al. (this volume) discuss the methodology they designed to link people to their disperse plots across the landscape.

In the absence of cadastral maps, it can be very difficult to define the location of household plots (note that cadastral maps indicate only who owns the parcel, not who uses it), thereby complicating the linkage of satellite data to lands used or owned by the household. Turner and Geoghegan (this volume) established this link by collecting field-based GPS measurements of plot locations for individual households. These locations were then linked to land treatments such as planting histories or fertilizer application patterns referenced on a parcel-by-parcel basis. The GPS can also be used to navigate to site locations that have previously been established as part of extant social surveys or biophysical measures and observations. Obtaining GPS measurements of individual plots is difficult work, particularly for large sample sizes and/or for households using or owning a large number of parcels that may further be broadly distributed.

Often too, administrative boundaries of villages (arranged in nuclear patterns or otherwise) are lacking or do not effectively describe the "functional" use of land at the household level and the geographic distribution of households across the landscape. There are a number of spatial approaches for setting village territories (see Rindfuss et al. 2002 for a review). Fox et al. (this volume) walked village boundaries with villagers and used a GPS to record the boundaries. Lambin (this volume) used radial buffers of different sizes to ascribe isotropic boundaries around villages. Dimensions can be set by estimating the normative time people are willing to commute to their fields and then translating this into distance by taking into account the mode of transportation and the nature of the road system. Theissen polygons can also be used to construct village territories, producing nonoverlapping, crisp boundaries set by the distribution of the village centroids. Region growing using Fuzzy set theory has been used to set village boundaries (Evans 1998). Finally, some have developed least-cost surfaces using natural features such as roads, water, and land-use/land-cover types, as well as reported demographic characteristics of villages, to either effectuate or constrain the movement of people across the landscape, as they travel from their household to their land parcels associated with functional village territories (Crawford 2000). In short, developing functional use boundaries around villages requires some combination of reasonable assumptions, empirical data, and spatial analysis.

Another set of problems faces researchers studying common property land-use systems where a one-to-one correspondence cannot be established between people who use the land and land-use/land-cover change. BurnSilver et al. (this volume), for example, study livestock grazing practices and face the problem of how to associate individual households with the land they use—or more specifically, the land their cattle graze. Liu et al. (this volume) study firewood collection patterns and the implications of these patterns on panda habitat. Their problem was identifying the parcels from which people collected firewood and quantifying the amount of wood collected. In common property systems like these, it is necessary to understand the rules and regulations of the group, and this may best be accomplished through group interviews. It may also be useful to conduct household interviews to gain insights into how individual households respond to common property rules, as well as violations of those rules.

3.4 Ephemeral Households and Boundaries

It is our expectation that a substantial number—perhaps the majority—of decisions influencing land use are decisions made at the household level. Even though any given household might have a dominant person who makes decisions, such decisions are made in the context of the household's needs and strengths. Issues such as household size, household wealth, available labor within the household, the household's age and sex composition, and circular migration of household members all influence how household land is used and whether to acquire (or sell) additional land. However, defining and following households over time is difficult because households are changeable—indeed, they are ephemeral. Likewise, we want to link household decisions to plots of land. The boundaries of these pieces of land—plots, villages, administrative units—are equally ephemeral, changing in response to population pressure, village growth, new administrative polices, and other events.

Consider individuals first. Individuals are discrete and identifiable over time. There is general consensus on the definition of the beginning and end of an individual's life. It is relatively straightforward to speak of the education, age, marital status, and so forth of an individual; and for many of these individual-level variables, there is general agreement on their measurement.

Now contrast individuals with households. The U.S. census definition of a household is: "A household consists of all the people who occupy a housing unit" (http://www.census.gov/population/www/cps/cpsdef.html). This definition, which is similar to other proposed definitions, contains numerous ambiguities that the researcher must resolve and for which there is little guidance in the research literature. For example, how long does a person have to be in the housing unit before that person is a member of the household? Circular migration is frequently associated with an agricultural season that occupies only part of the year, coupled with short-term employment opportunities in other parts of the country. Is an individual who is present in a dwelling unit three months a year and absent the other nine months a household member? Is an individual a member if the combination is four and eight months, five and seven months, and so forth? Can an individual be a member of more than one household?

The power of longitudinal designs is well established in the social sciences, providing numerous analytical and statistical advantages to cross-sectional designs. What does it mean to follow a household over time? For example, if all the individuals move out of a housing unit and another group moves in, does one follow the people or the dwelling unit? If one follows the people, what happens if the original group fissures? For example, a son might get married and move out with his bride. Or a husband and wife might separate and divorce, forming two collections of individuals living in two separate housing units. In each of these examples, which group is the successor household? If a person is picked to define the successor household, what are the theoretical arguments that guide how that person is chosen? What constitutes the death of a household?

The basic problem of following households is determining continuity rules for situations when members enter or leave the household (Duncan and Hill 1985; McMillan and Herriot 1985; Citro and Watts 1986; Keilman and

Keyfitz 1988). From the perspective of land-use/cover change, we need to develop an understanding of the best procedures for defining households longitudinally. Further, we need to ask empirically, Does it matter? How robust are our findings to alternative definitions of households over time?

Now let us consider plots of land. Walsh et al. (this volume) describe a longitudinal survey of farms in the Ecuadorian Amazon that was initially surveyed in 1990 using a probabilistic sample of farms selected from a list maintained by the government land office. Upon returning in 1999, many farms had been subdivided through sale and kinship ties. What was predominately a one-household to one-farm relationship in 1990 (similar to the situation described by Moran et al. in this volume) had been transformed into a many-household to one-farm relationship in 1999. But what constitutes a farm? For survey purposes, a farm in 1999 was defined by the farm boundaries that constituted the 1990 farm. Within that definition, farm boundaries were maintained regardless of the number of households that were spatially associated to the 1990 farm extents. From a functional perspective, new farms had been created when considered from the perspective of new households and new dwelling units, which were spatially confined to the boundaries of the 1990 farm layout.

Similarly, Rindfuss et al. (this volume) describe a longitudinal survey of households in 1984 and 1994. When they started the survey in 1984, they selected households from each of the fifty-one villages in the district. In 1994 the original villages had split administratively into seventy-six villages. Their approach to shifting administrative boundaries was to track all of the original component boundaries through time. This made it possible for them to reconstitute earlier administrative units. While this process had administrative and analytical advantages, it posed problems in terms of interviews. They worried that—no matter how much they reminded informants that they were talking about the village in terms of its 1984 historical boundaries, even showing them maps of these boundaries—changing boundaries might have caused some confusion in the minds of respondents and informants.

3.5 Temporal Depth: Prospective or Retrospective

The land-use and land-cover change research community is now in the fortunate position of having temporal depth on remotely sensed data for most regions of the world. This data richness comes from multiple sources, including aerial photographs, early U.S. and Russian espionage satellite images and U-2 aircraft overflights, Landsat, AVHRR, and SPOT systems, and recent MODIS and IKONOS data. For some sites, problems such as cloud cover, failure to have archived the data, failure with the archive's ability to retrieve the data, cost of the images, and other factors can limit the remotely sensed data available to researchers. On the social science side, we are fortunate to have a large number of longitudinal or panel studies, though few are from developing countries and few link social and spatial data. Their coverage varies by country, and this usually is for reasons of historical accident. With the available longitudinal data sets, the question is the extent to which they contain information that would be useful to those examining land-use and land-cover change. While we clearly are not experts on all extant longitudinal data sets, it is our distinct impression that the overwhelming majority do not contain the types of information that would be of use in examining land-use and land-cover change—at least given the current state of research.

In starting a new study, the analyst desiring temporal depth on the relevant social science variables needs to decide whether to use a prospective design, a retrospective design, or some combination of the two. A *prospective* design is a study in which sample households are interviewed at time 1 and then interviewed in subsequent time intervals with a periodicity determined by the researcher's substantive questions and available budget. A *retrospective* design asks respondents to recall previous events, activities, or intentions. Thus in a retrospective design, temporal depth is provided by the respondent's recall. The advantage of a prospective design is that there is no need to worry about recall error. The disadvantage is that one has to wait for time to elapse and multiple rounds of data collection to occur before the advantages of the prospective design are realized—and researchers and policy makers can be notoriously impatient. Further, prospective studies tend to be more costly.

With the retrospective design, the principal question is: What kinds of information retrospect well and under what circumstances? Within social demography, Westoff, Mishler, and Kelly (1957) showed that the correlation between recalled fertility preferences and actual childbearing was high and that the correlation between recalled fertility preferences and the previously measured fertility preference was low. Ever since, it has been conventional wisdom—repeatedly reaffirmed by empirical evidence—that preferences, attitudes, and intentions can change over time and do not retrospect well. People simply do not remember what their preference was five, ten, or fifteen years ago.

On the other hand, there is now considerable evidence that certain types of events are well remembered, such as moves (migrations), marriages, jobs, schooling, and residence arrangements (Baumgarten et al. 1983; Bradburn et al. 1987; Henry et al. 1994; Rindfuss et al. 1988). For land-use/cover research, do people remember how they used a given parcel of land last year, the year before that, and so forth? Do they remember why, such as a change in governmental policy? Are there ways the researcher can help trigger the memory of respondents, perhaps by showing a series of satellite images or aerial photographs or referring to special land-related events such as floods and droughts? To the best of our knowledge, the quality of retrospective data on land use has not been empirically examined.

What about the possibility of "retrofitting" longitudinal social science surveys to make them spatially explicit and hence make it possible to link to remotely sensed data? Walsh et al. (this volume) and Rindfuss et al. (this volume) have retrofitted longitudinal surveys that were begun in 1990 and 1984 respectively—prior to the availability of GPS technology. The Ecuador and Thailand projects, respectively, each sought a spatial perspective and organized their questionnaires and protocols to "retrofit" their surveys and to formally incorporate a spatial perspective. During subsequent data collections, GPS coordinates were collected for households and villages, and survey data previously collected were linked to their geographic positions through database management systems involving the creation of key fields that associated the "where" of things to the "what" and "when" of things contained in the surveys.

4. LINKING ISSUES AND METHODS

4.1 Start with Households or Land Parcels?

As we stated in the introduction, one fundamental question scientists studying land-use and land-cover change face is: Do we link human activities to land or do we link land to people? The perspective has major implications for project design, the questions that can be asked and answered, and the types of data collected and how they are stored and analyzed. For ease of discussion, consider the district of "Studyville," shown in panel A of Figure 2. It has two towns and eight villages. Assume that remotely sensed data are available for the entire area and for a reasonable buffer around Studyville. Thus, from the perspective of the remotely sensed land-cover data, there is no need to sample.

Imagine that the investigator has the resources and the theoretical rationale to obtain complete coverage of Studyville, linking households to plots. Except under very unusual circumstances, it is likely that some households residing within Studyville will have plots outside of Studyville, and that some plots within Studyville will be associated with households that reside outside Studyville (see panel B of Figure 2). If one starts with households with the intent of linking to all the land with which they are associated, some of the parcels of land will be outside the boundaries of

Studyville and some of the parcels of land within Studyville will not be attached to any household within Studyville. Assuming no refusals from households within Studyville, starting with complete coverage of households will lead to patchwork coverage of the land, both within and outside Studyville. Of course, this patchwork of land coverage will be exacerbated to the extent that there is land not associated with any household. Examples include land controlled by governments, businesses, churches, and so forth, as well as land not controlled by anyone.

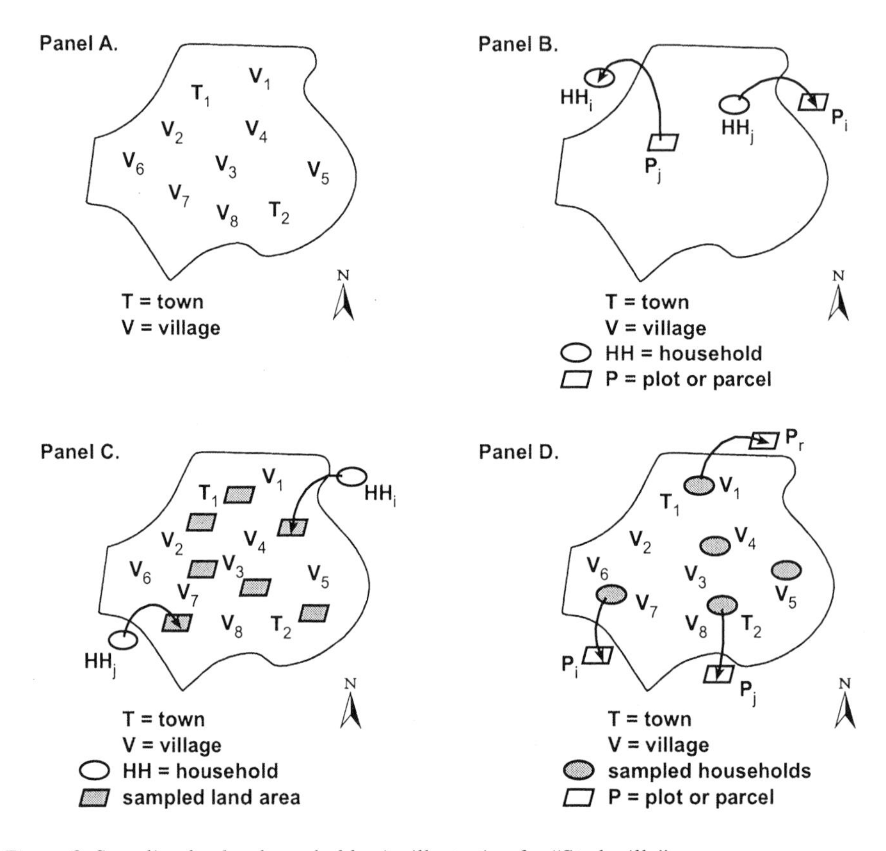

Figure 2. Sampling land or households: An illustration for "Studyville"

Conversely, if the investigator wants to start with the land and link it to all households that are associated with it, then the investigator will have a collection of some households residing within Studyville and some residing outside. Either way, starting from the land or starting from households, there may be a selectivity problem of the type that Heckman (1979) discusses with his original work on women's wage rates. To the extent that the households (or land parcels) not included in the data set are systematically different from those that are included, analytical results may be biased. Assuming one has the resources, this might argue for a strategy that blends starting from both land parcels and households.

In reality, one rarely has the resources to attempt complete coverage of either land parcels or households. The solution is to sample, again raising the question of whether to start from households within Studyville or land parcels within Studyville. Choosing one or the other starting point will yield samples with different properties—and perhaps yield different substantive results. Panels C and D of Figure 2 illustrate these two approaches. Starting with sampling land, some households will be outside Studyville; starting with sampled households, some land parcels will be outside Studyville. Further, numerous permutations are possible depending on the exact details of the study design. Note that the selectivity issue is a concern with both of these approaches.

The two different starting points can also yield quite different field strategies, as well as associated practical problems. These are discussed below. They involve issues such as land use versus ownership and whether you go to each parcel of land with a household member or whether you obtain the household-parcel linkage in some other manner.

In this volume, the Ecuador (Walsh et al.) and the Brazil (Moran et al.) studies represent situations in which land was sampled. In both cases, at the start of the studies the users/owners of the land also lived on the land parcels, hence giving results that would have been essentially similar to a household sampling strategy. The Ecuador study experienced land subdivisions over time and the development of lots used by households that commuted to work in nearby towns, which increased the number of households on a property that was formerly a single-household contiguous farm. The Brazil study sample was stratified by the date when there was visible deforestation on the parcel to examine cohort-effects hypotheses. Sometimes they approached a household living on one of the sampled land parcels and discovered that the original family that started deforesting the parcel had sold it and moved away. To preserve their goal of examining cohort effects, they then had a replacement strategy that involved going to a neighboring parcel. This resulted in a sample suited to their hypotheses, but one where the effects of a parcel changing ownership could not be investigated. The Mexico (Turner and Geoghegan this volume) and the Thailand (Rindfuss et al. this volume) cases are examples of sampling households that result in a patchwork of land surfaces covered (see Figure 6 in Rindfuss et al.).

The Ecuador and Thailand studies are longitudinal and suggest complications that arise when temporal depth is introduced. One is the question of what constitutes the successor household, an issue discussed above. Another is the question of how to handle subdivisions when land is sampled. In the Ecuador case this presented difficult study design, as well as budgetary and data management issues. A third issue is how to handle people or households that move out of the study area. In the Thailand case, migrants from some of the villages to selected destinations were followed, but little is known about their ownership and use of land back in their village. In the Ecuador case, out-migrants were not followed.

4.2 Use versus Ownership

There are numerous ways in which a household could be connected to a specific land parcel. They might live on the parcel. They might own it. They might have the right to use it. They might rent it, short or long term. Researchers who link households to land parcels need to decide what connection(s) will be operationalized. Presumably, such decisions need to be based on both theoretical grounds and practical grounds. The theoretical issues revolve around expectations of the effects of different levels of titling and land tenure policies, models of investment in land improvements, and the extent to which the land connection affects behavior such as migration or employment decisions.

One set of practical considerations also involves ethical overtones. Frequently, household interviews occur with multiple members of the household present and—in many parts of the world where much of everyday life takes place outdoors—neighbors, friends, and relatives are observing the interview. There might be joint ownership, involving multiple members of the household, or the owner could be a single member of the household. Whether or not there *should* be joint ownership may be a contentious issue within the household, and inquiring about it might generate conflict. In some places land ownership is a matter of public record and in other places it is not. Even if it is a matter of public record, it might not be a matter of common knowledge, and the household may not want others to know what they own. In short, the researcher needs to know a lot about the local situation prior to making a decision on use versus ownership.

Another set of practical considerations arises when researchers start with land parcels and plan to interview households that own or use the land. Once the land is sampled, how do you know which household owns or uses the land? If there is a dwelling unit on the land parcel, presumably the residents of that dwelling unit would be the starting point. If the residents are not the owners, they might be able to tell you the names of the owners. If there is no dwelling unit but ownership of land units is public information, then that might be a starting point. If ownership is not public information and no one is actually doing something on the land, what mechanisms are available for tracking and interviewing the owner or user? The problem is compounded if the owner or user lives outside the research site. What distance are you willing to travel to find the owner or user? Such issues are less of a problem when the information from households is linked to land parcels at the community level (e.g., Lambin this volume, and Fox et al. this volume) rather than the household level.

4.3 Land Parcels without Dwelling Units

It is a common situation in many parts of the world for households to own/use multiple plots of land, while their dwelling unit is on only one of them. In such a situation, should the other land parcels—those not containing the household's dwelling unit—also be included in the attempt to link households to their land parcels? If the answer is no, what might be the theoretical rationale? Perhaps the most basic decision is whether to take one or more household members to the land parcels (and obtain locational information with a GPS or other methods as is the case in Turner and Geoghegan this volume) or to attempt to do the linkage some other way (Rindfuss et al. this volume). Taking household members to their lands may be the least error prone, because they *should* know where their lands are located. Conversely, such an approach is likely to entail the greatest respondent burden, which increases as the number of parcels owned or used by the respondent household increases. Further, taking respondents to their fields can be expensive in terms of interviewer time and perhaps transportation costs. A likely outcome is that the sample size obtained is small.

If you do not go to the land parcels with the respondent, then some method of linking is needed that is essentially a "remote" link. Here, a variety of possibilities exist. Such linkages might already be available in land record offices in the form of cadastral maps and/or relevant documents in a local land office. Knowledgeable individuals in the community might be able to supply the necessary information. Map products could be brought into the household and household members could identify the parcels that they use.

4.4 Plot and Pixel Size Mismatch

In a number of environments, the size of plots associated with a single household may be below the spatial resolution of remote sensing systems. In such situations, digital or analog aircraft data might be considered, where the user can set the required minimum mapping unit for inter- and intraplot mapping. Also, digitization of ground-based videography and/or the use of aerial-based videography are also possible remote sensing solutions to high spatial resolution mapping requirements.

Linear mixture modeling (Adams et al. 1995) has been used to define land-use/land-cover proportions within a defined pixel in an attempt to decompose pixel response patterns to their component parts relative to the concept of an integrated pixel. In other words, this approach maps subpixel information. New remote sensing systems having higher spatial resolutions (e.g., IKONOS and QUICKBIRD) also are on-line to render detailed (to 1 x 1 m spatial resolutions) land-cover information. Note that these high spatial resolution sensors have high data volumes and costs associated with them. These high data volumes might, in turn, create data management and budget issues for the researcher, as well as design considerations about how best to use the high spatial resolution data—as a continuous data set for broad area mapping or for only a subset or sampled region, using models to extend the effects to broader areas. Regional views afforded by MODIS, for example, may be used to set context, while systems such as IKONOS detail plot-specific variability.

5. PRACTICAL ISSUES

5.1 Data Quality

Data quality is central to science, and that is clearly the case for those examining land-use and land-cover change. There are long literatures on the quality of survey data and remotely sensed data. Here we address only data quality issues at the interface between the two—specifically the quality of the link and whether anything in the linking operation has an adverse effect on the quality of the linked data components.

What methods are available for measuring the quality of the linking operation between households and the parcels of land they own or use? This is relatively uncharted methodological territory, and an area in need of further development. One question has to do with match rates: How accurately did we match the household to its plot of land? The first issue is: Should the denominator of the rate be based on land use/land cover or on information provided by the household? There is no reason why the two should necessarily be the same—and indeed there are likely to be substantial differences. But there are situations in which they might be the same, with the Brazilian case (Moran et al. this volume) being an example.

Ideally, to check the quality of the match, some independent method of matching should be used for a sample of the household-land combinations. For example, one might have ownership information from a household interview and a land office. To the extent that the independent methods agree, we will have greater confidence in the data. We are not aware of any published reports that have provided estimates of match quality between households and the pixels they use/own. If they do not agree, then we will have less confidence in the data, but we may not know the sources of error.

GIS technology may make it possible to use standard remote sensing "ground truth" accuracy assessment techniques to assess not only the accuracy of the remote sensing classification but also the land-use history/socioeconomic profile developed for each plot of land. Remote sensing procedures compute a root-mean-square-error (RMSE) term as part of the georeferencing procedure to indicate the degree of correspondence in the location of an earth feature that is referenced in the satellite data set through path-row coordinates and through map-based and/or GPS–based coordinates of the same feature. The RMSE indicates the +/– difference, generally in meters, between the two measures of feature location. Generally, analysts seek an RMSE of less than 0.5 pixel (Jensen 2000). Matches between attributes (e.g., land-cover/land-use categories) sensed through remote sensing systems and compared against referenced data of presumed higher spatial and/or attribute accuracy are also commonly applied. Theoretically, we could seek matches between attributes (land-use histories/socioeconomic profiles) developed through household and community surveys and compared with ground control information collected from randomly selected points.

5.2 Confidentiality Issues

It has become accepted social science practice that, when we collect information from individuals or households, we promise to protect their identities and not release data gathered from them that could be linked back to them. (There are a few exceptions, such as when public figures are being interviewed in their public capacity.) There are multiple reasons for protecting the confidentiality of our respondents; perhaps the most compelling is that, in the opinion of the social science community, it is the ethically correct thing to do. Further, in the long run, protecting respondent confidentiality is critical to insuring that respondents will cooperate with social scientists.

A countervailing force has been the pressure to release data sets to other users. Again, there are a variety of arguments, but perhaps the simplest is that contemporary surveys are too expensive to justify their use only by the investigator(s) who obtained the grant to collect the data. Within the United States, both NIH and NSF have been urging their grantees to make their survey data available to the broader scientific community.

A common strategy with social surveys to simultaneously protect confidentiality and release data has been to strip away names, addresses, and all other geographic identifiers. Instead, a unique (and fictional) ID is attached to each case. In more recent years, some contextual data have been made available under special arrangements with individual researchers. However, such contextual data have been at a sufficiently coarse scale, such as the respondent's county or district, that it would be very difficult to identify with certainty a respondent given the added information that context provides.

Linking survey data to the actual parcels of land a household owns or uses makes it extremely difficult to simultaneously protect the confidentiality of the respondents and make the data available to the broader research community. If the identifying information is stripped away from the household information, then other researchers will not be able to link household data with remotely sensed land-cover information. If the linkage data are not stripped away, then releasing it is tantamount to releasing the addresses of respondent households, because if you know the location of the respondent's dwelling unit and/or the land parcels used by that household, determining the identity of the respondent is straightforward.

The solution of aggregating the household data up to a larger geographic or political unit while protecting the confidentiality of respondent households is problematic because it precludes analysis at the household level by other researchers. Spatial transformations of the mapped location of the respondent's land parcels might be used to protect confidentiality. Such transformation could include the rotation and deformation of the study area so that location does not translate to easy respondent identification, but the procedures for doing so have not yet been fully developed or standardized. Even if they were available, it is the spatial relationships and position of land parcels linked to households that investigators seek to maintain, so such cartographic attempts at maintaining confidentiality often are limited to the creation of teaching data sets, test data sets, or demonstration data sets.

Another alternative would be to extract key aspects of the household's land parcels (e.g., size, land use, soil type) and attach this information to the household's survey data, strip away identifiers, and release the data. While this might provide for the needs of some researchers, it would not meet the needs of those who are more spatially oriented and want to understand such issues as neighborhood effects, the implications of fragmentation (see Malanson this volume), or the effects of household land-use decisions on watersheds (see Fox et al. this volume).

A different aspect of confidentiality involves the acquisition and use of remotely sensed data in conjunction with map products. Such maps and figures can frequently be easily grasped, understood, and used by a variety of groups who have an interest in a particular landscape. Such information can reveal things that are being done to the landscape that might not otherwise have been apparent. This information can be used to "police" those who are using the land in ways that others see as inappropriate or to identify valuable resources that others would like to "appropriate." Particularly in contested landscapes, such information can alter preexisting power relationships. When such information is linked to information on household economics, demographics, and land ownership, it can become even more problematic in terms of what group has access to the information generated and how such information might be used by different stakeholders. Judging from the extensive discussions on this topic by the coauthors of this introduction during its preparation, it would appear that consensus does not exist on how to share remotely sensed information with local stakeholders. Given the potential importance of this issue, it is a topic that ought to be discussed more widely in the land-use/land-cover research community.

5.3 Sample Size and Related Issues

There is a set of practical issues that researchers face at the confluence of available budget (both financial and time), number of interviewers available, and preferences for sample sizes. Given that "socializing the pixel" and "pixelizing the social" are relatively new scientific activities, there is limited experience about the expenses involved, the skills needed by interviewers, and the sample sizes needed to be able to answer relevant research questions. Further, there are different standards or "tastes" regarding sample size issues in the diverse research communities involved. For example, demographers are used to working with large samples, and anthropologists have been adept at maximizing information obtained from small samples. The lack of consensus and research base regarding sample size impacts the research community, beginning with the process of planning a budget. It makes the peer review of budgets difficult, and it can create a problem for foundations and governmental funding agencies because they do not know how much funding to make available. If the available amount is too low, then appropriate studies cannot be effectively planned. Because researchers so far have relied on extant aerial photos or satellite data, we limit the present discussion primarily to social survey type data.

Various factors affect the preferred sample size for a household or community survey. While there are formulae for calculating the statistical power of sample sizes associated with a given sample design, most surveys that would be designed to link households or communities to remotely sensed data would likely be designed for multiple types of analyses, some of which might not have been anticipated when designing the study. So instead of considering the formulae here, we will mention some of the principles underlying them.

As the level of statistical significance one is willing to tolerate moves from .10 to .01 to .001, other things being equal, the sample size required to detect a difference when one is present increases. If a dependent or independent variable of interest is categorical, the lower the percentage of the population in the smallest category, the larger the sample size needed. As one moves from trying to detect a difference between two populations in the distribution of some variable to multivariate analyses with numerous predictor variables, increased sample sizes are required. In general, when doing statistical analyses, most researchers would prefer larger sample sizes to smaller sample sizes because of statistical power and analytical flexibility.

In social surveys, scientific probability sampling is preferred. A probability sample is one in which the units are selected with a known, nonzero probability. Use of nonprobability sampling methods such as purposive sampling and quota sampling might provide reasonable estimates, but such methods cannot provide the necessary confidence in results. What kinds of sampling schemes are most appropriate for examining human impacts on land use and land cover? This volume contains a diversity of approaches.

While we frequently have a preference for larger sample sizes, practical realities such as cost and the available interviewer pool may push in the opposite direction. Based on our experience, personnel items are the largest single item in any budget for surveys. To the extent that this is true, what level of training is needed to carry out the various tasks? Clearly, to the extent that tasks can be standardized and that field workers can be quickly trained, the field workers can be drawn from a larger and lower-wage labor pool, and adding to in-country capabilities can be achieved. This would also reduce the costs related to transporting field staff from outside the region or country. Further, much of the extant literature on land use and social science has been set in developing countries. To the extent that interviewers are from those countries, wage rates are likely to be lower. Finally, from a practical standpoint, larger samples are more difficult to manage and supervise.

Questions that the land-use and land-cover change community needs to address are: What sample sizes are needed to address some of the more pressing theoretical and practical research issues? How should samples be selected to address questions of spatial autocorrelation? What cost savings can be achieved through standardization of procedures and using lower-wage personnel? Are there lower cost designs that can be every bit as effective as more expensive designs? What challenges do researchers face in finding appropriate local institutions and personnel? Can fieldwork lessons learned in one study area be generalized to other areas?

From the perspective of survey research designs, many of the interesting research questions have to do with change over time, which in turn pushes us to have temporal depth in the variables of interest. If land-cover data are coming from one of the frequently used sensors like Landsat TM, then temporal depth should be available. On the social survey side, however, there are many sampling issues involved in obtaining temporal depth.

One issue is whether to use a cross-sectional or longitudinal design. Cross-sectional designs are easier to administer because we do not need to follow sample households over time, which might move away from the study area or can change in size, composition, or character. Moreover, in a longitudinal design, repeated visits to the same household may affect the quality of household responses and, depending on the circumstances, the effect could be positive or negative. On the other hand, cross-sectional designs suffer from lack of comparability between sampled households at two time points, especially if the sample size is not large enough.

6. SUMMING UP

Our objective in preparing this introduction was to highlight some of the issues that underlie the linking of remotely sensed and social science data at the local or micro level. We sought to raise questions on methodological and practical issues that affect how researchers are linking across thematic domains and across space and time scales where the household and the community are the social units of organization and where the pixel and the landscape are the respective spatial and biophysical units. This discussion was motivated by the common concern shared by land-use and land-cover change researchers of how to relate population and the environment when social, biophysical, and geographical data are subject to differences in cartographic structures, space and time scales, units of measurement, fixed versus dynamic location, and so on.

As one reads the following essays, it becomes clear that the link between people and land has to be designed differently in different settings. No overarching theory or formula determines that the link needs to be established at any particular level (i.e., individual, household, community) or done in any one direction (land to people or people to land). Methods for linking humans to land should be responsive to the needs and conditions of different settings. This perhaps is most striking in the case of pastoralists, who continuously move across the landscape, versus settled farmers. It is also clear that in some settings, farm households do not make decisions about land use, but rather decisions are made at a more aggregated level, such as villages or corporate kinship groups. At the local level, researchers need the flexibility of deciding what links need to be examined and choosing or designing the appropriate tools for making these links.

It is also clear from the case studies that, given the variability in the type of groups making land-use decisions, prior knowledge of a study area is essential. This prior knowledge can be obtained as part of an ongoing study that seeks to add a land-use component by collaboration with knowledgeable local researchers, through image analysis involving multiresolution (and perhaps multitemporal) remotely sensed data, through field surveys of test sites, or through prior qualitative research. From a funding perspective, it highlights the importance of having funds available for "seed" grants.

Perhaps the most important conclusion that can be reached from these studies is that standards protocols have not yet emerged in the group of people doing land-use change studies—especially those who attempt to link to human data. Indeed, even though we have frequently heard the term *LULC community,* we remain doubtful that there is a true community, at least in the sociological sense of that term. Rather, there are people studying land-use and land-cover change from a variety of disciplinary and interdisciplinary perspectives, and they bring the standards of their own discipline to their work. Perhaps nowhere was this more evident than in the sample sizes of the various groups. Other examples include a lack of consistency in reporting on the accuracy of remote sensing classifications, survey response rates, or percent of land parcels in which an accurate link was established between a household and its land parcel.

This introduction and volume highlight numerous difficulties in linking social and remotely sensed data at the micro level. Yet in acknowledging these difficulties, we should not overlook the broader picture, which is quite uplifting. Within a remarkably short time period, the land-use/land-cover change research community has produced a variety of examples linking households, directly or indirectly, to the land parcels they own/use to the characteristics of those parcels as depicted from remotely sensed data. The examples contained in this book show that the research community (or communities) has been remarkably creative, resourceful, and resilient. Exciting data sets have been put together, and substantive research using these data sets is ongoing. Interdisciplinary and multidisciplinary research teams are emerging and functioning. Even though such teams are costly and can take a while before they function smoothly, they are necessary at this stage in the development of land-use change research. In looking forward, we are upbeat and optimistic that significant progress will be made in linking changes in global land cover with decisions about land-use practices that are made at the household and community levels.

REFERENCES

Adams, J. B., D. E. Sabol, V. Kapos, R. Almeida, D. A. Roberts, M. O. Smith, and A. R. Gillespie. 1995. "Classification of Multispectral Images Based on Fractions of End-Members: Application to Land-Cover Change in the Brazilian Amazon." *Remote Sensing of Environment* 52(2): 137–154.

Baumgarten, M., J. Siemiatycki, and G. Gibbs. 1983. "Validity of Work Histories Obtained by Interview for Epidemiological Purposes." *American Journal of Epidemiology* 118: 583–591.

Booth, A., and D. R. Johnson. 1985. "Tracking Respondents in a Telephone Interview Panel Selected by Random Digit Dialing." *Sociological Methods and Research* 14(1): 53–64.

Bradburn, N. M., L. J. Rips, and S. K. Shevell. 1987. "Answering Autobiographical Questions: The Impact of Memory and Inference in Surveys." *Science* 236: 157–161.

Call, V. R. A., L. B. Otto, and K. I. Spenner. 1982. *Tracking Respondents: A Multi-Method Approach.* Lexington, MA: Lexington Books.

Chambers, R., A. Pacey, and L. A. Thrupp, eds. 1989. *Farmers First: Farmer Innovation and Agricultural Research.* London: Intermediate Technology Publications.

Chase, T., R. Pielke, T. Kittel, R. Nemani, and S. Running. 1999. "Simulated Impacts of Historical Land Cover Changes on Global Climate in Northern Winter." *Climate Dynamics* 16: 93–105.

Citro, C. F., and H. W. Watts. 1986. *Patterns of Household Composition and Family Status Change.* Washington, D.C.: U.S. Bureau of the Census.

Clarridge, B. R., L. L. Sheey, and T. S. Hauser. 1978. "Tracing Members of a Panel: A 17-Year Follow-Up." In K. F. Schuessler, ed., *Sociological Methodology* (San Francisco: Jossey,-Bass).

Colfer, C., M. Brocklesby, C. Diaw, P. Etuge, M. Gunter, E. Harwell, C. McDougal, N. Porro, R. Porro, R. Prabhu, A. Salim, M. Sardjono, B. Tchikangwa, A. Tiani, R Wadley, J. Woefel, and E. Wollenberg. 1999. *The BAG: Basic Assessment Guide for Human Well-Being.* Bogor, Indonesia: Center for International Forestry Research.

Crawford, T. W. 2000. "Human-Environment Interactions and Regional Change in Northeast Thailand: Relationships between Socio-Economic, Environment, and Geographic Patterns." PhD dissertation, University of North Carolina.

Duncan, G. J., and M. Hill. 1985. "Conceptions of Longitudinal Households: Fertile or Futile?" *Journal of Economic and Social Measurement* 13(3-4): 361–375.

Evans, T. P. 1998. "Integration of Community-Level Social and Environmental Data: Spatial Modeling of Community Boundaries in Northeast Thailand." PhD dissertation, University of North Carolina.

Fox, J., J. Krummel, M. Ekasingh, S. Yamasarn, and N. Podger. 1995. "Land Use and Landscape Dynamics in Northern Thailand: Assessing Change in Three Upland Watersheds." *Ambio* 24: 328–334.

Geist, H., and E. Lambin. 2001. "What Drives Tropical Deforestation? A Meta-Analysis of Proximate and Underlying Causes of Deforestation Based on Subnational Scale Case Study Evidence." LUCC Report Series No. 4. University of Louvain, Louvain-la-Neuve, France.

Gibson, C., E. Ostrom, and T. Ahn. 2000. "The Concept of Scale and the Human Dimensions of Global Change: A Survey." *Ecological Economics* 32: 217–239.

Guyer, J. I., and E. F. Lambin. 1993. "Land Use in an Urban Hinterland: Ethnography and Remote Sensing in the Study of African Intensification." *American Anthropologist* 95(4): 839–859.

Heckman, J. 1979. "Sample Selection as a Specification Error." *Econometrica* 47(1): 153–161.

Henry, B., T. E. Moffitt, A. Caspi, J. Langley, and P. A. Silva. 1994. "On the 'Remembrance of Things Past': A Longitudinal Evaluation of the Retrospective Method." *Psychological Assessment* 92(101).

Houghton, R., J. Hackler, and K. Lawrence. 1999. The U.S. Carbon Budget: Contribution from Land-Use Change. *Science* 285: 574–578.

Jensen, J. R. 2000. *Remote Sensing of the Environment: An Earth Resource Perspective.* Upper Saddle River, NJ: Prentice Hall.

Kasperson, J., R. Kasperson, and B. L. Turner II, eds. 1995. *Regions at Risk: Comparisons of Threatened Environments.* Tokyo: United Nations University Press.

Keilman, N., and N. Keyfitz. 1988. "Recurrent Issues in Dynamic Household Modeling." In N. Keilman, A. C. Kuijsten, and A. Vossen, eds., *Household Formation and Dissolution* (New York: Clarendon Press), 254–286.

Lambin, E., X. Baulies, N. Bockstael, G. Fischer, T. Krug, R. Leemans, E. Moran, R. R. Rindfuss, Y. Sato, D. L. Skole, B. L. Turner II, and C. Vogel. 1999. "Land-Use and Land-Cover Change (LUCC): Implementation Strategy." IGBP Report 48 and IHDP Report 10. Stockholm: IGBP Secretariat, Royal Swedish Academy of Science.

Lambin, E., B. L. Turner II, H. Geist, S. Agbola, A. Angelsen, J. Bruce, O. Coomes, R. Dirzo, G. Fisher, C. Folke, P. George, K. Homewood, J. Imbernon, R. Leemans, X. Li, E. Moran, M. Mortimore, P. Ramakrishan, J. Richards, H. Skanes, W. Steffen, G. Stone, U. Svedin, T. Veldkamp, C. Vogel, and J. Xu. 2001. "The Causes of Land-Use and Land-Cover Change: Moving beyond the Myths." *Global Environmental Change* 11: 261–269.

Liverman, D., E. F. Moran, R. R. Rindfuss, and P. C. Stern, eds. 1998. *People and Pixels: Linking Remote Sensing and Social Science.* Washington, D.C.: National Academy Press.

McMillan, D. B., and R. Herriot. 1985. "Toward a Longitudinal Definition of Households." *Journal of Economic and Social Measurement* 13(3-4): 349–360.

Meyer, W., and B. L. Turner II. 1992. "Human Population Growth and Global Land-Use/Land-Cover Change." *Annual Review of Ecology and Systematics* 23: 39–61.

Moran, E. F., E. Brondizio, P. Mausel, and Y. Weu. 1994. "Integrating Amazonian Vegetation, Land Use, and Satellite Data: Attention to Differential Patterns and Rates of Secondary Succession Can Inform Future Policies." *BioScience* 44(5): 340–349.

Ojima, D., K. Galvin, and B. L. Turner II. 1994. "The Global Impact of Land-Use Change." *BioScience* 44(5): 300–304.

Ribisl, K. M., M. Walton, C. Mowbray, D. Luke, W. Davidson, and B. BootsMiller. 1996. "Minimizing Participant Attrition in Panel Studies through the Use of Effective Retention and Tracking Strategies: Review and Recommendations." *Evaluation and Program Planning* 19(1): 1–25.

Rindfuss, R. R., S. Morgan, and G. Swicegood. 1988. *First Births in America.* Berkeley, CA: University of California Press.

Rindfuss, R. R., B. Entwisle, S. J. Walsh, P. Prasartkul, Y. Sawangdee, T. W. Crawford, and J. Reade. 2002. "Continuous and Discrete: Where They Have Met in Nang Rong, Thailand." In S. J. Walsh and K. A. Crews-Meyer, eds., *Linking People, Place, and Policy: A GIScience Approach* (Boston: Kluwer Academic Publishers), 7–37.

Robinson, W. S. 1950. "Ecological Correlations and the Behavior of Individuals." *American Sociological Review* 15(3): 351–357.

Sala, E., F. Chapin, J. Armesto, E. Berlow, J. Bloomfield, R. Drizo, E. Huber-Sanwald, L. Huenneke, R. Jackson, A. Kinzig, R. Leemans, D. Lodge, H. Mooney, M. Osterheld, N. Poff, M. Sykes, B. Walker, M. Walker, and D. Wall. 2000. "Biodiversity: Global Biodiversity Scenarios for the Year 2100." *Science* 287: 1770–1774.

Skole, D. L., W. H. Chomentwoski, W. A. Salas, and A. D. Nobre. 1994. "Physical and Human Dimensions of Deforestation in Amazonia: In the Brazilian Amazon, Regional

Trends Are Influenced by Large Scale External Forces but Mediated by Local Conditions." *BioScience* 44(5): 314–322.

Turner, B. L. II. 2001. "Land-Use and Land-Cover Change: Advances in 1.5 Decades of Sustained International Research." *Emergent Sustainability Science* 4: 269–272.

Turner, B. L. II, and W. B. Meyer. 1991. "Land Use and Land Cover in Global Environmental Change: Considerations for Study." *International Social Science Journal* 130: 669–679.

Vitousek, P., H. Mooney, J. Lubchenco, and J. Melillo. 1997. "Human Domination of Earth's Ecosystems." *Science* 277: 494–499.

Walsh, S. J., T. P. Evans, W. F. Welsh, B. Entwisle, and R. R. Rindfuss. 1999. "Scale Dependent Relationships between Population and Environment in Northeastern Thailand." *PE & RS* 65(1): 97–105.

Westoff, C. F., E. G. Mishler, and E. L. Kelly. 1957. "Preferences in Size of Family and Eventual Fertility Twenty Years After." *American Journal of Sociology* 62(5): 491–497.

Chapter 2

LAND-COVER AND LAND-USE CHANGE (LCLUC) IN THE SOUTHERN YUCATÁN PENINSULAR REGION (SYPR)

An Integrated Approach

Billie L. Turner II
Graduate School of Geography and George Perkins Marsh Institute, Clark University
bturner@clarku.edu
Jacqueline Geoghegan
Department of Economics and George Perkins Marsh Institute, Clark University

Abstract The southern Yucatán peninsular region (SYPR) project seeks to demonstrate the value of "integrated land-change science" approaches to understanding tropical deforestation and land-use and land-cover change in the region in question. Ecological, social, and remote sensing–GIS sciences are joined in an examination of political economic forces driving land change in this "hot spot" of tropical deforestation, the impacts and feedbacks of these changes on forest and household dynamics, and the ability to advance spatially explicit land-change models of these dynamics linked to TM Landsat imagery. Emphasis here is placed on the methods used to collect and analyze household data in regard to a regional model that addresses both the magnitude and location of deforestation in the region. While exploratory in kind, the effort illustrates the potential to improve land-change models in ways that resonate with the interests of the three research communities involved in the project. The project indicates that integrated land-change science is possible, holds promise to improve understanding across a wide range of research questions, but invariably involves large start-up costs and entertains other requirements that make its practice a long-term research investment.

Keywords: Yucatán, tropical deforestation, land change, household survey, models

1. INTRODUCTION: THE SYPR PROJECT

The SYPR project began in 1997 as part of a three-year grant from NASA's LCLUC program, with sustained support from the NSF's Center for Integrated Studies of the Human Dimensions of Global Environmental Change (CIS–HDGEC, Carnegie Mellon University). Its theme during that phase of work was "spatially explicit probability approaches for modeling and projecting deforestation and land conversion linked to remotely sensed imagery." Subsequently, a second phase of the project began in 2001 (three years) under the auspices of LCLUC and CIS–HDGEC, focusing on "refining models and projections of deforestation with application to the carbon cycle, biotic diversity and regeneration capacity, sustainability and vulnerability."[1] Four overarching goals combine the two phases of the project: (1) to advance *integrated land science,* coupling ecological, remote sensing and GIS, and social science as a robust means of addressing land-use/cover change, (2) by developing a sound understanding of deforestation and agricultural change in the southern Yucatán peninsular region, and (3) placing this understanding into a suite of spatially explicit models that can explain and project land change, (4) with special references to forest carbon flux, loss in biotic diversity, and the vulnerability of the coupled human environment system to perturbations and stresses of various kinds. In addition, SYPR has joined a comparative LCLUC study on land-change effects on biodiversity near parks and reserves. Given the complexity of this large interdisciplinary and multi-institutional project (involving more than twenty-five researchers), attention here is given to phase one of project.[2]

2. BACKGROUND

The southern Yucatán peninsular region is the last tropical forest-agriculture frontier in Mexico. Encompassing about 22,000 km^2 (as defined by the project—see Figure 1) and spanning the states of Quintana Roo and Campeche, the region is a karstic upland, ranging from 100 to 300 m amsl (above mean sea level).[3] Precipitation varies north-south from 900 mm to 1,400 mm per annum, with a pronounced dry season during the winter. The karstic terrain promotes subsurface drainage of water with deep water tables (in excess of 200 m). Surface water flows only along the edges of the region, usually after prolonged rains or during extreme rainfall events (hurricanes). The rolling hills and ridges are dominated by rich but thin mollisols (a soil type). Dispersed among the uplands are large solution features *(poljes),* locally known as *bajos,* that infill with clay-rich vertisols (soil) and hold water during the wet season. The uplands and *bajos* (about 20 percent and

80 percent of the region, respectively) maintain seasonal tropical forests distinguished not so much by different species but by their relative abundance and structural appearance. Locally the forest on the uplands is known as *bosque mediano* and those in *bajos* as *alkache;* here we distinguish them as upland forest and *bajo* forest, respectively. Save for special development projects, most cultivation takes place on lands formerly under upland forest—the same lands possessing the better drained soils, as opposed to the poorly drained and massive clay soils of the *bajos*.

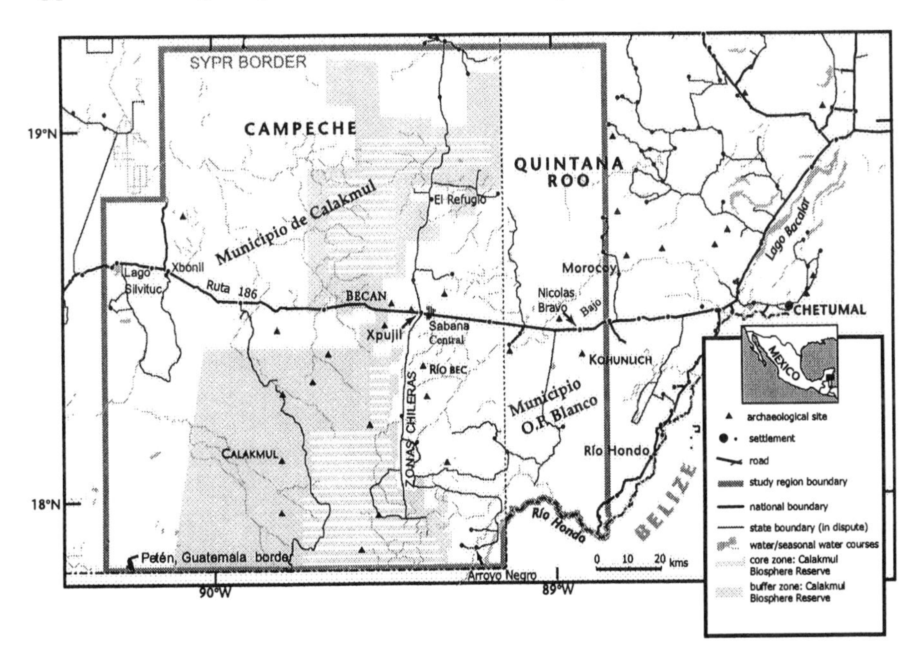

Figure 1. Summary of land areas and population densities

This region was once the northern part of the central Maya lowlands, and during the Late Preclassic and Classic periods, a large majority of the upland forests was obliterated for occupation and cultivation, as were parts of *bajo* forests in the east (Turner 1983). With the Classic Maya collapse (ca. A.D. 850–1000 locally), the region was abandoned and the forest returned, albeit modified in species abundance, remaining relatively untouched until the twentieth century. In the early part of that century, various logging schemes were planned but poorly implemented. By World War II, however, migrant camps of *chicleros* dotted the landscape to collect chicle (resin from the chicle tree) for American chewing gum. With the end of the war, this activity dissipated and was overtaken by serious logging of mahogany and Spanish cedar, which was abundant in the region (Lundell 1934). By the

1960s, much of the prime timber had been taken, and today it is difficult to find either species of substantial size in upland forests (Turner et al. 2001).

Major settlement in the region awaited the completion of Highway 186 (fully paved in 1972), crossing the heart of the forest and connecting Chetumal, the capital of Quintana Roo, with the remainder of Mexico (Klepies and Turner 2001). This "mini-Amazonian" highway (various parts paved between 1967 and 1972) opened the region to significant immigration from land-hungry peasants elsewhere in Mexico, and a large number of *ejidos* (communally controlled lands given to a group of people known as *ejidatarios*) were established in the region to accommodate this demand. As a result, the regional population grew from about 2,500 people in the early 1960s to over 28,000 people in the mid-1990s (Turner et al. 2001), not including an estimated 5,000 squatters located in the extreme southern part of the region (Stedman-Edwards 1997) (Table 1). The estimated population for agricultural *ejidos* for the year 2000 is 38,100 (Klepeis 1999). Smallholders initially pursued swidden cultivation in the uplands, locally known as *milpa* (i.e., maize, beans, and squashes). The federal government, however, had grander plans during the petro-boom of the late 1970s to mid-1980s and began to implement large-scale deforestation of the *bajos* Nicolas Bravo and Morocoy (in the east) and elsewhere to undertake mechanized wet rice cultivation, eventually felling 10,000 ha of *bajo* forest. During this time, *ejidatarios* and planned projects reduced the total forests of the region at an annual rate approximating 0.6 percent (Turner et al. 2001).

With the subsequent debt crisis of 1982 and the failure of the rice projects, the cleared *bajo* forests were planted in pasture for cattle experiments.. The enlarged population, however, maintained its attention to upland cultivation (swidden), while state-sponsored activities reversed their agricultural orientation and began to envision the forest itself and the Maya treasures within it as possessing value. A large portion (7,232 km^2) of the western portion of the region became the Calakmul Biosphere Reserve (UNESCO Man and Biosphere site, est. 1993) (Primack et al. 1998), and most of region fell under the interests of El Mundo Maya (The Maya World, an international archaeo-ecotourism program) (Garrett 1989). The former mantra of forest clearance for development has been replaced with forest preservation and conservation for development. Farmers, therefore, are embroiled in the dichotomous needs for household production via cultivation and the larger state aims to reduce deforestation, complete with a new GEF (Global Environmental Fund) program to make the region compatible with the aims of the Mesoamerican Biological Corridor to facilitate the movement of species across Middle America (Miller, Chang, and Johnson 2001).

NGOs have sponsored various "forest friendly" programs, but these have not stopped deforestation, which continued at 0.28 percent per annum from 1987 to 1997 (Klepeis and Chowdhury n.d.; Turner et al. 2001). Indeed, absent any federal, state, or NGO support, farmers have turned to the production of chili for the national market; by the late 1980s, a majority of those households with lands were engaged to different degrees in this activity (Keys n.d.; Turner et al. 2001). This phenomenon is critical for deforestation and forest in recovery in several ways: On average, mechanized and nonmechanized chili farmers cultivate, respectively, 7 ha and 3 ha more land than "pure" swidden farmers; and chili is added upfront in the normal swidden routine, typically reducing the crop-fallow cycle from 3–4: 12–15 years (crop: fallow) to 5–6: 10–12. More land, therefore, is being cultivated more frequently, impeding the fallow period of successional growth. The impacts on land degradation are not known, save that one invasive species (bracken fern) has increased more than fourfold from 1987 to 1997, spatially distributed where swidden and/or chili production has been undertaken the longest.[4] In addition, PROCAMPO, a federal program aimed at neoliberal reform in the rural sector, provided subsidies designed to intensify agriculture on extant cultivated lands. The funds were used to cut upland forest for pasture—largely unstocked, the rationale for which is under study (Klepeis and Chowdhury n.d.; Klepeis and Vance n.d.).

Table 1. Summary of land areas and population densities

Unit	Area [km^2]	Pop./km^2	Comments
Demarcated SYPR (Figure 1)	22,000	na	Rolling uplands (above 100 m) extensively occupied after 1967
Tenure units	18,703	2.0	*Ejidos*, private & government lands fully within demarcated SYPR
Agricultural *ejidos*	8,659	4.4	Does not include *amplicaciones* (Figure 2)—forest extension *ejidos*
Calakmul Biosphere Reserve	7,232	na	Estimated 5,000 squatters in southern part
Core	2,483	na	Technically, no agriculture permitted
Buffer	4,749	na	Agriculture where *ejido* intersects; also new squatters in southwestern part

Note: Population densities based on project estimate of 38,100 occupants of the "tenure units" in 2000. This estimate does not include squatters.

The impacts for pasture creation and former rice projects notwithstanding, the largest portion of deforestation in the region is driven by the combined consumption and market production of households and mostly takes place within upland forest lands. Overall, households hold a land surplus. They claim on average more than 40 ha but annually cultivate between 5 ha and 12 ha, depending on land holdings and degree of market activity. On average, households claim to hold 43 percent of their lands in older growth (beyond fifteen years), 11 percent in cultivation, and 46 percent in fallow (early succession) and pasture (Vance et al. n.d.). Thus, a majority of the land parcels within an agricultural *ejido* should display some level of human disturbance.

3. PROBLEM, THEORY, AND DESIGN

The study region (Figure 1) was selected for its match to the overarching goals of the NASA–LCLUC program—assessment of the coupled human environment system and spatially explicit understanding and modeling of deforestation. These goals are also consistent with those of the international LUCC program (IGBP–IHDP 1995, 1999) and offer an integrated assessment of land change consistent with the goals of NSF's CIS–HDGEC. In addition, the region offered a unique history of previous occupation-deforestation and abandonment-regrowth, the PI (principal investigator) worked there in 1973, 1974, and 1976, and combined aerial photographs and TM imagery captured most of the modern deforestation from the mid-1970s to present. The spatial juxtaposition of the farmers and forest reserves and the relatively late date for significant modern deforestation offered a case study highly suited for the examination of deforestation linked to TM imagery, with significant implications for the usefulness of spatial explicit land-change models, both theoretically and practically. As well, the ancient Maya experience in the region offers opportunities to explore the question of past human impress on forests and its comparison to the modern impress—planned future studies. Finally, the PI had documented the kind and spatial scale of ancient Maya agriculture and land change in the region, providing a strong basis for understanding the long-term history of the region and its forests as well as various observations and data pertaining to land uses in the region in the middle 1970s.

Forest dynamics are partly driven by the nature of forest use (e.g., intensity and frequency of cultivation) and, of course, by the age, structure, and environmental attributes of the forest. The study region contains segments of old growth (parts minimally older than a hundred years) and two basic forest types (upland and *bajo*). Sets of species tend to cluster within the uplands, and the overall dominance of certain species as well as biomass tends to correspond to the north-south rainfall gradient (Lawrence and Foster n.d.). In addition, the social science survey performed during phase one of the project (discussed below) opened the door for the sampling of growth on farmer's fallowed land under different phases of the crop-fallow cycle. Ultimately, the survey sampling designs must account for these different biophysical conditions.

Owing to recent occupation and deforestation, this "frontier" region offers a relatively simple political economy by which to assess land-change processes. State visions of how to develop the region have led to three episodes of use-occupation from the second half of the twentieth century to the present (Klepeis et al. n.d.; Klepeis and Turner 2001). (1) From the 1950s through the 1970s, the region was designated for tropical hardwood extraction, and logging trails were constructed as were a few larger settlements. (2) With the petro-boom of the late 1970s, the large seasonally inundated *bajos* (*bajo* forest) on the eastern and western flanks of the region were denuded by the state in pursuit of the development of large-scale (more than 10,000 ha were cleared) mechanized rice projects. These efforts drew large numbers of immigrants to the region and increased the number *ejidos*. (3) The petro-collapse in 1982 and the failure of the projects, however, did not stem the flow of immigrants into the region, given land scarcity elsewhere in southern Mexico. Farmers returned to upland swidden cultivation for consumption crops and began to engage in the national chili market, as well as other experimental crops (Keys n.d.; Klepeis et al. n.d.).

With the increase in cultivation, the state has begun to implement its new "green" or ecological vision for the region, one in which the forest ecosystem and the Maya ruins among them will derive value from tourism and carbon sequestration (e.g., Primack et al. 1998). Thus the central and western parts of region have been designated as the Calakmul Biosphere Reserve (Figure 1) and the *ejidos* within it made part of the new "green" *municipio* of Calakmul. El Mundo Maya, officially begun in 1993, seeks to develop the infrastructure for superior tourism, complete with new hotels and roads to the ruins (Garrett 1989). In addition, the Mesoamerican Biological Corridor seeks to reconcile farmers' use of the landscape with the movement and dispersal of biota, seeking to ameliorate the environmental impacts of deforestation and cultivation (Miller, Chang, and Johnson 2001).

These episodes are significant to the theoretical base of the study. At the heart of the SYPR project is the exploration of linking or coupling forest and household dynamics in order to understand and model forest change. Household decision making, of course, differs by the degree to which farmsteads engage the market and is mediated by resource institutions, stocks of capital, land, and labor, and off-farm employment opportunities. These qualities of the household have changed significantly during the three episodes noted. Imagery and socioeconomic data limitations, however, restrict model development to the last political-economic episode, such as after the large *bajo* projects, the rise of commercial chili cultivation, and the emerging rules of conservation-preservation activities underway. Thus, farming decision making is undertaken in a quasi-market fashion in which only a portion of the household assets are placed in commercial production and the remainder in subsistence production. The consumer-labor ratio in commercial production, therefore, remains an important element in subsistence production and access to hired labor (Vance and Geoghegan 2001). As well, farming decisions appear to be affected by various Calakmul-related programs and neoliberal programs designed to further develop commercial activities. These programs are under study and are not included in the assessment that follows.

4. FORESTS-*EJIDOS*-HOUSEHOLDS TO PIXELS

The linkage between the household-*ejido* as units of observation and their land-cover impacts, especially forest cover, is central to the aims of the project. TM imagery and aerial photographs were used to gain an understanding of the spatiotemporal dimensions of deforestation in the region, thereby affecting the methods used to stratify and select the *ejidos* for study. A primary aim is to use the household as a means by which to assess not only *ejido* forest–open land dynamics, but regionwide dynamics as well. To accomplish this task requires information on forest ecology taken from plot and transect studies, as well as household-*ejido* and farming information—all information linked to the pixels of TM Landsat imagery. Of course, the household does not make decisions independent of the larger socioeconomic conditions in which it exists (e.g., village land institutions or government subsidies), and these conditions are accounted for by limiting assessments to the last political-economic episode, that of diversification and tourism, in which most *ejidos* had access to the same federal programs. All the *ejidos*, save one, have the same land-tenure system. Some *ejidos* had different levels of attention from NGOs (Klepeis and Chowdhury n.d.), but these differences are not considered in this assessment.

4.1 The Forest Dynamics: The Forest Ecology Design[5]

Forest plot and transect analysis were undertaken to determine the species composition, abundance, and structure of upland and *bajo* forests, as well as successional growth. Two locations, one each in the north and east, were examined through multiple plots and transects, demonstrating the distinction between the two basic forest types (see above) as well as early successional growth (Lawrence and Foster n.d.; Read, Lawrence, and Foster 2001). This work also indicates that, in terms of species abundance, successional growth becomes highly similar to *mediano* after twenty-five to thirty years of regeneration (Pérez-Salicrup n.d.). Likewise, litter and biomass studies in three sites, selected to capture mature to recent successional forests as well precipitation gradients, demonstrate that twelve-to-twenty-five-year-old successional growth matches peak litter fall with older growth upland forest, indicating that twelve years of fallow may constitute the lower end of fallow cycle to permit the recapturing of biomass lost in swidden cultivation (Xuluc-Tolosa et al. n.d.).[6] Presumably, reduced fallows not supplemented by other inputs would not recapture the biomass.

The plots used in the studies above were located by GPS (Geographical Positioning System) and used in classification training (see below). More importantly, once the subregional variations in litter are worked out, the various data can be used to extend across all pixels classified as upland and *bajo* forests as well as various stages of successional growth. Phase two research will include other land covers as well.

4.2 *Ejidos* and Households: Social Science Design and Additional Issues

Entering the middle of the twentieth century, the Mexican government claimed control of virtually all the land in the region, testifying to its sparse occupation. It subsequently used most of it to award *ejidos* to various groups of petitioners, initially from adjacent Maya areas and later from other parts of southern Mexico. Today the region is comprised of four basic landholding units, shown in Figure 2 (after Turner et al. 2001): (1) private holdings—a few small, unproductive ranches; (2) government forest lands; (3) forest-extension *ejidos*—mostly in the west of the region and on which cultivation is not permitted or highly restricted; and (4) agricultural *ejidos* (8,659 km^2). Our project focuses on the last unit because this is where the overwhelming majority of deforestation and other land change takes place.

The ejido is a product of the land reforms of the Mexican Revolution (1910–17), modeled on a simplification of pre-European land holdings in central Mexico and directed to "rebalance" land inequities for rural peasants (smallholders) (Sanderson 1986). Ejidos were established under the same federal act that declared their community structure (no private holders), but variation exists among ejidos in Mexico in terms of the rules of access to land. Ejidos in the study region are relatively new compared to those elsewhere in Mexico, and most of them operate under a system of usufruct (tenure) in which individual ejidatarios maintain more or less permanent use rights to the parcels that they cultivate. This system translates into a practice in which any given parcel (usually a set of contiguous parcels) is controlled by one farmer or household over an extended period of time (e.g., decades) in such a way that household decision making is predicated on the assurance that they control certain parcels. This kind of household control is evident even in those few ejidos in which "informal" recognition of parcel control is absent. Thus, previous and subsequent to recent changes in the Mexican constitution concerning land tenure, discussed below, the farmers of the region have held relatively strong usufruct rights to their parcels. This tie between household and parcels permits land-use decisions to be linked to the geographical locations in which they have impact.7 In addition, the recent changes in the constitution initiated a comprehensive survey and mapping of ejido boundaries, information that allows the linkage of census data at the ejido level with the specific TM pixels associated with that ejido (discussed more fully below).

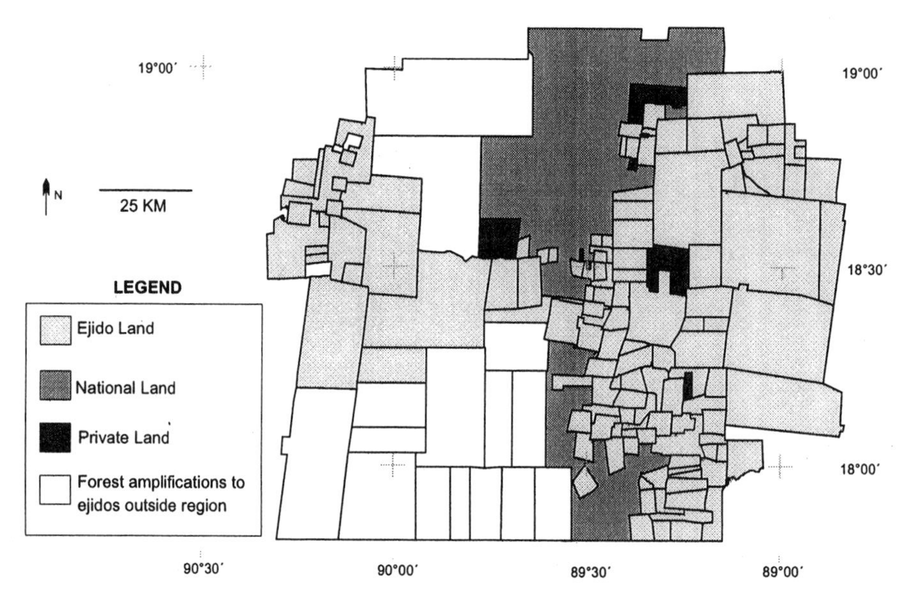

Figure 2. Types of *ejidos* in the study region

A typical *ejido* consists of a settlement cluster or village (in some cases a town), a forest reserve, and household-controlled lands (plots or parcels), creating a landscape of cultivated-fallow lands, a modest amount of pasture land, and *mediana* (older forest).[8] Most land change takes place primarily on household-controlled lands. The number of household parcels and their total land area varies among *ejidos*, typically related to the origins of the *ejido* (established with forestry in mind or not) and the length of tenure of the household within the *ejido*. The average farm household in the region maintains usufruct over 40 ha (land surplus), although this figure drops significantly for the smaller, more recently established *ejidos* (Figure 2).

Within the past several decades, federal rules have required that each *ejido* establish a forest reserve that cannot be cut (although examples of clearing on these lands have been observed). It is also noteworthy that *ejidatarios* are technically restricted from cutting forest growth of more than twenty years, even if it exists on their land. It has not been established by the SYPR project that this rule or that relating to the *ejido* forest reserves is strictly observed.

All households cultivate subsistence crops—primarily maize *(Zea mays L.)*—and do so with minimal capital input. A slash-and-burn or *milpa* strategy is used in which, initially, *bosque mediano* is cut, dried, burned, and cultivated for three to four years and then fallowed for twelve to fifteen years, during which substantial successional forest regrowth helps to replenish soil nutrients and reduce weed and pest disease concentrations (Klepeis et al. n.d.). In such a system, one unit of cultivated land is matched by four to five units in fallow. As noted, however, over 50 percent of the households in the region have begun to cultivate chili for the market. With the exception of mechanized experiments, most of this production is folded within the *milpa* system, save for the use of chemical pesticides, herbicides, and fertilizers during the chili cultivation phase.[9] In most cases, chili is cultivated for two to three years, followed by one to two years of *milpa* before fallowing the land for ten to twelve years, invariably reducing the crop-fallow cycle to 1:3 or less. Importantly, all chili households clear more land than *milpa*-only households, signaling that chili expands cultivation as well as intensifying it (Keys n.d.).

4.3 The Household-*Ejido* Sampling Design and Survey Questionnaire

The project began in 1997 with a Marsh-Clark, Harvard Forest, and ECOSUR workshop, one part of which was devoted to elaborating the institutional and cultural setting of the study region in order to fine-tune our hypotheses concerning land-use change. The choices of explanatory variables are informed by social science theory as to what elements are hypothesized to affect land-use change decisions: for example, in a von Thünenian model, accessibility is hypothesized to affect choice; in a Ricardian model, land quality; in a Chayanovian model, consumer-labor ratio; and in a Boserupian model, land pressures. Local information was essential to the identification of the best measures of these variables, as well as how they might be affected through local institutional conditions. In addition, this expertise was crucial in designing the survey instrument and sampling framework, especially in regard to which kinds of questions would elicit the control and other exogenous variables needed for modeling and testing of hypotheses and to determine the stratification scheme discussed below.

The survey data used in the first phase of the SYPR study were collected by the project over an eleven month period beginning in October of 1997. To accommodate the limitations of sampling frames and control data collection costs, selection of respondents proceeded according to a stratified, two-stage cluster sample, with the first stage unit being *ejidos* (approximately 125 in the region) and the second stage unit being the farm household (Warwick and Lininger 1975; Deaton 1997). Completion of the first stage selections involved compiling an enumeration of *ejidos* in the region with the aid of 1990 cartographic sources and census records, both published by the Mexican government's National Institute of Statistics and Geography (INEGI). Using these maps, the region was partitioned into eleven geographic divisions, or *strata*, and one *ejido* from each of the strata was randomly selected. These strata were chosen to get representation of different "types" of *ejidos* as a result of our initial workshop noted above: for example, wet versus dry, old versus new, and Maya versus migrant *ejidos*. Each *ejido* was assigned a probability of selection equal to the ratio of its population to the population of the stratum. The strata were drawn to ensure that *ejidos* from across the region were represented in the sample, thereby capturing variability in both ecological conditions and in the influence of market and road proximity. Ten out of the eleven *ejidos* selected have assigned parcels in which households have usufruct access to a specific area of land. Settlers from some twenty-three states in Mexico have colonized the southern Yucatán peninsular region, and the sample captures this regional and ethnic diversity.

In the second stage, the survey respondents themselves were randomly selected from each *ejido* after an inventory of households was enumerated. This inventory was secured after establishing contact with the *comisariado* (community leader) of each *ejido*, who, along with the *ejido* assembly, granted permission to proceed with the survey (all selected agreed to the survey). The number of households surveyed from each *ejido* was approximated so that the corresponding stratum was represented in roughly the same proportion as its share of the total population of the study zone. After eliminating twelve records collected during a pilot phase of the survey, the final sample size for phase of study was 188 observations.[10] Of the households selected, 93 percent agreed to participate in the survey.

A standardized questionnaire was used to elicit the socioeconomic and land-use data. The questionnaire was administered to the household head (in Spanish) and was comprised of two sections. The first section, completed at the individual's house, elicited information on migration history, farm production and inputs, off-farm employment participation, land acquisition, and the demographic composition of the household—both current conditions and conditions associated with different dates in the past. Many questions were designed as built-in "checks" on the data for internal consistency. From the survey, approximately 180 different variables were derived (for a brief overview of some of the data from the survey, see Table 2).

Completion of the second section involved a guided tour of the respondent's agricultural plot, a task undertaken by all the participating households. Using a GPS, the interviewer created a georeferenced sketch map detailing the distinct fields within the plot as the farmer-respondent provided an interpretation of the use of these fields.[11] By visiting the respondent's parcels to draw the sketch map, a more open dynamic between researcher and respondent was established than would have emerged by way of less intensive surveys at the farmstead. Via a process akin to "participant observation," the confidence of the respondent in the researcher was heightened, data from the in-house survey was revised, and new information on land-use dynamics was obtained. The sketch map not only documented the spatial configuration of contemporary land uses, including forested areas, but also the land-use transition histories of the principal areas of activity. This information was later used to calculate various indices of land-use change, such as the area deforested since acquisition of the plot, as well as to inform the classification and interpretation of the satellite imagery (Figure 3).

Table 2. Summary statistics from household survey: Socioeconomic characteristics of households in the southern Yucatán peninsular region

Variable	**Mean**	**Standard Deviation**	**Min**	**Max**
Household demographics				
Household size	6.43	3.60	1.00	31
Number of children of head	5.30	3.79	0	27
Age of male head	46.28	15.10	18.00	86.00
Age of female head	40.60	12.90	16.00	80.00
Years of education of male head	2.80	3.16	0	14.00
Years of education of female head	2.93	3.05	0	14.00
Years of education of oldest son if older than 17	6.61	4.12	0	16.00
Years of education of oldest daughter if older than 17	6.26	3.48	0	14.00
Household characteristics				
Percent households having electricity	71	45	—	—
Percent households having cement walls	11	32	—	—
Percent households having cement floor	53	50	—	—
Percent households having potable water	66	48	—	—
Percent households having a radio	27	44	—	—
Percent households having a television	37	48	—	—
Household origin				
Percent originating from Tabasco	16	36	0	1

Table 2. Summary statistics from household survey: Socioeconomic characteristics of households in the southern Yucatán peninsular region (continued)

Variable	Mean	Standard Deviation	Min	Max
Percent originating from Yucatán	8	25	0	1
Percent originating from Chiapas	14	35	0	1
Percent originating from Veracruz	22	41	0	1
Percent originating from Michocan	5	19	0	1
Percent originating from Quintana Roo	6	25	0	1
Percent originating from Campeche	29	45	0	1
Years of occupancy at interview location	19.45	11.38	2.00	58.00
First language of household head				
Percent Spanish	0.70	0.46	0	1
Percent Maya	0.22	0.41	0	1
Percent Tzetal	0.02	0.14	0	1
Percent Totonaco	0.06	0.25	0	1
Farm capital and inputs				
Percent with access to tractor	0.11	0.32	0	1
Percent with access to chain saw	0.37	0.48	0	1
Percent owning vehicle	0.16	0.37	0	1
Percent applying fertilizer	0.46	0.50	0	1
Percent applying insecticide	0.51	0.50	0	1
Percent applying herbicide	0.43	0.50	0	1
Total expenditure for hired labor (pesos)	4,173	12,352	0	120,000
Area in hectares of land endowment	88.23	50.45	10.00	390.50
Ecological indices				
Percent of average plot in upland soils	77	42	—	—
30-year average annual rainfall (mm)	1,093	103	931	1,275
Elevation (meters amsl)	169.46	78.50	34.30	290.57
Incidence and area of principal land uses				
Percent of households with forest cover on plot	100	0	—	—
Hectares forest	51.58	52.25	1.00	297.00
Percent of households planting milpa	98	13	—	—
Hectares milpa	4.51	4.04	.50	30.00
Percent of households planting chili	53	50	—	—
Hectares chili	1.33	1.36	0.06	9.00
Percent of households planting pasture and owning grazing animals	26	44	—	—
Hectares pasture: w/ grazing animals	25.36	28.43	0.50	144.50
Percent of households planting pasture and not owning grazing animals	23	42	—	—
Pasture: w/out grazing animals	11.03	17.17	0.40	95.00
Indicators of access to farm plot and to market				
Distance from *ejido* to nearest market (km)	26.66	21.56	0	82.63
Distance from household to plot (km)	6.59	6.16	0	28.60
Percent accessing plot by foot	42	50	—	—
Percent accessing plot by bike	19	39	—	—

Table 2. Summary statistics from household survey: Socioeconomic characteristics of households in the southern Yucatán peninsular region (continued)

Variable	Mean	Standard Deviation	Min	Max
Percent accessing plot by horse	22	41	—	—
Percent accessing plot by hitchhiking	6	24	—	—
Number of domestic animals				
Cows	3.38	11.12	0	91.00
Turkeys	3.57	5.91	0	30.00
Chickens	14.89	16.76	0	100.00
Pigs	3.56	5.36	0	32.00
Horses	0.87	1.33	0	8.00
Sheep	1.22	5.17	0	45.00
Nonfarm sources of income				
Percent practicing apiculture	25	43	—	—
Percent extracting chicle	8	27	—	—
Wage income (pesos)	10,328	22,313	0	200,000
Income from government credit/subsidy				
Credit (pesos)	2,329	2,236	0	14,850
PROCAMPO subsidy (pesos)	2,050	1,968	0	13,076

4.4 Regional Land Cover: Imagery Analysis and Spatial Design

Landsat TM images from several dates between 1988 and 1997 were used for this study. Given major problems of regional cloud cover, phase one of the project sought to minimize cloud cover in the images rather than to match precisely the chronology of the images. Four scenes have been processed for the project, variously dating from the mid-1980s to the late 1990s (Chowdhury and Schneider n.d.).12 The individual images were first georeferenced to a latitude-longitude projection (Datum NAD27, Mexico) to under 0.5 RMS positional error (Figure 4). Following image registration, the TM scenes were subjected to haze removal using ERDAS Imagine's Tasselled Cap transformation, reducing the haze-related atmospheric noise in each of the dates processed. Using the dehazed red and infrared bands, NDVI images were produced for each date for subsequent analysis. The dehazed images were then processed using Principal Components Analysis (PCA) to complete noise removal and reduce data redundancy. PCA was run separately on the visible and near-infrared bands, and the first three principal components from the separate analysis were retained for further processing. The higher-order components captured random as well as systematic noise, such as striping, in the data and were dropped from further analysis.

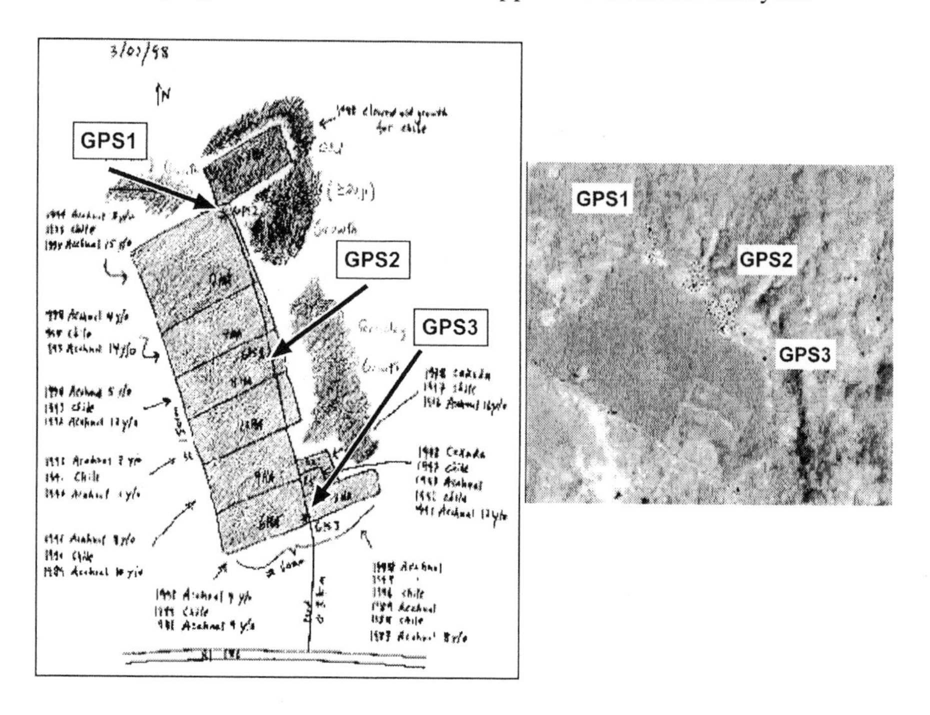

Figure 3. Example of a field sketch map of parcel usage for one household linked to TM imagery (Chowdhury and Schneider n.d.)

The next steps consisted of performing texture analysis on the three-band principal components image. By generating measures of spectral variance in immediate spatial neighborhoods of individual pixels, texture analysis produced three additional bands of information based on the original three-band PCA image. The three texture bands were stacked with the three PCA bands to produce a six-band image. Next, the NDVI image produced from the originally dehazed red and infrared bands was added to the PCA and texture bands to generate a final seven-band image for signature development and classification. Signature development involved the creation of training sites for land-cover classes of the target classification scheme. Ground truth data were derived from a variety of sources: GPS assisted field visits, topographic maps, vegetation and land-use maps, and detailed sketch maps on recent land-use history collected during field research. After the development of training sites, the corresponding land-cover signatures were evaluated using accepted measures of separability—for example, Euclidean distance, divergence, transformed divergence, and Jefferies-Matusita distance—and then further refined. The final signatures were then used in a maximum-likelihood supervised classification to produce categorical land-use/cover maps for change detection and modeling.13

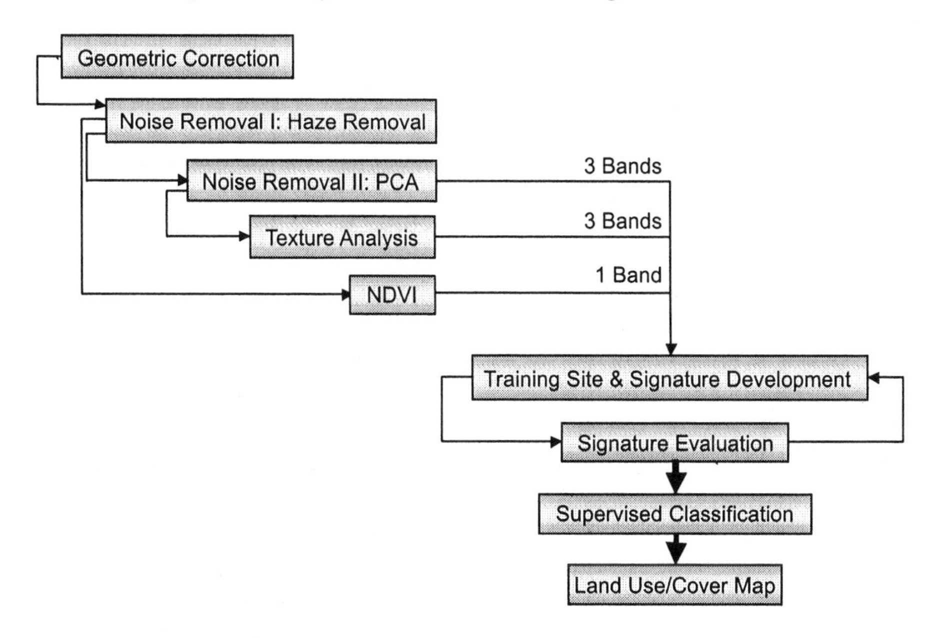

Figure 4. SYPR project image analysis methods (Chowdhury and Schneider n.d.)

A two-stage approach was used in the satellite imagery database, with an attempt to register land-cover classes critical to questions related to land-use decisions (e.g., age of successional growth in the swidden cycle). This effort was assisted strongly by the parcel-level, historical land-use maps generated in the household surveys (Figure 3). The details of land-cover changes in the "sketch" maps permitted intensive training on particular sets of pixels and their transitions from scene to scene. This detail in turn permitted individual household- or parcel-level classification of ten land classes: seasonally inundated semievergreen *bajo* forest *(bajos);* semievergreen upland forest; herbaceous (one to three years secondary growth); shrubby (four to seven years secondary growth); arboreus (seven to fifteen years secondary growth); cropland; pasture; bracken fern; inundated/semi-inundated savannas; and water.

Given the use of several scenes to create regional coverage, the regional variation in pixel signals, and the chronological mismatch of the scenes, the fine-tuned classifications generated at the parcel level could not be replicated at the regional level. The regional or mosaic database, therefore, was reduced to seven land-cover classes: seasonally inundated semievergreen bajo forest; semievergreen upland forest; secondary vegetation (four to fifteen years); agricultural use (cropland, pasture, and one to three years secondary growth), bracken fern, inundated/semi-inundated savannas; and water.

4.5 Additional Data Sources for Spatial Analysis

Other spatial data collected or created for use in the various models include: elevation and slope from a digital elevation model; soil types digitized form a 1:250,000 Mexican government (INEGI) map; digitized road network from INEGI 1:50,000 topographic maps; rainfall data (interpolated to cover the region) from the Mexican government (Secretary of Agricultural and Hydrological Resources); and sociodemographic data from Mexican government censuses. The soil maps for the region are complex, especially in regard to the implications for soil moisture, and offer many geographical "clusters" of soils based on taxonomic order and type, plus other characteristics. The uplands are dominated everywhere in the region by rendzinas or rendolls (mollisol order) of various qualities; this base type forms over limestone, is high in organic matter and soft when dry, and maintains a base saturation of greater than 50 percent. Exceptionally thin and rocky soils, often near the top of hills, are classified as litosols and are largely avoided for cultivation. The *bajos* or seasonal wetland soils are almost everywhere of the order vertisols; such soils are heavy "Maya" clays (in excess of 35 percent clay) that display various degrees of shrinking and swelling according to seasonal moisture flux. For modeling purposes, the many types and subtypes have been aggregated into three that reflect the basic conditions facing farmers: rendzina and other mollisols, litosols, and vertisols.

Only four precipitation-temperature stations exist in the region. To these, sixteen others lying just outside the region were added to create twenty stations or precipitation points and used to create an interpolated map (krigging analysis) of region rainfall (average annual, 1980–98). A federal population census is taken every ten years in Mexico, with the unit of observation being the *ejido*. The data include information on number of households, language and literacy, and structural characteristic of houses, such as electricity and water service (see Table 2). While the level of detail is not ideal, the data do span the entire spatial extent of our study region and constitute the only current source of demographic data for *ejidos* not covered in our survey. Using the GIS map of *ejido* boundaries described previously, the census data could be linked to the pixels associated with each of the *ejidos*. Using the road layers, the distance from each pixel to roads and markets could be calculated. The data also permit other spatial information to extracted and analyzed, including the diversity and fragmentation of land uses around each individual pixel and the distance from each pixel to the nearest agricultural pixel. Such data permit assessments of land-use change according to particular spatial patterns of swidden on the landscape.

5. PROBLEMS OF DATA, DATA COLLECTION, AND MODELING

5.1 Data and Data Collection

For the most part, the SYPR project started from ground zero in data collection, encountering extremely high start-up costs. Despite various projects elsewhere using aerial photography and satellite imagery to create land-cover classifications of parts of the region, no intensive work on forest structure had taken place in the region for over fifty years, and no sustained work of any kind has taken place on forest function and dynamics.[14] This meant that all forest ecology work began more or less from scratch, focused on sampling strategies for intensive plot and transect data. The resulting information was then matched to the TM imagery to create regionwide assessments (e.g., what type of forest occurs where), and the initial "classification" results were laboriously ground-checked by our ECOSUR colleagues. Problems of other environmental data were noted above—primarily the paucity of meteorological stations within the region. Soil maps do cover the entire region, but these were produced largely by linking an unspecified number of ground truth observations and analysis to aerial photography, with the result that interpreted vegetation cover constitutes the source of the soil maps.

Surprisingly, even base boundary information for *ejidos* and other land units was not fully developed and available, in part because the actual surveying of *ejido* boundaries began only recently. RAN (Registro Agrario Nacional), obtained through ECOSUR, provided a first-version map of these boundaries, but the SYPR project had to expend considerable effort correcting various errors (e.g., overlapping *ejido* boundaries) to create a "political unit" map of the region. Likewise, parcel surveys within *ejidos* began only over the past decade, and some *ejidos* have avoided them. As a result, *ejido*-wide parcels that can be linked to households are not available. This project, therefore, had to create sketch maps (located by GPS) of the parcels farmed or controlled by each of the households surveyed in phase one of the study (see above and Figure 3).

Infrastructure information and maps for the region are less than ideal, and the project had to construct these as well. With the exception of the several major improved roads (paved or large but unpaved) in the region, maps depicting the routes and quality of unimproved roads are inaccurate. Owing to canopy coverage, some known roads are not even visible in the TM imagery. Unimproved roads were digitally mapped by way of "best observation"—routes actually traveled by the project members.

Table 3. Summary statistics from 1990 Mexican government census: Average household statistics per *ejido* in the southern Yucatán peninsular region

Variable	**Mean**	**Standard Deviation**	**Min**	**Max**
Total population	274.45	386.70	20	2,916
Male population	144.71	203.91	9	1,529
Female population	129.74	183.17	11	1,387
Alphabetic population older than 15	97.57	168.33	6	1,322
Analphabetic population older than 15	39.87	42.34	2	267
Population at age 5+ not speaking Spanish	5.33	12.57	0	69
Population at age 5+ speaking Spanish	50.12	90.80	0	636
Total number of inhabited houses	53.43	76.27	3	579
Number of private houses with piped water	22.39	65.50	0	486
Number of private houses with sewers	2.39	7.57	0	47
Number of private houses with electricity	27.38	68.67	0	509
Population at age 5 attending school	4.20	5.30	0	29
Population at age 5 not attending school	4.18	4.70	0	25
Population age 6–14 attending school	49.45	56.98	0	334
Population age 6–14 not attending school	18.37	19.69	2	101
Population age 15+ never attending school	36.92	37.28	0	220
Population at age 15+ that attended post-elementary school	10.60	20.14	0	109
Population literate between 6–14 years	48.29	59.82	0	352
Population age 15+ not completing school	50.67	56.84	2	298
Population age 15+ that completed elementary school	17.38	24.27	0	114

Save for the sparse data from population (ejido-level) and agricultural (municipio level) censuses (e.g., Table 3), minimal social data suitable for modeling efforts of the SYPR project exist for the region. This situation required that the project produce these data in phase one through sample-survey methods, a procedure that we hope to replicate in phase two of the project. One major shortfall in the regionwide analysis is the paucity of spatial fine-tuned socioeconomic data on which to assess trends and test hypotheses. The fiscal and time costs of project generation of these data are large.

5.2 Analysis and Modeling

The project is ambitious in its aims, requiring fine-tuned data and imagery classification, as well as analysis across different spatial and temporal scales (Figure 5). Of course, few of the data used in the study are arrayed continuously region-wide such that they can be matched to the 30 x 30 m coverage of the TM imagery. In most cases, therefore, point source (contiguous) data are interpolated or extrapolated to continuous coverage (as for rainfall), or data, such as the population census, was distributed over the unit of data collection (the *ejido*).[15] Note, however, that individual household and parcel analysis encountered less of these problems, but did not provide region-wide coverage. Therefore, the modeling effort up to this point has taken a dual approach: regional modeling using the aggregated data and parcel-level modeling using the survey data (Geoghegan et al. 2001). In addition, the project is undertaking an integrated assessment model for projecting near-term land change under different sets of assumptions, but we defer discussion of this part of the project (Manson 2000, n.d.).[16]

As the land allocation mechanism is not market-based, but rather politically/institutionally-based in Mexico, (i.e., *ejido* land grants from the central Mexican government) the classic spatial von Thünen model of land use - where market equilibrium results from a trade-off of farm-gate prices (agricultural returns minus transportation costs) and agricultural land prices, leading to concentric rings of land use around a market center - is mostly irrelevant as a theoretical modeling structure because there is no land market in this part of Mexico.[17] Therefore, distant-related concepts cannot be relied on as the sole theoretical framework for understanding the location of the agricultural activity of an individual household. Other hypothesis similar to von Thünen, however, can be tested (Chisholm, 1965); for example, are older *ejidos* afforded lands with better access to markets, or were initial members of an *ejido* allocated the "best" lands, or does the location of individual household parcels affect land-use decisions owing to access and transportation costs? Clearly, if a household is engaged in the market, its decisions will be made on his expectation of farm gate prices; if it is a strictly subsistence farmer, then the distance from residence to agricultural plot will likely affect cultivation decisions.

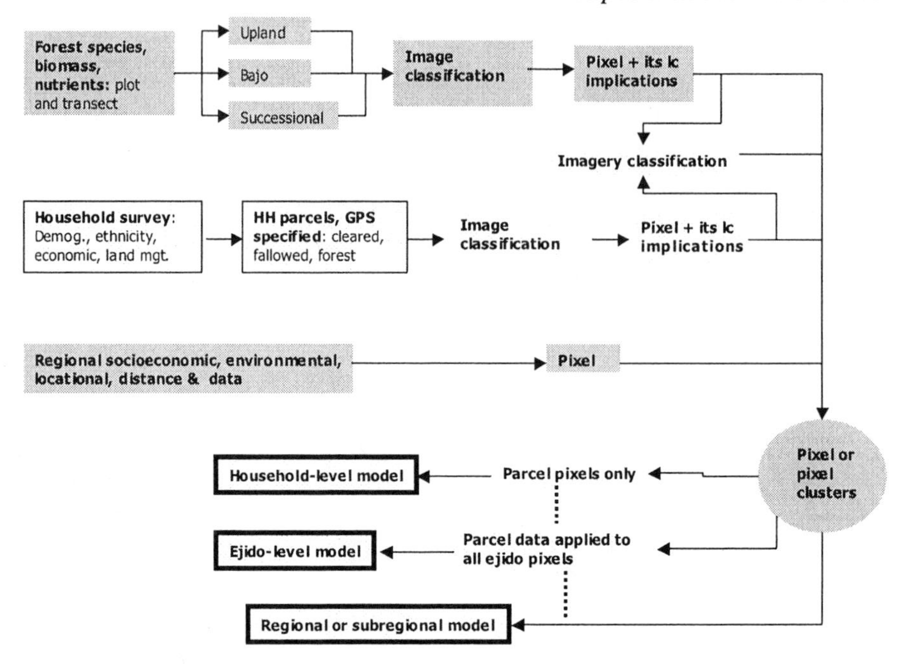

Figure 5. Relationships among SYPR research components, imagery classification, and pixel-level analysis and models

For the empirical specifications of the theoretical models, the largest challenges remain regarding how best to use these spatial data. While GIS has been used to create a number of spatial variables, as previously mentioned (e.g., distance measures and landscape ecology measures), there are likely to be many more imaginative measures that could be created and used in the statistical models. More importantly, whenever spatial data are used, there is a large potential for spatial statistical problems to arise, such as spatial autocorrelation in the residuals, spatial heterogeneity (e.g., nonstationarity over space of the estimated coefficients), spatial dependency (such as a spatial autoregressive process), and so forth. However, for modeling at the pixel level, as has been the main approach taken in this project, then discrete choice models are used (e.g., logit/probit for the aggregate regionwide model and hazard models for the parcel-level models). As the dependent variable is the change in a pixel from one discrete state (e.g., forest cover) to another discrete state (e.g., cropland), the theoretical statistical techniques for these models are not as developed as they have been for regression models; as a result, many of the potential spatial econometric problems cannot be "fixed" as readily as in regression models (Anselin 1988; Anselin and Florax 1995).

6. OPERATIONAL AND PRACTICAL ISSUES AND PROBLEMS

Operational problems invariably abound in an undertaking as large and ambitious as the SYPR Project. In phase one, three institutions were brought together for the first time: Marsh-Clark University, Harvard Forest, and ECOSUR. As the impetus for the NASA LCLUC grant rested in Marsh-Clark, the other units had to be brought on board in terms of a strong structural design. Two major problems were encountered in this regard: (1) the highly integrated character of the research that required specialists in forest ecology, socioeconomic survey and analysis, imagery and GIS, and land-change modeling to come together in special ways; and (2) the extremely high start-up costs of such a group, as well as the high cost of sustained work and analysis.

In regard to the first issue, various specialists had to "learn" the overall project, appreciating how the synergies among the many research parts not only enhanced each part but built toward the overall synthesis of the project—the land-change models. Fortunately, the project had outstanding participants who were open to such integration and synthesis, although the level of involvement needed to reach a "working understanding and integration" of the entire project, varied by individual and unit. Minimally, three years were needed to get all institutions and participants fully integrated and all units marshalling the right set of skills toward the aims of the project. This issue alone constitutes a high start-up cost, but others include the development of linkages to the myriad of federal and state agents and NGOs working in the region as well developing the good will of *ejidos* and *ejidatarios* who are beset by a bevy of researchers in the region, largely focused on activity friendly to the Calakmul Biosphere Reserve and El Mundo Maya landscape preservation. As well, after three years of research, the project is only now confident that it has a nuanced understanding of the human environment dynamics driving deforestation and land-change in the region; that is, we understand what we know well and what we do not know well (and must improve) as we continue to improve existing and develop new project models.

These start-up costs reduced significantly the size of the household survey that the project was able to undertake in phase one of the study. Essentially, two field workers (as part of their dissertation work), with modest assistance, spent nearly a year of intensive household survey activities to generate information on 180 sketch-mapped households. Ideally, the size of survey teams should have been larger, increasing the number of *ejidos* and households sampled. In addition, the survey was time-consuming because the project had to meet with each *ejido* council in question, describing the project and seeking permission to survey the *ejido*. The survey was also encumbered by the need for the farmer to take the surveyor to the household's parcels, in some cases several kilometers or more from the homestead. This requirement meant that special times had to be set up with *ejidatarios* in which it was convenient for them to take the project members to their plots; many appointments were missed or cancelled after the surveyors traveled hours to reach the household.

The availability of remote sensing data was surprisingly less than had been expected at the outset of the study. No cloud-free scenes were available, and those with minimal cloud coverage were few. This circumstance forced the project to select images that were not precisely congruent temporally in order to provide "generalized" coverage of the region for three time frames. Of course, this circumstance increases problems of signature differences between scenes and, thus, classifications of the mosaic pieced together for the region. Phase two of the project is designed to overcome some the problems encountered in this respect. The project was able to obtain aerial photography classifications of the Calakmul Biosphere Reserve from ECSOUR to match against the TM imagery classifications. Importantly, the TM imagery missed the phase of most rapid deforestation in the region—that of the large government projects.

Perhaps more importantly in regard to the remote sensing work, the project began without a senior researcher with remote sensing experience. The addition of such an expert—now on board with the project—would have made the initial imagery assessment more efficient and would have directed the research in ways that certain errors need not be relearned. As well, the project should have begun with attention to multiple sensor data (e.g., AVHRR) by which to add additional information and help fill gaps in the temporal dimensions of the regional assessment.

The project had to create almost all other data in spatial form; few digitally mapped data exist for the region, despite "records" indicating data were on file in various government agencies. These records proved difficult to obtain in some instances. The project also had to create all forest ecology ground data from scratch.

7. SUMMARY, CONCLUSIONS, AND THE FUTURE

The aims of the first phase of work were to combine socioeconomic, ecological, and remote sensing data and analysis (1) to create an understanding of deforestation and agricultural dynamics from the late 1960s onwards, where the socioeconomic data were derived from both government census data and individual household original survey work; (2) to link these dynamics to their spatial outcomes in terms of land covers detected by TM imagery analysis; and (3) to create spatially explicit land-cover change models (Geoghegan et al. 2001; Klepeis and Turner 2001; Turner et al. 2001; Turner, Geoghegan, and Foster n.d.). These aims were ambitious given that the project was essentially starting from scratch datawise and doing so by combining three research units that not only had not worked together previously but whose "land" expertise and interests differed considerably.

The major lesson learned in this effort should be transparent to those who have attempted such programs of study. Nevertheless, we elaborate these below, inasmuch as the global-change, land-change, and sustainability science communities increasingly call for projects of the kind undertaken here.

(1) "Pre–start-up" costs are large and not typically funded. Organizing several units that have not worked together previously is time-consuming, and it is difficult to gain a full measure of interinstitution cooperation previous to the grant application. The kinds of units needed for such research are typically deep into their own research activities and, absent grant preparation funds, getting all units on board intellectually is no easy task. Valuable research time is subsequently expended "reinventing" the project once the award is obtained and in locating extant data. The critical point here is that "integrated land science" of the kind promoted by LUCC (IGBP–IHDP 1995, 1999) and NASA's LCLUC program is complicated and requires considerable upfront investment. SYPR phase two research, of course, is now well positioned to take on a wide range of research questions.

(2) Projects that attempt to link the human, environmental, and remote sensing sciences require a lead investigator for each of the major domains of study. While this observation constitutes common sense, the SYPR project phase one has not been (and will not be) the only project missing a lead expert in one of its domains. As long as this circumstance exists within the integrated land-change research projects, the overall quality of the results—or the efficiency to research robust results—will suffer.

(3) Such projects by their very nature require attention to three different data domains and, when carried out in the developing world, they require that much of the "human" and environmental data be generated by the project. In short, such projects are time and labor intensive and fiscally expensive. Few if any agencies, foundations, or program sources will provide sufficient funding individually to cover these expenses adequately. A major objective, therefore, is to design the project to achieve high levels of efficiency, to reduce the overall aims, or to seek additional funding from other sources. While this observation is obvious, it is worth noting because the breadth and depth of the objectives of land-change science are

incommensurate with the single-source funding involved—and in some cases with the "RFPs" calling for projects.

(4) Almost all of the technical issues that created difficulties in phase one of the SYPR project, which would be approached differently in hindsight, are identified above. It is noteworthy, however, that almost all these difficulties were a product of the absence of "pre–start-up" funding to insure proper institutional input and coordination in the project design (item #1 above) or the limitations to data acquisition and analysis that follow from the programmatic issues raised in items #3 and #4 above.

(5) Despite these obstacles, integrated land science among multiple units and researchers is possible and can produce synthesis or integrated results that are beyond the capacity of a single researcher or research unit to produce. This production is especially acute for complex land-change models, a central element for global environmental change studies, as well as the numerous questions and issues that the sustainability science initiatives, including biocomplexity, seek to address.

NOTES

[1] Reviews and details of the first phase of the project can be found in Turner and colleagues (2001), Geoghegan and colleagues (2001), and Turner, Geoghegan, and Foster (n.d.).

[2] The first phase of research was a collaboration of the George Perkins Marsh Institute (Clark University), Harvard Forest (Harvard University), and El Colegio de la Frontera Sur (ECSOUR, Mexico). The second phase of the work adds the University of Virginia as a collaborator, researchers at the Instituto de Ecología, Universidad Nacional Autonoma de Mexico (UNAM). The comparative parks study is led by Montana State University and the University of Maryland.

[3] Karst is a terrain, usually limestone dominated, in which the principal erosional process is solution: water penetrates the limestone, giving rise to subterranean cavities and surface sinks. As a result, minimal surface water is typical (Jennings 1985).

[4] No studies address the contemporary carrying capacity of swidden cultivation in the region (see Turner, Klepeis, and Schneider n.d.), but Turner (1976) reviewed the literature (past and present cultivation) for the Maya region. He found that estimates for long fallow (forest) did not exceed six people/km^2 and that bush fallow ranged from eleven to twenty-four people/km^2 (Turner 1976: 80). No sustained bush fallow has been observed in the region, however, and various studies of land degradation under fallow have focused on soil nutrient depletion, not weed and pest invasion, which are critical factors in the region. The issue, however, is not strictly defined carry capacities but the agroecological strategies that are employed to offset the impacts of reduced fallow. These strategies may well take the system beyond "classical" swidden.

[5] The SYPR project addresses issues both of ecological and biophysical impacts of land change as well as biophysical impacts on land-use decisions. Given that the workshop and volume for which this study is prepared focuses on social science issues related to coupled human-environment work linked to imagery, we provide information on the ecological issues only as they inform the volume's objectives.

[6] Litter refers to the deadfall of leaves, twigs, stems, bark, and other vegetative materials that fall from the forest and is one indicator of biomass and nutrient cycling. Standing biomass is

registered by measures of the trees themselves (e.g., trunk size, crown height, root mass, and leaf area).

[7] Article 27 of the constitution was amended in 1992 (1) to permit lands formerly held in usufruct under the *ejido* system to be bought and sold, (2) to open the possibility for joint ventures between *ejidos* and private interests, and (3) to terminate the continued distribution of land to peasant communities. The impacts of these changes on land access in the southern Yucatán area are not yet clear and are the subject of ongoing study. As elsewhere in Mexico, *ejidatarios* may receive certificates specifying their rights to the use of *ejido* land, rights which cannot be transferred to another family member or sold to another member of the same *ejido* without permission of the *ejido* assembly. Many if not most of the *ejidos* in the study region grant this land by designating specific parcels to the *ejidatario*, who must use these lands and no others. This "parcelization" is an *ejido* function, not part of the Mexican certification process. If given assembly permission, an *ejidatario* may receive title to the said parcels, in which case the property ceases to belong to the *ejido* and can be bought and sold outside the *ejido* structure. To our knowledge, no privatization by *ejidatarios* has taken place within the southern Yucatán region, although many if not most *ejidatarios* have obtained certificates to their *ejido* lands (Klepeis and Vance n.d.). Informal "renting" takes place, but the project has yet to address this issue in detail.

[8] While we do not deal with pasture here, it is noteworthy that land-surplus households may convert forest to pasture after a few years of cultivation, absent any livestock. The apparent reason rests in a PROCAMPO (government program) subsidy for this purpose. The household assumes that yet another subsidy may be forthcoming from which they will benefit for making this conversion (Klepeis and Vancen.d.).

[9]. Mechanized land is that which is disked by tractor and purportedly undertaken for more or less permanent cultivation. The system is too new to establish the sustainability of this method.

[10] This figure represents as large a sample as could be obtained given the labor and time extended to the first-phase survey and the fact that each survey involved an extensive period of trekking to the household's cultivated parcels where sketch maps were prepared. The survey was undertaken by C. Vance (economist) and P. Klepeis (geographer), both addressing their dissertation research through the project. They were assisted in a few cases by R. Keys and B. Schmook, also doctoral students on the project. Subsequently, another survey of households in the *chilero* zone (the area of high chili activity in the south) was undertaken by Keys (n.d.). Still other surveys are planned by different parts of the project.

[11] Given the remote sensing and ecological work of the project, the different stages of successional growth on abandoned swidden plots were well understood. The enumerators, therefore, could assess the accuracy of the farmers' comments about which parcels were cultivated when.

[12] The scenes and dates are: path 20/row 47 (4/1/87, 4/27/88, 5/8/92, 10/29/94, 5/22/95, 2/5/96); path 19/row 47 (1/14/85, 1/4/87, 11/7/94,1/31/97); path 20/row 48 (12/7/88, 12/13/93/ 2/5/96); path 19/row 48 (11/11/84, 2/21/93, 1/31/97).

[13]. A full validation assessment of the classification is underway in the second phase of the project, as are the uses of other imagery.

[14] See Galendo-Real (1999) and a map of the vegetation communities of the region (http://jasper1.stanford.edu/~jfay/sandler.jpg).

[15] Currently the census data are simply spread over each *ejido* as a uniform distribution. The project currently has point locations of *ejido* population centers, so that future research includes "draping" the census data in some fashion to give more population "weight" to these centers.

[16] The SYPR IA (integrated assessment) model is designed to project short-term land-use/cover change scenarios in the region. Its prime rationale lies in the principles of

integrated assessment: to aid policy development and advance understanding of global change through identification of interrelationships among socioeconomic and biophysical factors. To this end, underlying the model is a conceptual framework based on land manager decision making in relation to socioeconomic institutions and the environment. This conceptual framework is mapped onto a simulation environment based on the dynamic spatial simulation methodology, recognized as "one of the most advanced modelling approaches for a complex, dynamic and spatial problem such as LUCC" (Lambin 1994: 92). The model expands on this method by combining it with other dynamic simulation methods, agent-based modeling, and cellular automata techniques. Regardless of its conceptual and methodological underpinnings, this integrated assessment simulation model is intended to complement other land-use/cover change modeling techniques. It is therefore calibrated with data compatible with other approaches. In the case of SYPR, this includes expert knowledge, socioeconomic information and research, remotely sensed data, and spatial socioeconomic and biophysical data. Key among model outputs is simulated land-use/cover change, which can be validated against other models and fed into other endeavors such as estimating carbon regimes or biodiversity.

[17] The reforms in the Mexican Constitution on land tenure have had little if any impact on the SYPR (Klepeis and Roy Chowdhury n.d.). "Informal" renting of land does take place, but little is known about the pervasiveness of this activity or its details.

REFERENCES

Anselin, L. 1988. *Spatial Econometrics: Methods and Models.* Dordrecht: Kluwer Academic Publishers.

Anselin, L., and R. J. G. M. Florax, eds. 1995. *New Directions in Spatial Econometrics.* Berlin: Springer-Verlag.

Chisholm, M. 1965. *Rural Settlement and Land Use: An Essay in Location.* London: Hutchinson University Library.

Chowdhury, R. R., and L. Schneider. n.d. (forthcoming) "Land-Cover and Land-Use in the Region: Classification and Change Analysis." In B. L. Turner II, J. Geoghegan, and D. Foster, eds., *Integrated Land-Change Science and Tropical Deforestation in the Southern Yucatán: Final Frontiers* (Oxford: Clarendon Press).

Deaton, A. 1997. *The Analysis of Household Surveys: A Microeconometric Approach to Development Policy.* Baltimore: Johns Hopkins University Press.

Galindo-Leal, C. 1999. "The Greater Calakmul Region: Biological Priorities for Conservation and a Proposal to Modify the Biosphere Reserve." Final Report to World Wildlife Fund. Mexico, D.F.

Garrett, W. E. 1989. "La Ruta Maya." *National Geographic* 176: 424–479.

Geoghegan, J., S. C. Villar, P. Klepeis, P. M. Mendoza, Y. Ogneva-Himmelberger, R. R. Chowdhury, B. L. Turner II, and C. Vance. 2001. "Modeling Tropical Deforestation in the Southern Yucatán Peninsular Region: Comparing Survey and Satellite Data." *Agroecosystems and Environment* 85: 25–46

IGBP–IHDP. 1995. "Land-Use and Land-Cover Change Science/Research Plan. IGBP Report No. 35 & HDP Report No. 7. Stockholm: IGBP and Geneva: HDP.

———. 1999. "Land-Use and Land-Cover Change (LUCC) Implementation Strategy." IGBP Report No. 48 & IHDP Report No. 10. Stockholm: IGBP and Bonn: HDP.

Jennings, J. N. 1985. *Karst Geomorphology.* Oxford: Basil Blackwell.

Keys, E. n.d. (forthcoming) "Jalapeño Pepper Cultivation: Emergent Commercial Land-Use of the Region." In,B. L. Turner II, J. Geoghegan, and D. Foster, eds., *Integrated Land-*

Change Science and Tropical Deforestation in the Southern Yucatán: Final Frontiers (Oxford: Clarendon Press).

Klepeis, P. 1999. "Deforesting the Once Deforested: Land Transformation in Southeastern Mexico." Ph.D. dissertation, Clark University, Worcester, MA.

Klepeis, P., and R. R. Chowdhury. n.d. (forthcoming) "Institutions, Organizations, and Policy Affecting Land Change: Complexity within and beyond the *Ejido*." In B. L. Turner II, J. Geoghegan, and D. Foster, eds., *Integrated Land-Change Science and Tropical Deforestation in the Southern Yucatán: Final Frontiers* (Oxford: Clarendon Press).

Klepeis, P., and B. L. Turner II. 2001. "Integrated Land History and Global Change Science: The Example of the Southern Yucatán Peninsular Region Project." *Land Use Policy* 18: 272–309.

Klepeis, P., and C. Vance. n.d. "Neoliberal Reform and Deforestation in Southeastern Mexico: Land Reform, PROCAMPO, and Land Use Change." Manuscript.

Klepeis, P., C. Vance, E. Keys, P. M. Mendoza, and B. L. Turner II. n.d. (forthcoming) "Subsistence Sustained: Swidden or *Milpa* Cultivation." In B. L. Turner II, J. Geoghegan, and D. Foster, eds., *Integrated Land-Change Science and Tropical Deforestation in the Southern Yucatán: Final Frontiers* (Oxford: Clarendon Press).

Lambin, E. 1994. *Modeling Deforestation Processes: A Review*. Luxembourg: European Commission.

Lawrence, D., and D. Foster. 2001. "Determinants of Regional Variability in Litter Production of Forests in the Southern Yucatan: Environmental Gradients or Human Legacy?" *Eos. Trans. AGU* 82 (20), Spring Meet. Suppl. Abstract B22A-01.

———. n.d. (forthcoming) "Recovery of Nutrient Cycling and Ecosystem Properties following Swidden Cultivation: Regional and Stand-Level Constraints." In B. L. Turner II, J. Geoghegan, and D. Foster, eds., *Integrated Land-Change Science and Tropical Deforestation in the Southern Yucatán: Final Frontiers* (Oxford: Clarendon Press).

Lundell, C. L. 1934. "Preliminary Sketch of the Phytogeography of the Yucatan Peninsula. *Contributions to American Archaeology* 12 (436). Washington, D.C.: Carnegie Institute.

Manson, S. M. 2000. "Agent-Based Dynamic Spatial Simulation of Land-Use/Cover Change in the Yucatán Peninsula, Mexico." Paper presented at Fourth International Conference on Integrating GIS and Environmental Modeling (GIS/EM4), Banff, Canada.

———. n.d. (forthcoming) "The SYPR Integrative Assessment Model: Complexity in Development." In B. L. Turner II, J. Geoghegan, and D. Foster, eds., *Integrated Land-Change Science and Tropical Deforestation in the Southern Yucatán: Final Frontiers* (Oxford: Clarendon Press).

Miller, K., E. Chang, and N. Johnson. 2001. *Defining Common Ground for the Mesoamerican Biological Corridor*. Washington, D.C.: World Resources Institute.

Pérez-Salicrup, D. n.d. (forthcoming) "Forest Types and Their Implications." In B. L. Turner II, J. Geoghegan, and D. Foster, eds., *Integrated Land-Change Science and Tropical Deforestation in the Southern Yucatán: Final Frontiers* (Oxford: Clarendon Press).

Primack, R. B., D. Bray, H. A. Galletti, and I. Ponciano, eds. 1998. *Timber, Tourists, and Temples: Conservation and Development in the Maya Forests of Belize, Guatemala, and Mexico*. Washington, D.C.: Island Press.

Read, L., D. Lawrence, and D. Foster. 2001. "Litter Nutrient Dynamics in the Secondary Dry Tropical Forests of the Southern Yucatan." *Eos. Trans. AGU* 82 (20), Spring Meet. Suppl. Abstract B32A-13.

Sanderson, S. E. 1986. *The Transformation of Mexican Agriculture: International Struggle and the Politics of Rural Change*. Princeton, NJ: Princeton University Press.

Stedman-Edwards, P. 1997. "Socioeconomic Root Causes of Biodiversity Loss: The Case of Calakmul, Mexico." Mexico, D.F.: World Wildlife Fund. Unpublished report.

Turner, B. L. II. 1976. "Population Density in the Classic Maya Lowlands: New Evidence for Old Approaches." *Geographical Review* 66: 73–82.

———. 1983. *Once Beneath the Forest: Prehistoric Terracing in the Rio Bec Region of the Maya Lowlands.* Dellplain Latin American Studies No. 13. Boulder, CO: Westview Press.

Turner, B. L. II, S. C. Villar, D. Foster, J. Geoghegan, E. Keys, P. Klepeis, D. Lawrence, P. M. Mendoza, S. Manson, Y. Ogneva-Himmelberger, A. B. Plotkin, D. Pérez Salicrup, R. R. Chowdhury, B. Savitsky, L. Schneider, B. Schmook, and C. Vance. 2001. "Deforestation in the Southern Yucatán Peninsular Region: An Integrative Approach." *Forest Ecology and Management* 154: 343–370.

Turner, B. L. II, J. Geoghegan, and D. Foster, eds. n.d. (forthcoming) *Integrated Land-Change Science and Tropical Deforestation in the Southern Yucatán: Final Frontiers.* Oxford: Clarendon Press.

Turner, B. L. II, P. Klepeis, and L. Schneider. n.d. (forthcoming) "Three Millennia in the Southern Yucatán Peninsular Region: Implications for Occupancy, Use, and 'Carrying Capacity.'" In A. Gómez-Pompa, M. Allen, S. Fedick, and J. Jimenez-Osornio, eds., *The Lowland Maya Area: Three Millennia at the Human-Wildland Interface* (New York: Haworth Press).

Vance, C., P. Klepeis, B. Schmook, and E. Keys. n.d. (forthcoming) "The Ejido Household: The Current Agent of Change." In B. L. Turner II, J. Geoghegan, and D. Foster, eds, *Integrated Land-Change Science and Tropical Deforestation in the Southern Yucatán: Final Frontiers* (Oxford: Clarendon Press).

Vance C., and J. Geoghegan. n.d. "Semi-Subsistent and Commercial Land-Use Determinants in an Agricultural Frontier of Southern Mexico: A Switching Regression Approach." Manuscript.

Warwick, D. P., and C. A.Lininger. 1975. *The Sample Survey: Theory and Practice.* New York: McGraw Hill.

Xuluc-Tolosa, F. J., H. F. M. Vester, N. Ramirez-Marcial, J. Castellanos-Albores, and D. Lawrence. n.d. (forthcoming) "Leaf Litter Decomposition of Tree Species in Three Successional Phases of Tropical Dry Secondary Forest." Forthcoming in *Forest Ecology and Management.*

Chapter 3

HOUSEHOLD DEMOGRAPHIC STRUCTURE AND ITS RELATIONSHIP TO DEFORESTATION IN THE AMAZON BASIN

Emilio F. Moran
Anthropological Center for Training and Research on Global Environmental Change, Indiana University
moran@indiana.edu
Andréa Siqueira
Anthropological Center for Training and Research on Global Environmental Change, Indiana University
Eduardo Brondizio
Anthropological Center for Training and Research on Global Environmental Change, Indiana University

Abstract The greatest challenge to theory in human ecology has been how to define the unit of study so that it can be reasonably well studied, while at the same time not losing sight of the larger whole or ecosystem within which human beings interact with their biophysical surroundings. Some relevant theoretical approaches used to address particular questions include those aiming at explaining patterns of agriculture intensification (e.g., Boserupian and Von-Thunenian models), and household development cycles (e.g., Chaianovian models). In this project we ask what the role of household demographic structure on observed rates of deforestation might be. The study of human impacts on land cover can follow any number of approaches. What we have found most useful is to take a multi-scaled approach that examines at each level of aggregation both biophysical and socioeconomic variables. We have used a variety of methods of data collection and mined a variety of data sources: time-series Landsat satellite data; survey research; stratified random sampling of properties; registering property boundaries onto satellite image time series in a geographic information system; carried out soil and vegetation stand sampling with precise coordinates using GPS; examined the reproductive histories of women and their decisions using survey research at the household level; and obtained time series price data and other economic statistical time-series. A key goal of this study was to understand whether trajectories of deforestation could be better understood knowing the age and general structure of households through time, rather than just in aggregate number.

We think that our study does show the technical feasibility of examining land use and land cover change at the level of households and properties—and that the insights are worth the effort and investment required to achieve it.

Keywords: Population and environment, Amazon, household demographic structure, deforestation, land use, land cover

1. INTRODUCTION

Our research on population and the environment evolved from earlier interests some of us have had in government-directed colonization into the Amazon Basin, the process of adaptation by migrants to a new biophysical and social environment, and on trajectories of deforestation (Moran 1976, 1981, 1987, 1990, 1993). Theoretically we have been guided by a set of theories generally referred to as human ecology (Moran 1979, 2000; Moran and Brondizio 2001), and more recently as environmental social science (Moran in preparation). Its antecedents are the work of geographers and anthropologists such as Julian Steward (1955), Robert Netting (1968, 1981), Carl Sauer (1958), William Denevan (1976), Karl Butzer (1980), Roy Rappaport (1967), and others. The greatest challenge to theory in human ecology has been how to define the unit of study so that it can be reasonably well studied, while at the same time not losing sight of the larger whole or ecosystem within which human beings interact with their biophysical surroundings. Several solutions have been offered in the past: the use of a cultural area as the equivalent of a biogeographical area (Kroeber 1939); the use of social organization for subsistence as a core set of variables (Steward 1955); and the ecosystem as a unit of analysis (Rappaport 1967; Moran 1990, for a review of the ecosystem literature). From the onset we realized that no "monolithic" theory could account for human decisions and land use change in the region. However, some relevant theoretical approaches used to address particular questions include those aiming at explaining patterns of agriculture intensification (e.g., Boserupian and Von-Thunenian models), and household development cycles (e.g., Chajanovian models).

In this particular project we were driven to ask what the role of household demographic structure on observed rates of deforestation might be. This question was suggested by earlier work in the region (Moran 1976, 1981) in which it was observed that younger households pursued very different land use strategies than middle aged and aging households. In a frontier region, where labor is generally scarce, the number of working age members might reasonably be inferred to play a key role in how much labor a household can muster for farm work and thus which strategies are likely to be chosen. Yet, it was also observed that households pursued a more intense process of deforestation at the outset, when they had least labor. The result of these observations in the field, in the course of studying issues of adaptation and trajectories of secondary succession, led to our current project on population and environment.

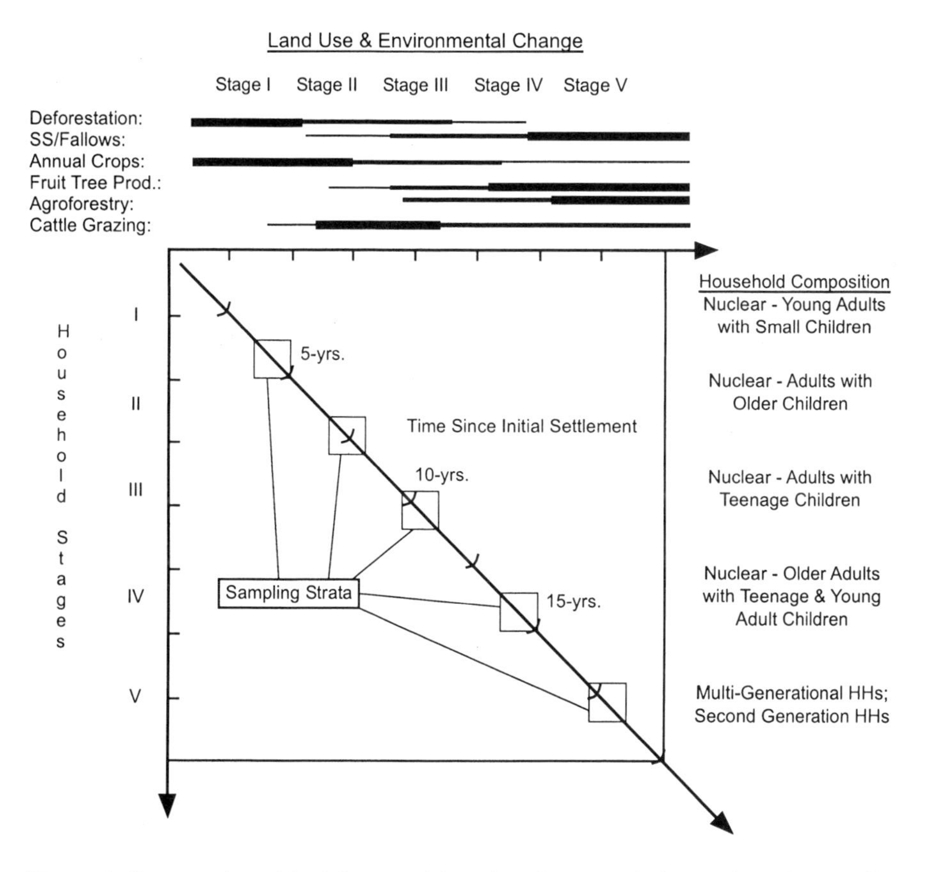

Figure 1. Conceptual model of demographic and environmental change (Brondizio et al. In press)

Figure 1 illustrates the conceptual model that guided this research linking household demographic structure to the deforestation and land use behavior of households. The model posits that there is a developmental cycle (Goody 1962, 1976) resulting from the changing age and gender composition of the household over time. The timing and magnitude of these reproductive decisions are expected to affect how households relate to environment and economy. The model proposes that younger households, with very young children and low supplies of capital, will focus on annual crops in the frontier as a way of building up their capital stock, and as a way to transform the biomass of the forest into fertilizer for fast growing crops. Over time, we expected these households to shift from annual crops to pasture and more cash and labor demanding activities such as cash crops and permanent tree crops, such as mahogany, cocoa, pepper, and sugar cane. As the young people in the household reach marriageable age, we expect the household heads to shift again towards less labor demanding crops, and to crops, which ensure them of regular cash flow. We expect with another generation taking over the farm that the process will be repeated, but limited by the much smaller supply of forest and the presence of permanent crops and pastures as initial conditions.

The study of human impacts on land cover can follow any number of approaches. What we have found most useful is to take a multi-scaled approach that examines at each level of aggregation both biophysical and socioeconomic variables. We have used a variety of methods of data collection and mined a variety of data sources: time-series satellite data such as Landsat; survey research; stratified random sampling of properties; registering property boundaries onto satellite image time series in a geographic information system; carried out soil and vegetation stand sampling with precise coordinates using GPS; examined the reproductive histories of women and their decisions using survey research at the household level; and obtained time series price data and other economic statistical time-series. We will examine these design questions in the pages that follow.

2. RESEARCH DESIGN

The project began with one great advantage in that the PI and co-PIs had years of experience in the region, and had already collected substantial biophysical and social data over a period of several years. We developed detailed measurements to characterize land cover structure and composition to inform spectral data analysis. We emphasized the most dynamic land cover classes -- stages of secondary succession and pasture. These are key land cover classes that inform the analysis of land use trajectories associated with agropastoral cycles. Detail vegetation inventories of 25 areas representing land cover classes in the area were undertaken. In each vegetation area, we used nested plots (n=10) of 150 square meters, with subplots (n=10) of ten square meters randomly distributed. These plots were used to measure (height, DBH, number of individuals) and identify trees, and saplings and seedlings, respectively. Using these inventoried areas to provide structural parameters defining types of land cover, more than 300 training samples were collected to be used during image classification (see Moran and Brondizio 1998).

Thus, we already had a reasonably good idea of the trajectories of deforestation at the landscape level between 1985 and 1991. We had good survey data on histories of land use in the regions where vegetation and soil sampling had been carried out. This work gave us substantial insights into the role of soils and land use on the rate of secondary succession or regrowth following deforestation, and we were able to demonstrate the feasibility of monitoring stages of secondary succession in the Amazon using a combination of field studies and satellite remote sensing (Mausel et al. 1993; Moran et al. 1994; Brondizio et al. 1994; Moran and Brondizio 1998, 2001). This work was at the landscape level and provided important insights. However, we felt that to understand decisions about deforestation we had to focus more on household structure and household decision-making and thus our project grew to incorporate this dimension.

2.1 Study Area

The original government colonization scheme proposed to build a road across the Amazon Basin, the Transamazon Highway, and it also included building planned communities/villages (agrovilas) every 10 kilometers along the main road, and also 10 km inside the feeder roads. The initial plan was to have in each village, a community center where basic services (a primary school, a health post staffed by a nurse, an ecumenical church, a water tower to ensure supply, and some agricultural storage) were provided to farmers and their families. Ideally, the farm plots were to be located at a distance no greater than 4 km from the village within which the farmers resided. However, this idea's scheme was subverted by the decision to let farmers choose their own properties, resulting in many cases, in farms ending up as much as 20 km from the village. In addition, the communities lacked any basis for social cohesion: families came from very different regions of Brazil, belonged to very different religious denominations which were unfriendly to each other and unwilling to be ecumenical or planned for, the health and education posts often went unfilled, and the water towers failed to guarantee a supply of potable water. Over time, families began to abandon these poorly-serviced villages and began to settle on the farm properties to make their commute easier.

The conception of community in any frontier site must by its very nature be fluid. Church membership and region of origin can help define a sense of community. But in an area with so many people from different regions and religions, the sense of community emerged slowly. Few things acted to bring members of the villages together into a "community." Over time, most of these villages have been abandoned as farm communities and they have been occupied by single workers or landless relatives. They are, in fact, dorm communities with very little social cohesion. Many of them have been entirely abandoned, especially those in the side roads.

A key element in our research strategy was an early methodological decision. We had obtained in the past a property grid used by the colonization agency in keeping track of the land allotments to settlers. Since in this colonization project the colonization agency gave each household one property in the settlement scheme, there was a one-to-one relationship between households and properties, so that by observing changes in a property using satellite images, we could begin to observe the outcome of household decisions over time in a spatially explicit fashion. So one important first step in our research was figuring out a way to overlay the property grid onto the satellite images (see Figure 2a, b) (which themselves were overlaid on each other in a time series) with sufficient accuracy that we could examine change in land cover at both the landscape and the property level. The property grid added an entirely new capability to the understanding of land cover changes over time. Observation of the Landsat image provides a good idea of changes but these cannot be tied to any particular household. With the overlay of the grid, and of an ID for each property, it becomes possible to query land cover change on any particular property.

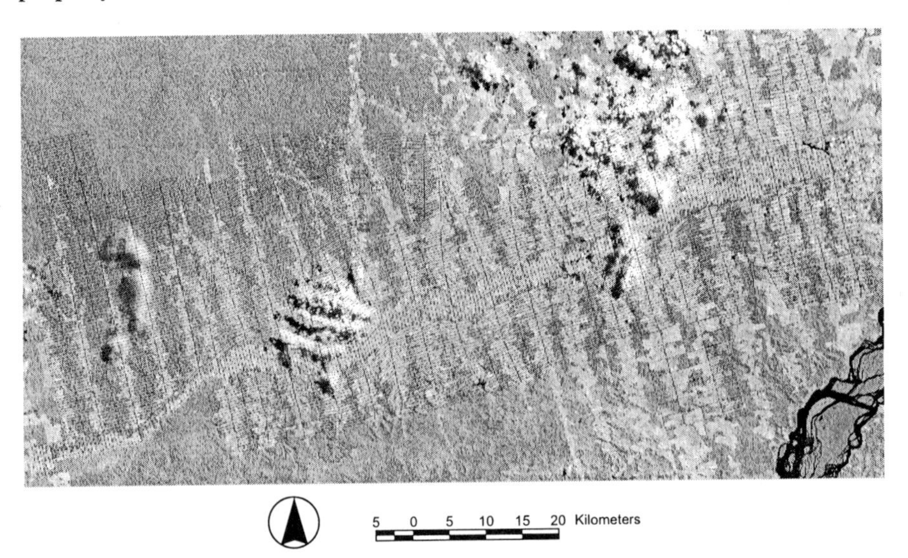

Figure 2a. Property grid and landscape (after McCracken et al. In press [b])

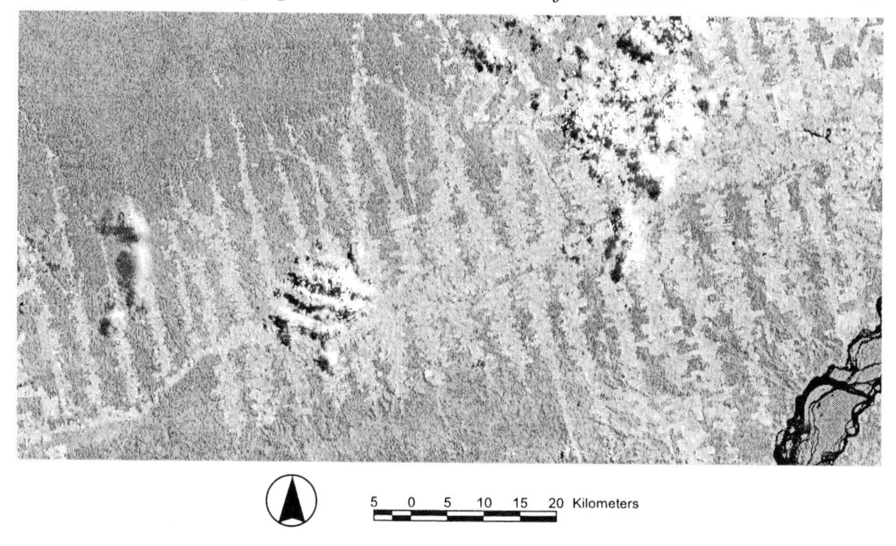

Figure 2b. Landscape (1996 Landsat TM Image, Bands 5, 4, 3) (after Boucek and Moran In press)

This process proved to be far more time consuming than we originally imagined, but this was not the result of technical failures, or of taking the wrong direction in the work. The property grid we had obtained proved to be an "ideal" view of the landscape and the properties. It did not take into account topographic features, rivers, and other natural features. In reality, as land was occupied, owners and neighbors negotiated changes in the precise boundaries of many properties when this made sense for their farm production. The "ideal" properties, imagined to be exactly 100 hectares, have, over time, become variable in size from 84 to 123 hectares, and their perfect rectangular shape has experienced realistic rounding (see Figure 3). Thus we were trying to fit a grid that was not the real grid. We solved this problem through very large investments of GPS work and visual on-screen modifications based on how land use evolved in the time series to bring out the real boundaries between properties, so that the final grid we have at present is well over 90% accurate. This is no mean feat, considering that this is an area of 3,784 properties or a total of 3,800 square kilometers. We have detailed the technical details followed in another paper (McCracken et al. 1999 and McCracken et al. in press [a]).

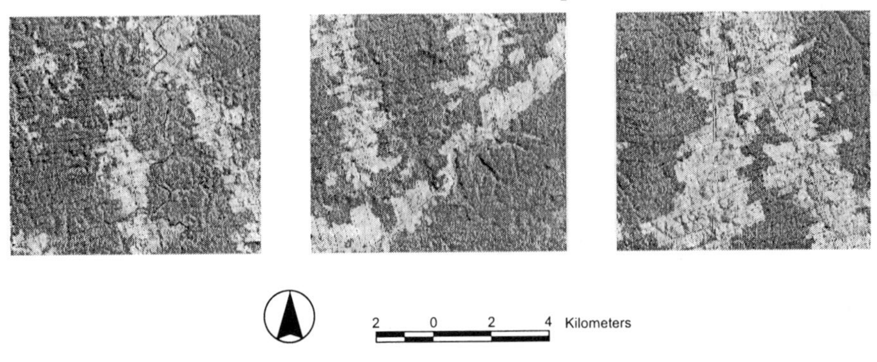

Figure 3. Examples of atypical properties

The approach we took can not be used in places that lack a one-to-one relationship between household and property, given the all too common reality of people living in villages and walking to their fields, the uneven size and commonly fragmented nature of landholdings in many other cases with a household having many very small holdings, and the presence of very diverse systems of land tenure in the same area. In our study area we benefited from a system based on individual private land tenure (although titling is a problem), of most households having only one property and living right on the property. But our study area is hardly unique. We have used the same methods to study a two settlement region in western Amazonia in Rondônia (Batistella 2001), a traditional peasant region that has combined common and private property (Futemma 2000), and in a region of south Central Brazil at Alta Floresta, Mato Grosso (Oliveira Filho 2001). The lessons learned in Altamira were applied and speeded up the work in these other regions.

Ideally, a time series should (a) cover the duration of the settlement; (b) cover consistent intervals to observe dominant land use systems; and, (c) capture inter-annual agropastoral cycles to allow observation of crop decisions following deforestation events.

The temporal scope of the study was extended backwards to 1970 before the land settlement scheme began to bring colonists, and forward to 1996 (covering therefore a period of 25 years). We were unable to obtain cloud-free images of the study area for 1997 or 1998 when field research took place. Similarly, the deterioration of Landsat MSS image archives has limited the availability of digital data covering the period of 1972 to 1984. Fortunately, photographic print outs of Landsat MSS products were available at the regional development agency (SUDAM).

Aerial photos for 1970 and 1978, together with Landsat MSS images for 1973, 1975,1976, and 1979, and Landsat TM images for 1985, 1988, 1991 and 1996 provided a dramatic time series of the trajectory of deforestation (see Figure 4).

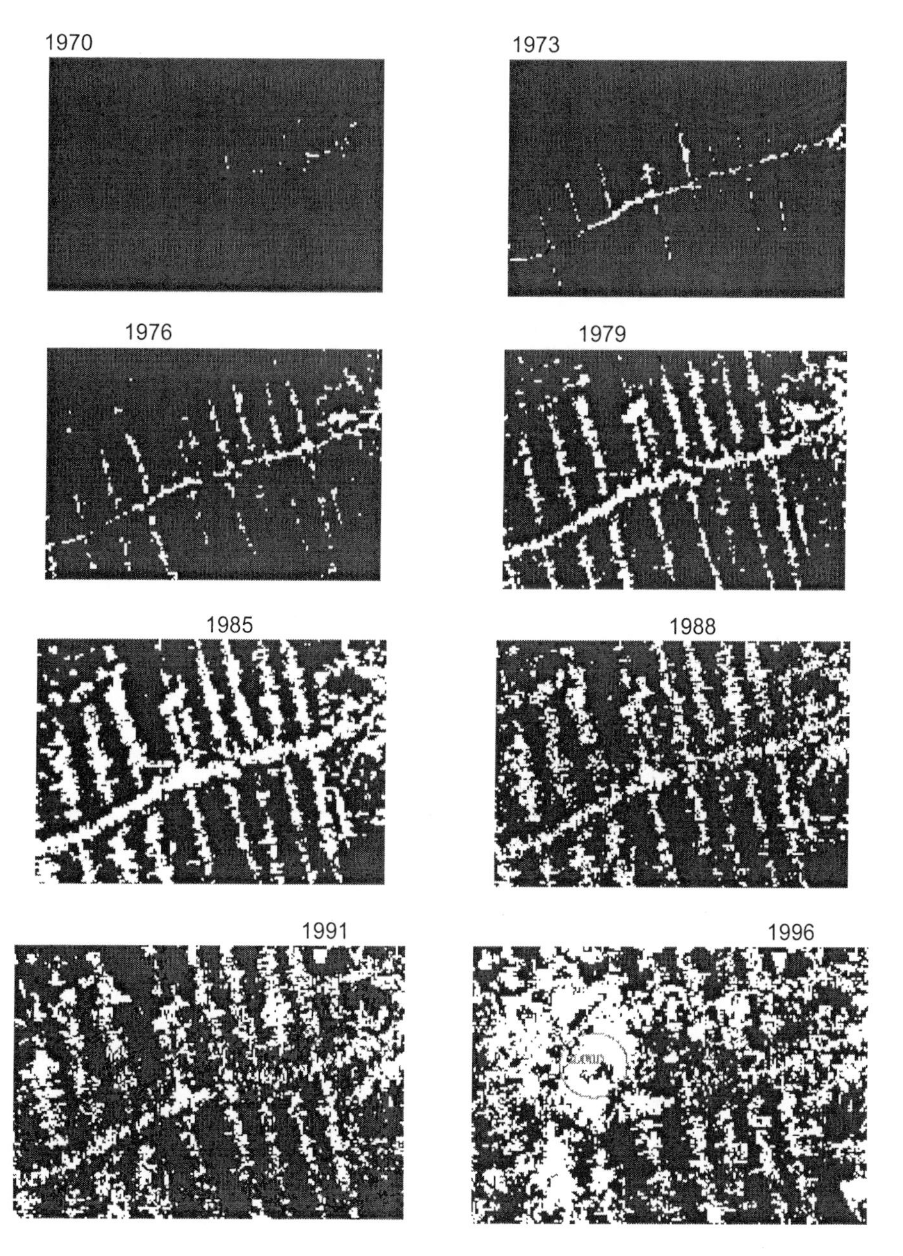

Figure 4. Time series of the trajectory of deforestation in Altamira (from Moran and McCracken, submitted manuscript)

Training samples representing land cover classes were evaluated using statistical and spectral analysis (e.g., separability analysis) to ensure proper aggregation of samples and accuracy during classification. Image classification was based on a hybrid approach including pre-field unsupervised classifications and incorporation of training samples and unsupervised spectral signatures. We relied on a spatial-spectral classifier ECHO and Maximum Likelihood algorithms (see Mausel et al. 1993, Moran et al. 1994, Brondizio et al. 1996, McCracken et al. 1999). It was possible to distinguish among three stages of secondary succession representing initial, intermediate, and advanced phases of regrowth. Initial classification aiming at distinguishing early stages (e.g., one-two year-old regrowth from three-five year-old regrowth) and advanced stages from mature forest proved difficult. To ensure a minimum classification accuracy of 85% across all classes we merged five classes of secondary succession into the three classes presented here. We used standard Kappa accuracy measures based on test fields. Test fields are areas initially collected as training samples, but not used during classification. Overall image classification accuracy is around 88% for the images of 1996 and 1991 and 85% for the images of 1988 and 1985. However, classification accuracy varied from 92% to 85% among land cover classes. Classification accuracy for our Landsat MSS (1973, 1975, 1976, and 1979) and aerial photography (1970, 1978) was not possible to calculate using test field procedures. For these dates, we defined four land-cover classes (forest, non-forest, roads, and water), which allowed us a higher mapping accuracy than otherwise possible with the inclusion of secondary succession stages. Accuracy was further assured during the development of transition matrices across dates. Inconsistent transitions (for instance from non-forest to forest in few years) were re-coded properly to correct for spatial mismatch.

The accuracy of our property grid varies within the region, with an overall rate of 90%. Accuracy was lower (around 80%) in areas of variable topography as well as in areas where the definition of lots was not closely followed by INCRA. In this area, lots and the road network tend to assume various shapes making if difficult to guarantee a perfect match between our vector layer and field reality. Property grid accuracy is higher ~95% for the farm lots where households members were interviewed (n=402).

Another key methodological decision made early on, which shaped the entire study was to try to distinguish between period and cohort effects. Period effects are those effects resulting from events that all actors experience equally, such as hyperinflation in the national economy or a major shift in world commodity prices relevant to a population. Cohort effects are those specific to a cohort but not necessarily experienced by other cohorts. Since the population migrating to the Amazon came and is still arriving, it is possible to distinguish cohorts of arriving settlers. Those arriving in 1971 experienced a very different set of conditions than those who came in 1985 or in 1996.

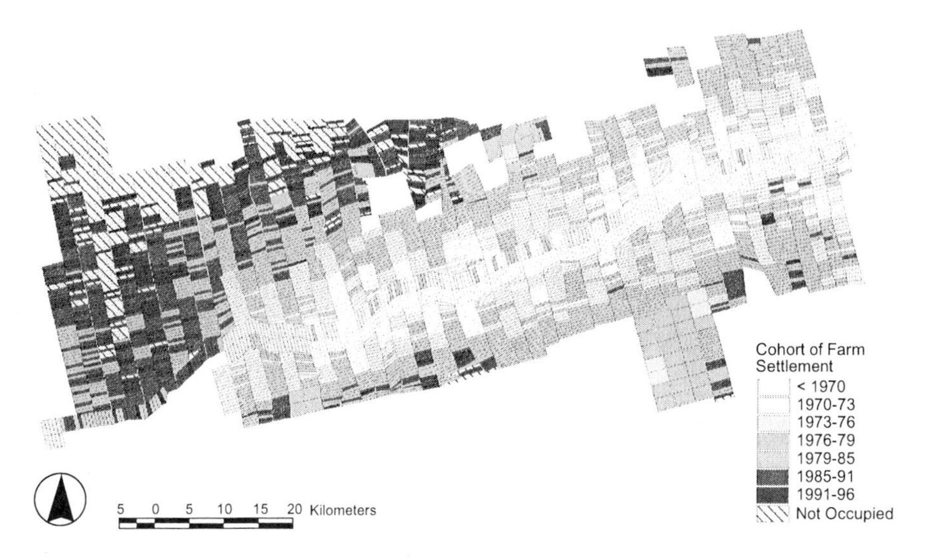

Figure 5. Farm property cohort of settlement (after McCracken et al. In press [b])

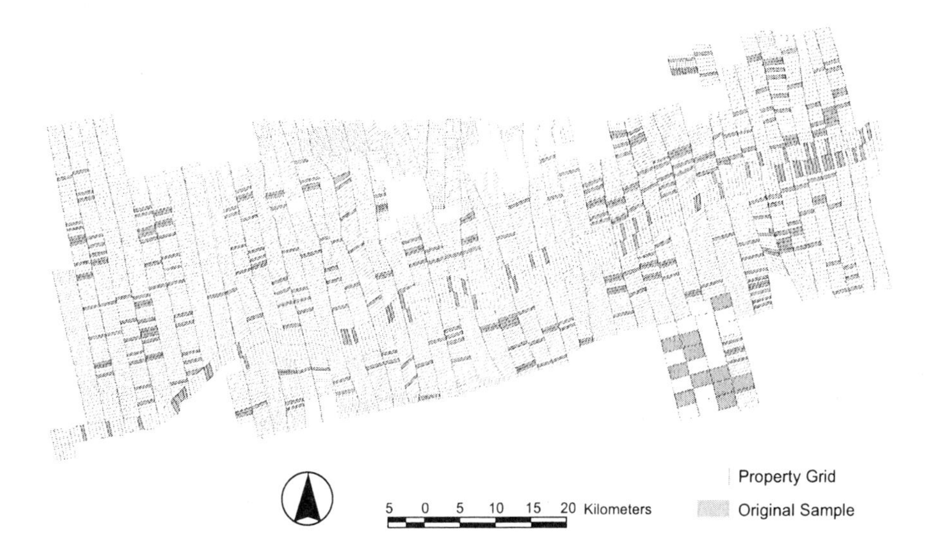

Figure 6. Property sampling strategy

For the purposes of our study, we decided to determine the membership of a household in a particular cohort by when we were able to observe on the satellite image the clearing of at least 5 hectares of mature forest. This was based on our earlier work in the Amazon wherein we had noted that most households cleared a minimum of 3 hectares per year. Since our satellite time series was roughly spaced in three year intervals, 5 hectares would indeed be a minimum that a household could be expected to clear, thereby announcing their arrival on the property—and membership in a given cohort. In this fashion, we were able to define 8 cohorts. Our stratified sample was based on the analysis of five land cover classifications and their respective land cover transitions (1970, 1978, 1985, 1988, 1991). This allowed for the development of a stratified sampling frame for selecting properties and households based on (a) timing of settlement based on period of initial forest clearing, and (b) extent of deforestation as of 1991. This method proved useful to allow a sampling frame compatible with our conceptual model of life cycle of households, agriculture strategies, and deforestation levels. The size of each cohort was quite variable because of the heavy involvement of government in the 1971-76 period in bringing colonists to the area at an accelerated rate, and due to shifts in the economy and in the presence of alternative areas for migrants to go to. Because some of the cohorts were small, we chose to draw a disproportionate sample to ensure a sufficient number of households from each cohort in the final sample (Figures 5 and 6).

A potential problem was the inclusion of "replacement households" where families that originally opened the lot sold the lot and moved away. To overcome this problem, we developed an alternate sampling frame consistent with the goals of our sampling design. The use of an "alternate" sampling frame proved very important to guarantee work continuity during the day (that is, to find neighboring farms that could be used for replacement during a interview trip).

Households in the study area are comprised of those sharing the same roof. We considered households as joint economic units and our survey did not attempt to grasp possible conflicts between household members related to access and allocation of resources and processes of decision-making. However, we included questions in our survey referring to outside members (usually grown up children and relatives) and their contribution in remittances, work, food, and gifts, if any, to the focus household in order to understand possible flows of labor, goods and resources among households. Keeping track of changes on household composition through time was possible with retrospective histories of household membership. One problem we faced while reconstructing household/family history was to match it with the farm's land use history. Besides being a region with already a significant settled population, property turnovers occur with relative frequency.

The size of our sample was based on our past experience in the region, and on the past work of Pichon and Bilsborrow (1992) in the Ecuadorian Amazon. From long experience working in the Amazon, we know the best we can hope for is one or two interviews per day per team of interviewers given distances and transportation difficulties in the region. We used 6 teams of interviewers and completed interviews of 402 households in the Altamira region using a survey instrument of some 27 pages that inquired from the male head of household about land use and economics of the farm, and a women's questionnaire that examined household structure and composition, reproductive histories and decisions made by women. We focused our sample on the settlement scheme and we did not interview urban households except in those cases where the sampled household happened to live in town rather than on the farm. We are currently conducting a study in this region and another region which focuses on the question of whether there are discernable differences in the land use decisions of urban and rural dwellers. Our sample covers the landscape and allows us to examine other important variables such as distance to road and markets, to towns and other features. With the development of a digital elevation model we are moving towards addressing the role of topography in land use and improving our soil mapping (Figure 7).

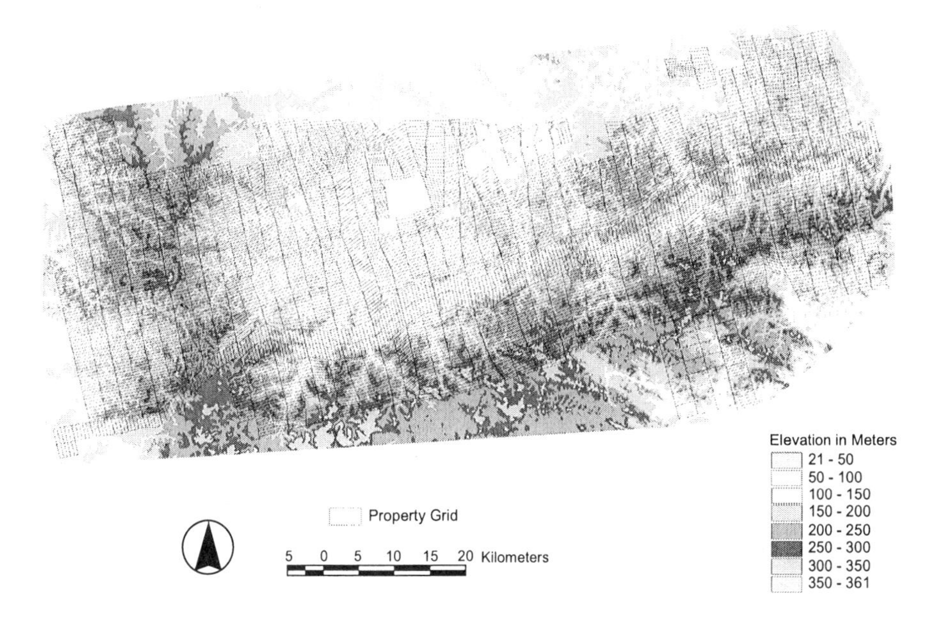

Figure 7. Digital elevation model with property grid overlay

Households in the region were on the whole very cooperative to the study and welcomed us. They answered our very long questionnaires with remarkable patience and interest. There was some suspicion that our team might be from the environmental protection agency (IBAMA) and thus discover some violation of regulations about deforestation and the use of fire. But in general we were able to convince them that our information was largely academic, and would be kept confidential, and that it would not be passed on to IBAMA or other enforcement agencies. We focused on the history of land use in the farm and highlighted the importance of understanding the farmer's experience, successes and failures during the life in the frontier; also, we ensure the academic nature of our research. Contrasting with the receptiveness of farmers was the reluctance of absentee landowners living in Altamira, usually professionals and wholesale merchants, to meet us for interviews. Interesting to note, however, is that absentee landowners, usually merchants, living in smaller urban centers, such as Brasil Novo and Medicilandia, were willing to answer our survey questions when requested. The satellite images were a source of continuous amazement to the population and recognizing that, we printed a copy of a time series of the property in question, and a couple of adjacent ones on which we made markings along with them on their land use through time. This was helpful in aiding memory of past events in land use. We left a clean copy of this same image with each household interview as a gift. This was deeply appreciated, and in some cases where return visits have taken place, households still had the image after a couple of years, and they had obviously discussed it at some length after we left. We think this is a very useful tool in promoting thoughtful discussion by households of their land use strategies and the outcome of their decisions (Figure 8).

Despite this participative use of satellite images, households varied a good deal in their ability to recall land uses several years in the past. The images were helpful in capturing the larger changes, but we have found that smaller management decisions may be lost until better resolution such as IKONOS images are widely used (Figure 9). We encouraged our teams of interviewers to not only conduct a careful interview in the home of the farmers, but also to go out and walk around the property to better understand the spatial distribution of the maps created and the location of major types of land cover. This led in many cases to modifications of the maps to better reflect what was there at present. Remarkably, a very large percentage of farmers have a keen understanding of the spatial location of their plantations and are able to provide if not a perfectly scaled map, at least one that is reasonably accurate about the position of their various crops. Careful use of the reflectance data can help distinguish the different crops, should the images be concurrent with the field work. Of course, in the humid tropics this is not always possible due to a high frequency of cloud cover.

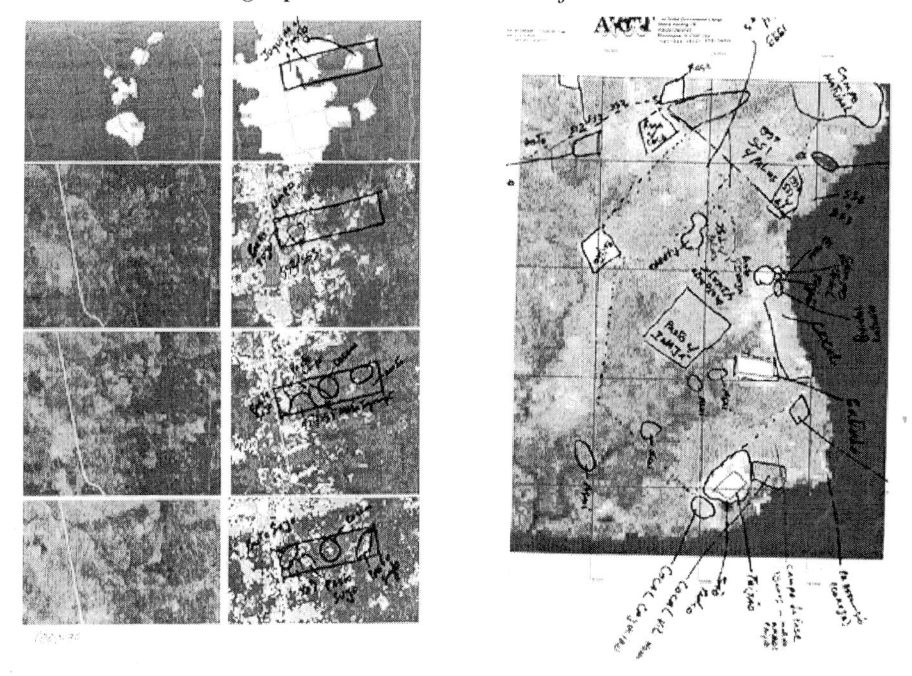

Figure 8. Using images in the field: farm and community level interview

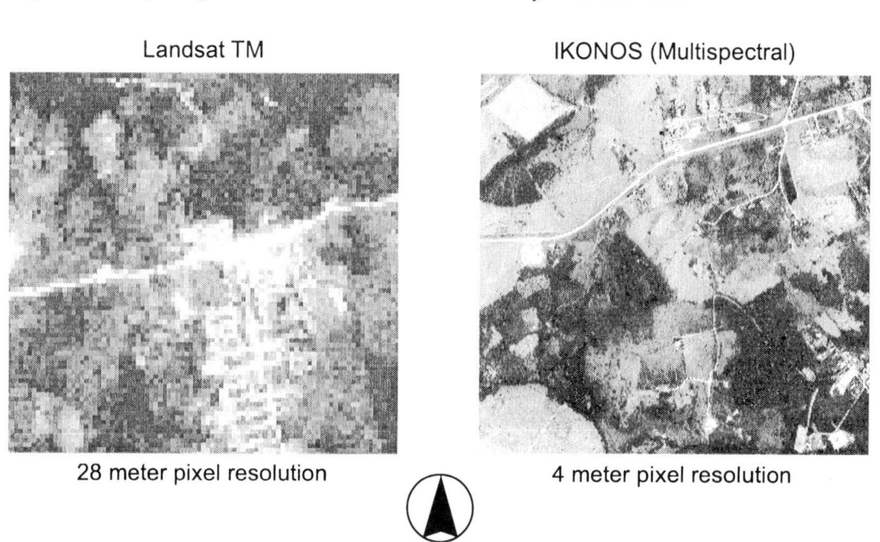

Figure 9. Comparison of Landsat TM and IKONOS Multispectral Data (The images provided are of two different locations within Altamira and are intended only to illustrate the differences in resolution.)

In this particular study, we did not face the difficult choices faced by others over frequent joint ownership of parcels, fragmentation of properties over time, and the reconstitution of properties through inheritance and land portioning. In this area, to date, land has been largely maintained as a unit, and commonly one heir takes over the property. More common has been the aggregation of multiple neighboring lots into large land holdings owned by absentee landowners. In a few cases more than one heir lives in the property if there are rich resources such as cocoa plantations, in which case they divide the number of trees that each owns and takes care of. Some households acquire additional properties so that their children will each be able to have one full one hundred hectare property. However, many children choose to live in the city and to leave the rural way of life.

The greatest source of frustration in this regard was the poor quality of government records on land titles and who owned what piece of land. In most cases the records were out of date and inaccurate. Thus we relied largely on our survey research but could not have much assurance that all the properties in the landscape studied were adequately recorded. The agencies seemed to be working with poor maps and records of ownership. According to Brazilian law, those receiving land from government colonization programs should return the land to the colonization agencies and are not allowed to sell it to third parties. As expected, this law is not enforced, but does not permit the government agencies to keep track of the changes in ownership. Given that, a remarkable number of farmers did not have a fully recorded title to their land, although they had some sort of document that proved ownership.

For the survey research, we relied on a combination of our USA-based team members and a largely local population (often children of settlers) who had pursued a high school education and some technical education. Each team was made up of a man and a woman, with the man usually asking the male head of household the land use and farm production questions, and a woman asking the female head of household and other women about demographic and other reproductive and child raising questions. Our research team includes anthropologists, geographers, sociologists, demographers, GIS/remote sensing specialists, and ecologists. The presence on almost each team of at least one child of the region, as they are called, facilitated rapport with the population who recognized them and who could converse on events with them in familiar terms. Even our USA-based team consisted of fluent and experienced Brazilian researchers, such as the co-authors, Andrea Siqueira and Eduardo Brondizio.

Teams were trained before beginning the interviews. Regular meetings to go over entries in the forms that were not clear to the supervisors took place regularly. This has always proven to be one of the most challenging aspects of this work, because the pace at which the information comes back daily requires that several knowledgeable staff be on hand to check each questionnaire and correct errors found within a day of the interview. This means having at least two people who are not going to the field to interview but who are just devoted to checking the surveys. Since most people prefer to be in the field, than in an uncomfortable office with minimum infrastructure, it is a thankless job from their point of view. We did not allocate that much personnel in previous studies, an error that we have corrected in our current study in the region, and this proved to increase the yield and reliability of the answers. We tried to rotate who stayed behind so that the spirit of the team was better and everyone had a chance to appreciate both the difficulty of the fieldwork, and the difficulty of ensuring accuracy from a large group of enumerators.

Each team also had to be trained to use a GPS and how to interpret a satellite image so that they could answer questions from farmers and use the image to help the interviewees recall past events. This is a very important step and we did not at first fully appreciate the amount of time required to train the enumerators. This is a fundamental step, otherwise the spatial data may be compromised and a great opportunity to link households to landscape can be missed.

Data entry, data clean up, and data analysis is, as we all know and recognize, an arduous and time-consuming processes. Because of the very large number of questions asked, it took a long time to go from the survey data to the writing of papers. The spatial analyses came first, with examination of the insights gained from having a property level analysis vis-à-vis one based purely at the landscape level (McCracken et al.1999). Then came a series of papers examining the role of soil fertility in farmers' choices of crops [1](Moran et al. in press, 2000, Figure 10); examination of trajectories of land use and deforestation at both household and landscape levels (Brondizio et al. in press and Figures 11 and 12); and examination of the role of time on the farm on the trajectory of deforestation[2] (Brondizio et al. in press, Figure 13).

A key goal of this study was to understand whether trajectories of deforestation could be better understood knowing the age and gender structure of households through time, rather than just their aggregate number. In a paper presented at Population Association of America 2001, and currently under review, we find that indeed the model proposed at the outset of the study is consistent with the empirical results of the study (Moran et al., submitted). Each cohort begins occupation of the land with an exponential rate of deforestation necessary to establish their rights to the land, and to establish the farm as a productive unit. This stage lasts approximately 5 years. After this period, there is a steady decline in the rate of deforestation as the household begins to manage the areas already cleared, and tries to control the aggressive regrowth of the native species.

Households begin to shift from largely annual crops to pasture and perennial crops. This requires more labor, now available to them through the aging of their children into teenagers. Each cohort experiences a less steep but still noticeable second rapid increase in deforestation that is more short-lived. This we refer to as a consolidation stage of the farm when the now aging household tries to put their property in order before either passing it on to their children or selling it (Brondizio et al. In press). Over this 20-year period, the deforestation has declined from 5 - 6% per year, to 3%, and settled around 1.2% per year. Projecting out to 2020, we expect that in the study area, 24 to 32% of forest will remain.

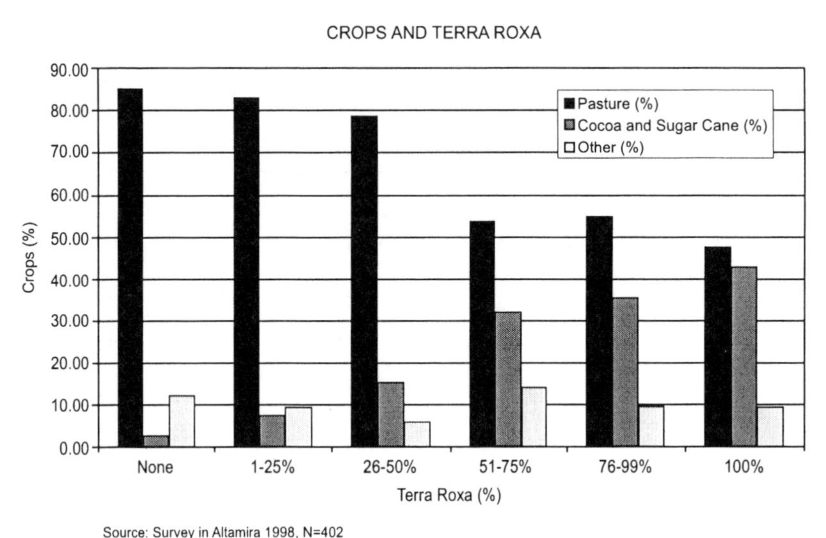

Figure 10. Crops and terra roxa (Moran et al. In press)

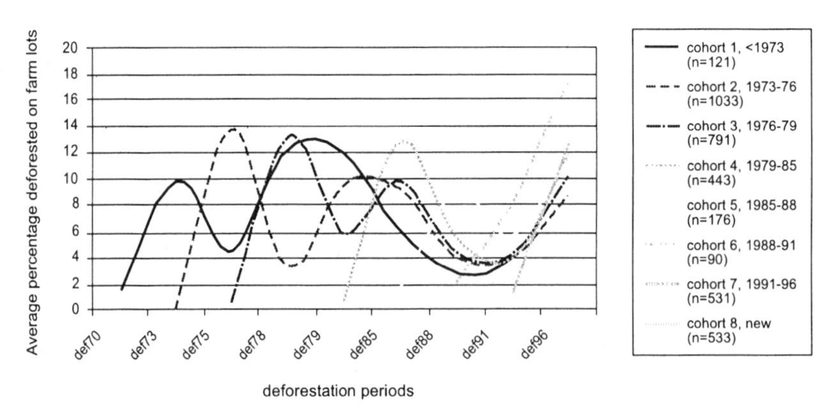

Figure 11. The colonist footprint: average deforestation trajectories across cohorts (Brondizio et al. In press)

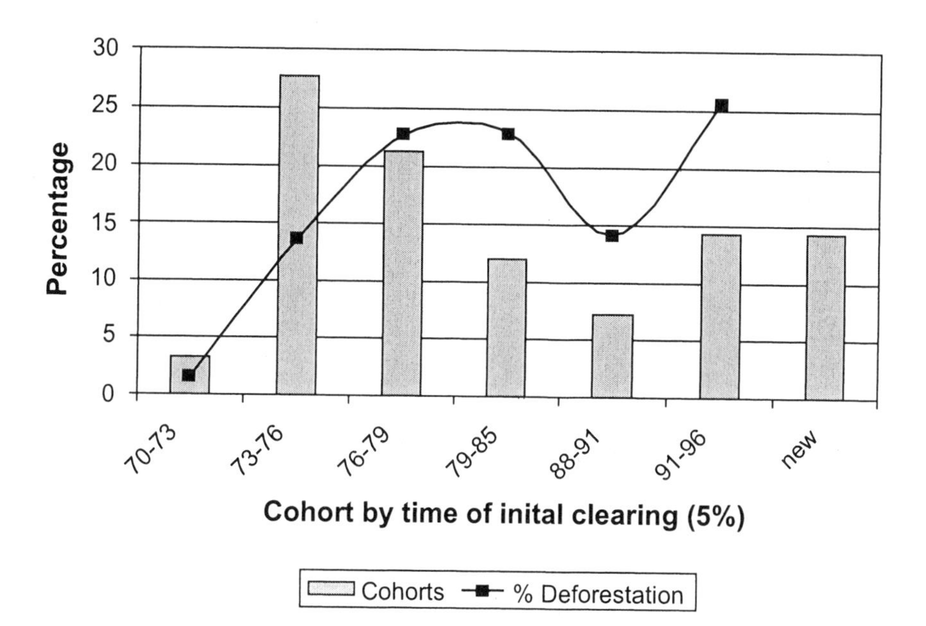

Figure 12. INCRA colonization: Altamira, Brasil Novo, and Medicilandia; distribution of colonization cohorts and percent deforestation (Brondizio et al. In press)

In a paper currently in press (McCracken et al. in press [b]), we suggest that there are at least three plausible scenarios depending on which set of variables is selected. To obtain the age-dependent pattern of deforestation at the farm level we fitted our mean annual rates of farm area deforested with ordinary least-squared regression (OLS) with a simple linear model based on age, and curvilinear models with the inclusion of age-squared and a quadratic equation with an age-cubed term. The results are presented in Table 1. In Figure 14 the observations for each interval between remotely classified images are presented by approximate age of the farm, and a line from the quadratic equation is used to illustrate the general age pattern associated with deforestation on the farms. We found considerable variation about the regression line. The simple linear regression suggests that farms begin by clearing about 4.8% of the forest during the first year, based on the y-intercept. In each subsequent year the area deforested declines by 0.1724. By the 12th year, the percent of the farm deforested fall to less than 3%, and to 1.5% per year by year 20. Applying these rates to a 100 hectare typical farm suggests that approximately 32% of the farm would remain in forest by 2020.

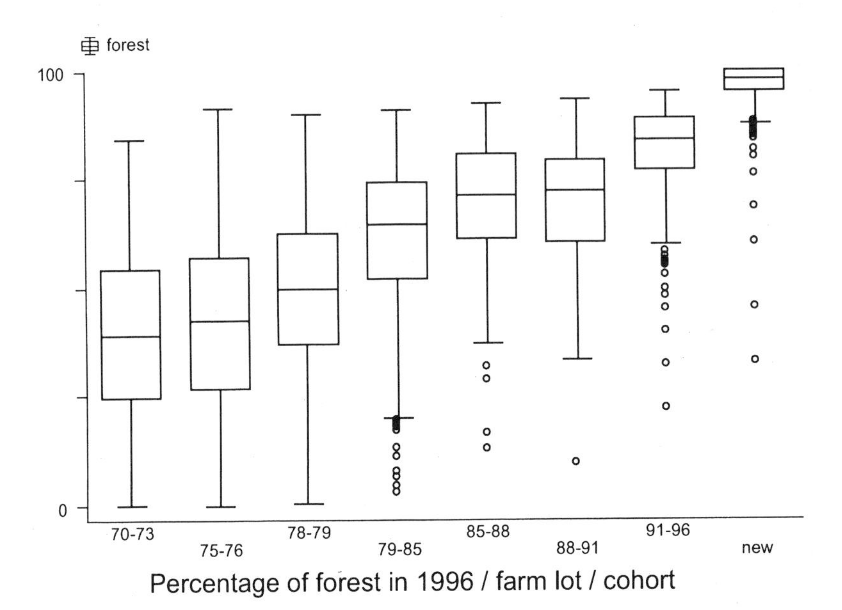

Figure 13. Percentage of forest in 1996 / farm lot / cohort (Brondizio et al. In press)

Table 1

OLS Regression on Mean Annual Percent of Farm Area Deforested

Mean Annual Percent of Farm Area Deforested	Linear	Curvilear-Sq.	Quadratic
Independent Vars			
Years on Lot	-0.1724	-0.4667	-0.7200
Years-Squared		0.0125	0.0374
Years-Cubed			-0.0007
Constant	4.7712	5.8458	6.3826
Number of Obs.	13383	13383	13383
F	2007.24	1269.45	870.43
Prob > F	0.0000	0.0000	0.0000
Adj. R-squared	0.1304	0.1594	0.1631

All results were significant at the 0.01 level

Table 1. OLS regression on mean annual percent of farm area deforested (McCracken et al. In press [b]

The use of age-squared and age-cubed models results in curvilinear patterns with very different consequences for the amounts of forest remaining. Both suggest that deforestation is greater in the beginning and declines to about 2 percent by the 13th year and begins to level off in year 20. After year 20 the simpler model using age and age-squared suggests that deforestation will increase by year 25 and then accelerate—resulting in no more original forest remaining on the farm by year 36. The use of a quadratic equation provides a very different portrayal: it begins with high rates of farm clearing, levels off by year 20 at about 2 percent, and then declines slowly after 30 years with no more deforestation after year 36. Applying these rates to a typical farm would result in 24 percent of the forest remaining on the property.

The linear regression in our view understates the amount of deforestation associated with initial farm creation, and overstates the amount of forest remaining in future years. The curvilinear resulting from use of a squared term for age results in complete deforestation. The quadratic model results are more in line with field observations and knowledge of the area. We have found frequent concern among settlers with preserving some portion of their farm as forest, and a preference to shift to clearing secondary vegetation rather than primary forest over time.

Our project is now entering its more productive phase as the data both spatial and survey is fully available for analysis, and allows us to ask many questions of relevance to population and environment. The region still presents a demographic pattern typical of a frontier[3], where there are more men than women in all age groups (McCracken et al. in press [b]). Women leave their parents' households at an earlier age than their brothers in order to marry, work or pursue education in nearby urban centers, while there seems to be a men's "farm labor retention" (Siqueira et al. in press). One notable question, which bears close examination, is the role of the precipitous decline in female fertility in Brazil, and even in the Amazon frontier. In our study, 43% of women aged 25-29 are already sterilized, with a sizable number using contraception (see Table 2) (McCracken and Siqueira, in preparation). Very few younger women are having more than one or two children, and are using sterilization to ensure that this is a permanent choice. What does this say about the future of land use and land cover in the Amazon frontier? Is this another indication that the process of urbanization is a major driver of land use and land cover change already, and will be more so in the future? As labor in farms declines with lower fertility, does this mean that the landscape will be inescapably mechanized? Or dominated by very large extensive cattle ranches? How best to study and capture these changes now and in the future?

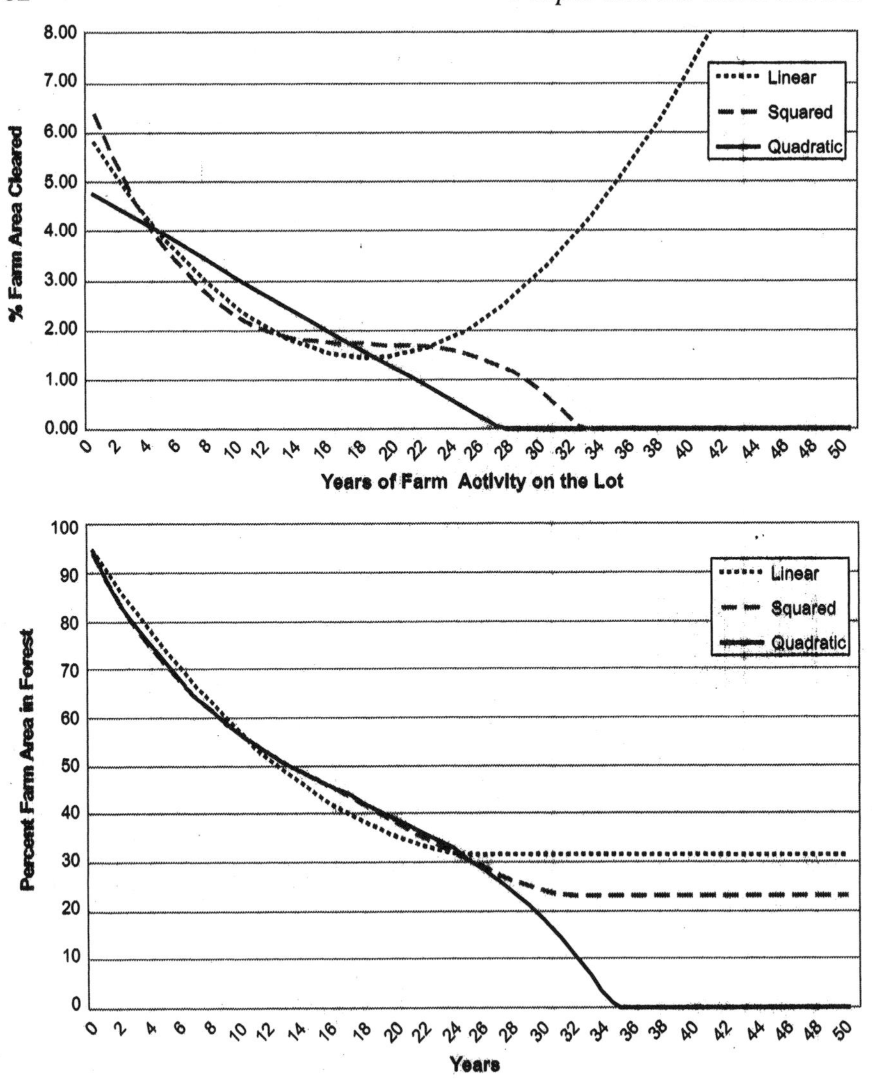

Figure 14. General age pattern associated with deforestation on farms

We think that our study does show the technical feasibility of examining land use and land cover change at the level of households and properties -- and that the insights are worth the effort and investment required to achieve it. We have been able to do it in Rondonia in a fraction of the time and effort it took in Altamira because the grid for Machadinho in Rondonia was almost perfect. It was developed 12 years later, with greater sophistication and resources, and made our work fairly easy by comparison. As more government agencies make use of GPS technology to survey and map, we can expect that this kind of precise linkage between households and land use/cover can be carried out with greater economy of effort and cost.

Age Group	Method Used Currently			Past Contraceptive Use		
	Pill	Other	Steriliz.	Past Use	Never	Total
15–19	15	1	3	18	3	40
	37.50	2.50	7.50	45.00	7.50	100.00
20–24	14	3	5	37	7	66
	21.21	4.55	7.58	56.06	10.61	100.00
25–29	6	0	27	28	1	62
	9.68	0.00	43.55	45.16	1.61	100.00
30–34	2	3	29	7	3	44
	4.55	6.82	65.91	15.91	6.82	100.00
35–39	0	1	43	10	3	57
	0.00	1.75	75.44	17.54	5.26	100.00
40–44	2	1	37	4	2	46
	4.35	2.17	80.43	8.70	4.35	100.00
45–49	0	0	30	6	5	41
	0.00	0.00	73.17	14.63	12.20	100.00
50–54	0	1	29	12	4	46
	0.00	2.17	63.04	26.09	8.70	100.00
55–59	0	0	20	11	13	44
	0.00	0.00	45.45	25.00	29.55	100.00
60–64	0	0	16	30	16	62
	0.00	0.00	25.81	48.39	25.81	100.00
Total	**39**	**10**	**239**	**163**	**57**	**508**
	7.68	**1.97**	**47.05**	**32.09**	**11.22**	**100.00**

Table 2. Contraceptive use among married women by age group, colonization area, Altamira (Moran and McCracken, submitted manuscript)

Our team has been nurtured in an interdisciplinary fashion and we have not experienced major difficulties working together. We have encouraged our team members to become reasonably familiar with GIS, GPS, and other spatial approaches so we can all communicate easily. There is a division of labor, wherein the demographers on the team focus more on those issues, while others focus on credit and the economy, soils and vegetation, or labor allocation and work effort in the farm by age and gender. However, we emphasized the role of "unifying questions" in fostering interaction across fields and disciplines. We encourage papers to be written across these boundaries of topics and disciplines. We try to encourage our younger team members to publish in their disciplinary journals, and not just in interdisciplinary ones, to ensure that they are credible in their own disciplines. This is sometimes a challenge given that many key disciplinary journals still resist papers that are interdisciplinary in nature and that are framed in ways that may be unfamiliar to narrowly trained reviewers in the disciplines.

3. FINAL REMARKS

We would not make any major changes in how we went about our study. We were challenged by the poor quality of the property grid, and this cost us years of man-labor to improve to the point we could trust the boundaries and thus the statistical output of land cover changes at the level of properties. An error of 100 meters could cause an error of some 25% in the changes in land cover obtained, and thus swamp any real differences resulting from decisions taken by households. In other study areas, we have obtained better property grids and the work has been much less time consuming and more rewarding. The use of satellite images and the beginning of deforestation to create cohorts and then to use these cohorts to draw a sample was very useful and we would encourage others to use this approach to sampling whenever possible. The results of our work suggest that we need to more closely examine the role of household structure on any number of changes taking place in the environment. Deforestation was just one clear example, but there are surely others. One of the greatest challenges to researchers is how we are to change our methods to capture the increasingly urban-driven nature of land use and land cover change.

ACKNOWLEDGMENTS

The authors wish to thank the many colleagues who have generously commented upon the design, field and lab work, and the substance of our papers. We thank NICHD for their support of this work through grant #9701386A, and the National Science Foundation through grants 9310049 and 9521918. We thank colleagues at the Fogarty International Center at NIH, and colleagues in Brazil at EMBRAPA, the Federal University of Para, CNPq, Fiocruz, and other institutions for their assistance in obtaining all the authorizations for this work. We appreciate the good humor and openness of the people in the study area who bore patiently with our endless questions. Any errors and interpretations are the responsibility of the authors.

NOTES

[1] Our research has found that soil fertility differences account for a significant proportion of the variance in observed rates of secondary successional regrowth (Moran and Brondizio 1998; Tucker et al. 1998) and even in the portfolio composition of crops cultivated by farmers (Moran et al. in press; 2001). In the more fertile sites we have studied, we find at least a two-fold difference in biomass over nutrient-poor sites, a difference that amplifies in the second and third decade of regrowth. We have also found significant differences in species composition, with the more fertile sites having greater tree species diversity because of greater canopy development but lesser total plant species diversity because of lesser understory development (Tucker et al. 1998; Moran et al. in press). We have also found that households with above average soils (specifically, households with terra roxa estruturada eutrofica, or alfisols) are consistently more able to hold on to their land than households who lack these soils (see Table 3). In Table 3 one can see that households who arrived early were able to choose the best soils and that farms with these soils have not entered the real estate turnover pool, otherwise one would see a greater number of farmers who came later having access to the better soils. Farmers prefer to buy properties that are already developed or partially deforested to virgin plots. There is a very high turnover rate in land holding, but these are disproportionally on the poor soils. Crop choice is affected as well. Households with poor soils had more than 80% of their cleared area in pastures. There is a steady decline in the proportion of the cleared land in pasture as the percentage of good soils increases (see Figure 10). For those households who have their entire property in good soils, there is a balanced portfolio of pasture and cash crops evident in Figure 10.

CHOOSING TERRA ROXA BY COHORT*

% terra roxa	Before 1971	1971-75	1976-80	1981-85	After 1985	Total
None	6.25	38.82	**62.69**	**62.12**	**72.62**	59.45
1-25%	6.25	10.59	14.93	19.7	13.1	13.68
26-50%	**62.5**	10.59	7.46	12.12	1.79	8.71
51-75%	18.75	7.06	4.48	3.03	5.95	5.97
76-99%	6.25	12.94	2.99	1.52	5.36	5.97
100%	0.00	20.00	7.46	1.52	1.19	6.22
Total	100.00	100.00	100.00	100.00	100.00	100.00

*cohort is based on the year of arrival in the lot

Source: Survey in Altamira 1998, N=402 households

(Moran et al In press, 2002)

Table 3. Choosing Terra Roxa by Cohort (Moran et al. In press)

[2] The combination of time series remote sensing data, property grid maps, and field surveys has provided us the opportunity to look at land use and cover change at several levels, such as the colonization landscape, groups of farm lots, and individual lots; this way, it provides tools to capture the arrival of colonization cohorts and the simultaneous process of farm consolidation and expansion.

Colonization landscape level: Despite of the high rates of deforestation during the period of settlement, 61% of the total colonization area remain in forest by 1996, whereas total deforestation adds up to 37%, and non-classified area represent about 2%. By taking into account areas of bare soil, pasture, and cocoa plantations (as estimated by Comisão Executiva do Plano da Lavorura Cacaueira - CEPLAC), our estimate shows that about a half of the deforested area from 1970 to 1996 remains in production by 1996, whereas half of it has been taken over by different stages of secondary vegetation.

Fluctuation in deforestation rates can be observed during this period. Frontier occupation is an on-going dynamic process where "old settlers" co-exist with new ones, the last being recent migrants or second-generation colonists taking over new lots. Colonization rates decreased after the withdrawal of government support in 1974 for about fifteen years, returning to an increased rate after 1991. Fluctuation of deforestation rates after 1985 coincides with the arrival of colonist farmers and with national-level economic indicators. Economic depression and inflation during the second half of the 1980's and withdrawal of cattle ranching incentives are potential explanations for the sharp decrease in deforestation rate seen between 1985 and 1991. (See Figure 12)

Cohort level: There is a close relationship between time of settlement and forest cover, that is, older cohorts have on average less forest cover than younger cohorts (Adj. r 2 = 0.58, significant at 95% conf. interval), however with strong internal variation. Three main groups can be distinguished according to forest cover and time of arrival. Cohort farm lots of the 1970's present similar average in forest cover, about 40% of forest in the farm, but with strong variation within cohort, that is from 0% to 90% of forest. Cohort farm lots of the 1980's show on average about 60% of forest cover in their lots, but ranging from about 30% to 90% forest cover in average. Cohort farm lots of the 1990's show in average more than 75% of forest cover by 1996, but ranging from about 60% to 95% of the farm lot.

The area in secondary succession and production, however, present less significant correlation with time of settlement. Whereas secondary succession (Adj. r 2 = 0.27, significant at 95% conf. Interval) and production (Adj. r 2 = 0.31, significant at 95% conf. Interval) increases with time of settlement, differences among cohorts are less notable. Older cohorts have larger variation of secondary succession and production areas. Cohorts of the 1970's have in average about 8% to 10% of the farm area in secondary succession, varying between 0% and 50% of the total property in fallow. Cohorts of the 1980's have in average 5% in secondary succession, varying within cohorts from 0% to 20% of the total property in fallow. Similar distribution is perceived in production areas. The 1970's cohorts have 6% of their farm lots in production, with variation among farm lots ranging from 0 to 25% of the property. Cohorts of the 1980's present on average 4% of the farm lot in production, however, they show smaller variation within cohorts, with production areas ranging from 0% to 8% of the farm lot. (See Figure 14)

Whereas positive significant correlation exists between time of settlement and deforestation, this is offset by the internal variability within cohorts, which is stronger than across cohorts. Such variability is even stronger in older cohorts suggesting variation in land use systems probably associated with different trajectories in household economic strategies, composition, and in farm production potential.

The data suggest that deforestation trajectories across cohorts are marked by cycles of farm lot formation characterized by deforestation pulses of different magnitudes--termed by Brondizio et al. (in press) as the colonist footprint (Figure 11). Independent of cohort group, frontier farms consistently present cycles of deforestation and agro-pastoral activities development associated with periods of establishment, expansion, and consolidation. These cycles are marked by pulses of deforestation followed by strategies of crop and pasture development, and secondary succession management. Whereas these cycles represent an "age effect," the magnitude of each pulse of deforestation is influenced by factors such as national economic conditions (e.g., inflation rates) and the availability of credit support programs.

In this sense, "the colonist footprint" is characterized by the co-existence of extensification and intensification of production strategies marked by cycles of expansion and consolidation of the farm operation. These processes, however, are characterized by high variation within farm cohorts resulting from differential rate, extent, and direction of land cover change across farm lots.

Understanding deforestation trajectories and the colonist footprint requires a combination of variables related to time of settlement (e.g., cohort effect), cohort and household dynamics (e.g., aging, household labor composition, experience, origin), and period effects (e.g., credit, inflation), underlined by environmental, market, and infrastructural conditions.

[3] The observed pattern of family size is much closer to the current Brazilian urban pattern, that is, it is nuclear and relatively small. The current household size is about 4.6 individuals. Marriage is the most common reason (about 67%) to leave a parents' household for both sexes, followed by search of off-farm jobs (~11%), schooling (~10%) and other reasons (11%). On the other hand, most of the incorporation into the household is through birth (about 72%), marriage representing only 10%. Usually, the young women between the ages of 15-24 are the individuals being incorporated into the household (McCracken et al. in press [a], Siqueira et al. in press). Figure 15 (on the CD-ROM) shows the current age and sex distribution of the study population. Overall, this figure illustrates two main processes taking place in the area, the general aging process of households in this by now 30 year agricultural frontier, and the general loss of labor from children as these become young adults and leave the households, especially women (McCracken et al. in press [a]). In all five-year age intervals there are more males than females. Women are more likely to leave their family household in earlier ages than men, and usually due to marriage and schooling. Young men are more likely to stay in the farm longer, which seems to be a household "labor retention" strategy. Nevertheless, the current sex ratio is more balanced than the initial migration flows to the region. Figure 16 (on the CD-ROM) illustrates what could be called the "gender selectivity process" at arrival. When arriving on the frontier, households are composed of predominantly young members, and slightly more males than females. This pattern of male dominated sex rations, even among infants and children and through the early 20s suggests selectivity in favor of male labor as families migrate to the frontier (ibid).

Perhaps the most important demographic change taking place at the level of household is related to women's reproductive behavior. For the past twenty years we observed a rapid fertility decline among the women in this frontier area. For the late 1960s and early 1970s, fertility for women was very high, on the order of 10-11 children per women. By 1980s fertility declined to about 4.3 children, and we observed that by age 29 more than 40% of women have been sterilized by tying their tubes, which is a common procedure elsewhere in Brazil but unexpected on a frontier area due to the reported scarcity of labor. The current fertility rate is still between 0.5 and 0.8 children above the rest of rural Brazil, but the observed decline in fertility is as important as the rapid fertility decline occurred in the country as a whole (McCracken and Siqueira, in preparation; Siqueira et al. in press).

The possible explanatory reasons for this behavior and the consequences for family labor availability still require further investigation, but it seems fair to point the role of institutional and economic changes taking place in Brazil since late 1950s. Rapid urbanization, the expansion of consumer society and social security coverage, increase in mass communication and better access to health care are possible causes/incentives to the changes observed on the reproductive behavior of these frontier women.

REFERENCES

Batistella, M. 2001. Landscape Fragmentation and Land-Cover Dynamics in Rondonia Brazilian Amazon. Ph.D. Dissertation, School of Public and Environmental Affairs, Indiana University.

Boucek, B., and E.F. Moran. In press. Inferring the Behavior of Households from Remotely Sensed Changes in Land Cover: Current Methods and Future Directions. In: *Best Practices in Spatially Integrated Social Science.* M. Goodchild and D. Janelle (eds.) University of California Press.

Brondizio, E.S., E.F. Moran, P. Mausel and Y. Wu 1994. Land use Change in the Amazon Estuary: Patterns of Caboclo Settlement and Landscape Management. Human Ecology 22(3): 249-278.

Brondizio, E., S. McCracken, E.F. Moran, A.D. Siqueira, D. Nelson, and Rodriguez-Pedraza. In Press. The Colonist Footprint: Towards a conceptual framework of deforestation trajectories among small farmers in Frontier Amazonia. In: *Deforestation and Land Use in the Amazon.* C. Wood and R. Porro (eds.) Gainesville, University of Florida Press.

Futemma, C.R.T. 2000. Collective Action and Assurance of Property Rights to Natural Resources: A case study from the Lower Amazon Region Santarem, Brazil. Ph.D. Dissertation, School of Public and Environmental Affairs, Indiana University.

Goody, J. 1962. *The Development Cycle of Domestic Groups.* New York: Cambridge University Press.

Goody, J. 1976. *Production and Reproduction: A Comparative Study of the Domestic Domain.* New York: Cambridge University Press.

Kroeber, A.1939. *Cultural and Natural Areas of Native North America.* Berkeley: University of California Press

Mausel, P., Y. Wu, Y. Li, E. Moran, and E. Brondizio. 1993. Spectral Identification of Successional Stages following deforestation in the Amazon. Geocarto International 8: 61-71.

McCracken, S., E. Brondizio, D. Nelson, E. Moran, A. Siqueira, and C. Rodriguez-Peraza 1999. Remote Sensing and GIS at Farm Property Level: Demography and Deforestation in the Brazilian Amazon. Photogrammetric Engineering and Remote Sensing 65 (11): 1311-1320.

McCracken, S. and A. Siqueira. nd. Fertility Decline in an Amazonian Agricultural Frontier in Brazil: New Evidence for Old Debates. Paper presented at the Annual Meetings of the Population Association of America, New York, March 25-27, 2000.

McCracken, S., A.D. Siqueira, E.F. Moran, and E.S. Brondizio. In Press [a]. Land-use Patterns on an Agricultural Frontier in Brazil: Insights and Examples from a Demographic Perspective. In: *Deforestation and Land Use in the Amazon.* C. Wood and R. Porro (eds.) Gainesville, University of Florida Press.

McCracken, S.D., B. Boucek, and E.F. Moran. In Press [b]. Deforestation Trajectories in a Frontier Region of the Amazon. In: *Remote Sensing and GIS Applications for Linking People, Place and Policy.* S. Walsh and K. Crews-Meyer (eds.) Kluwer Academic Publishers.

McCracken, S. and A. D. Siqueira. In preparation. Fertility Decline in an Amazonian Agricultural Frontier in Brazil: New Evidences for Old Debates.

Moran, E.F. 1976. Agricultural Development along the Transamazon Highway. Bloomington: Center for Latin American Studies, Indiana University, Monograph Series No. 1

Moran, E.F. 1979. *Human Adaptability: An Introduction to Ecological Anthropology.* N. Scituate: Duxbury Press. 404 pp.

Moran, E.F. 1981. *Developing the Amazon.* Bloomington: Indiana University Press

Moran, E.F. 1987. Social and Environmental Systems. In *Latin America: A Regional Perspective.* J. Hopkins (ed.). New York: Holmes and Meier. Pp. 3-18.

Moran, E.F. ed. 1990. *The Ecosystem Approach in Anthropology: From Concept to Practice.* Ann Arbor: Univ. of Michigan Press.

Moran, E.F. 1993. *Through Amazonian Eyes: The Human Ecology of Amazonian Populations.* Iowa City: University of Iowa Press, 1993. 230 pp.

Moran, E. F., E. Brondizio, P. Mausel, and Y. Wu 1994. Integration of Amazonian Vegetation, Land Use and Satellite Data. BioScience 44: 329-338

Moran, E.F. and E.S. Brondizio 1998. Land Use after Deforestation in Amazonia. In: *People and Pixels: Linking Remote Sensing and Social Science.* D. Liverman, E. Moran, R. Rindfuss and P. Stern (eds.). Washington DC: National Academy Press.

Moran, E.F., E. Brondizio, J. Tucker, M.C. Silva-Forsberg, S. McCracken, and I. Falesi 2000. Effects of Soil Fertility and Land Use on Forest Succession in Amazonia. Forest Ecology and Management. 139(2000):93-108.

Moran, E.F., and E.S. Brondizio 2001. Human Ecology from Space: Ecological Anthropology Engages the Study of Global Environmental Change. In: *Ecology and the Sacred: Engaging the Anthropology of Roy A. Rappaport.* E. Messer and M. Lembeck (eds.) Ann Arbor: University of Michigan Press.

Moran, E.F., E.S. Brondizio, and S. McCracken. In press. Trajectories of Land Use: Soils, Succession, and Crop Choice. In: *Deforestation and Land Use in the Amazon.* C. Wood and R. Porro (eds.) Gainesville: University of Florida Press.

Moran, E.F., and S.D. McCracken (submitted manuscript). The Developmental Cycle of Domestic Groups and Amazonian Deforestation.

Netting, R. 1968. *Hill Farmers of Nigeria.* Seattle: Univ. of Washington Press

Netting, R. 1981. *Balancing on an Alp.* New York: Cambridge Univ. Press

Oliveira Filho, F.J.B. 2001. Deforestation Pattern and Evolution of the Landscape Structure in Alta Floreste (MT). Master Thesis. Bioscience Institute; University of São Paulo; Brazil.

Pichon, F. and R. Bilsborrow. 1992. Land Use Systems, Deforestation and Associated Demographic Factors in the Humid Tropics. Paper prepared for the IUSSP Seminar on Population and Deforestation in the Humid Tropics. Campinas, SP, Brazil.

Rappaport, R. 1967. *Pigs for the Ancestors: Ritual and Ecology of a New Guinea People.* New Haven: Yale Univ. Press.

Sauer, C. 1958. Man in the Ecology of Tropical America. Proceedings of the Ninth Pacific Science Congress 20:104-110.

Silva-Forsberg, M.C. 1999. Protecting an Urban Forest Reserve in the Amazon: A Multi-Scale Analysis of Edge Effects Population Pressure, and Institutions. Ph.D. Dissertation, School of Public and Environmental Affairs, Indiana University.

Siqueira, A.D., S.D. McCracken, E.S. Brondizio, and E.F. Moran. In press. Women in a Brazilian Agricultural Frontier. In: *Gender at Work in Economic Life.* G. Clark (ed.). SEA Monograph Series, Maryland; University Press of America.

Steward, J. 1955. *The Theory of Cultural Change.* Urbana: University of Illinois Press

Chapter 4

INTEGRATION OF LONGITUDINAL SURVEYS, REMOTE SENSING TIME SERIES, AND SPATIAL ANALYSES

Approaches for Linking People and Place

Stephen J. Walsh
Department of Geography and Carolina Population Center, University of North Carolina
swalsh@email.unc.edu
Richard E. Bilsborrow
Department of Biostatistics and Carolina Population Center, University of North Carolina
Stephen J. McGregor
Carolina Population Center and Department of Geography, University of North Carolina
Brian G. Frizzelle
Carolina Population Center, University of North Carolina
Joseph P. Messina
Department of Geography, Michigan State University
William K. T. Pan
Department of Biostatistics, University of North Carolina
Kelley A. Crews-Meyer
Department of Geography, University of Texas
Gregory N. Taff
Department of Geography and Carolina Population Center, University of North Carolina
Francis Baquero
Ecociencia, Ecuador

Abstract Linkages between people and the environment are examined within a space-time context as part of population-environment research ongoing in the northeastern Ecuadorian Amazon. In this chapter, we consider how a longitudinal household survey, a satellite time series, field sketch maps and image products, GPS (Global Positioning System) coordinates for household farms and built structures, GIS (Geographic Information System) data management schemes, pattern metrics, image change-detections and pixel histories, and cellular automata and multilevel models can be used to assess LCLU (land-cover/land-use) dynamics and socioeconomic, biophysical, and geographical drivers of change.

Approaches, protocols, and philosophies are described for linking data types to represent historical, contemporary, and possible future characterizations of LCLU patterns for the study region. Challenges and opportunities for relating data across thematic domains and space-time dimensions are considered, with emphasis on strategies and rationales for data linking.

Keywords: remote sensing, GIS, longitudinal survey, pattern metrics, cellular automata, Ecuador

1. INTRODUCTION

Beginning in the early 1970s, roads were built by the petroleum industry into the Oriente region (the northeastern Ecuadorian Amazon) to explore for oil and subsequently to lay pipelines to extract it. Prior to that time the main inhabitants of the region were indigenous peoples who lived in small, remote, and isolated communities located primarily along rivers. The new roads dramatically changed the geographic accessibility of the region. People with little or no land living in other rural areas of the country poured into the Oriente and settled on plots of land made accessible by the newly built roads.

As a result, small-scale farmers have been the primary direct agents of land clearing or deforestation (Bromley 1989; Rudel and Horowitz 1993; Southgate and Whittaker 1994; Pichón and Bilsborrow 1999). Upon establishing themselves on plots, groups of farmers living in one area or sector sought to have the boundaries of their plots determined by the government land-titling agency, IERAC, resulting in household farms—or *fincas*—being established with an area of approximately 50 ha and measuring 0.25 x 2.0 km. Subsequently, farmers would apply to IERAC for provisional land titles and ultimately full land titles, or *escrituras*. The lands most sought after were along roads, but parallel rows of farms—*lineas*—extended back from the roads sometimes up to 16 km. Deforestation and agricultural extensification occurred on household plots and proceeded on *fincas* from lands adjacent to roads, moving deeper into the forest as additional lands were developed or made more accessible. Because of low soil fertility in most areas, the common cycle of land use has been land clearing to plant annual subsistence crops, followed by planting of perennial crops (mainly coffee), then pasture, the beginnings of secondary plant succession, and ultimately to land abandonment. Although land abandonment has been rare in Ecuador to date because of better soils and land scarcity, some land has been, at a minimum, temporarily abandoned through land conversion processes that have altered the landscape. Towns grew to supply the oil industry and, subsequently, to support the agricultural activities and consumption needs of the colonists.

It is within this context that we examine LCLU (land-cover/land-use) dynamics and the drivers of change in the Oriente. To do this, a comprehensive integrated GIS database has been developed comprising the following elements: (1) a remote sensing image time series (i.e., Landsat Multispectral Scanner (MSS), Landsat Thematic Mapper (TM), IKONOS Multispectral and Panchromatic, and Panchromatic Aerial Photography) to characterize LCLU patterns over a twenty-seven-year period from 1973 to 2000; (2) longitudinal household survey data for a representative sample of 418 migrant settler plots sampled in 1990 and revisited in 1999; (3) a community survey of sixty locations conducted in 2000; (4) preparation of geographic site and situation variables for households and communities within a GIS (Geographic Information System) environment to emphasize geographic accessibility, resource endowments, and the spatial organization of LCLU; and (5) GPS (Global Positioning System) coordinates, field sketch maps, and database management techniques to link people and environment by emphasizing spatially and temporally explicit characterizations of people, place, and environment.

The intent of this research is to characterize the spatial and temporal nature of deforestation and agricultural extensification in northeastern Ecuador and to assess the multi-thematic and scale-dependent drivers of landscape change. The goals of this chapter are to (1) describe the nature of our population-environment research relative to the collection of socioeconomic, biophysical, and geographical data and their integration in analyses; (2) describe products and protocols generated to support data collection and integration efforts; (3) examine mechanisms developed to impose spatial and temporal dimensions and to embed geographic, ecological, and socioeconomic principles into data collection protocols and analyses; and (4) describe approaches that we have implemented to link people, place, and environment in spatially and temporally explicit ways through associations characterized by (a) longitudinal household surveys, (b) sketch maps of household LCLU patterns, (c) GPS technology and GIS database management systems, (d) remote sensing systems of differing spatial and spectral resolutions, (e) image change-detection and change trajectories, and (f) LCLU simulation models developed within a CA (Cellular Automaton) approach, where rules, initial conditions, neighborhood relationships, and feedbacks are used to visualize future patterns of LCLU.

The organization of this chapter is as follows. We first describe the study area and forces that have combined to shape and alter the socioeconomic, biophysical, and geographical landscapes of the Oriente. Next, approaches and techniques used to collect the various types of data are described, with emphasis on the issues of linking people, place, and environment. Finally, the challenges and opportunities for building linkages between thematic domains and space-time dimensions are discussed, followed by concluding comments.

2. THEORETICAL CONTEXT AND RESEARCH QUESTIONS

2.1 Theoretical Arguments

A number of scholars have described the effects of the growing frontier population on the local environment as seen through political, human, and landscape ecology theory, as well as complexity theory (e.g., Bilsborrow 1987; Marquette and Bilsborrow 1994; Jolly 1994; Moran and Brondizio 1998; Wood and Skole 1998; Walsh et al. 2002). In this context, political ecology can be viewed as referring to the context within which household land use and related decisions are made, including sociocultural norms, the economic standard of living, technology, local and national government policies, commodity prices at local to global scales, and physical infrastructure, such as roads. Local decisions at fine scales (e.g., the household level) are framed within a broader set of issues at coarser scales (e.g., the community) and are ultimately subject to the influences of macro-level forces operating at the regional, national, and global scales (e.g., public policy and institutions, global markets and prices). Human ecology argues that people are active agents on the landscape that shape it and are shaped by it. Activities such as the gathering of fuelwood and the out-migration of young adults because of excessive land fragmentation are examples. Landscape ecology points to the interactions between landscape structure (spatial organization of landscape units), function (operative mechanisms), and change (temporal dynamics of pattern and process) (Forman and Godron 1986). Complexity theory states that systems contain more possibilities than can be actualized (Luhman 1985). The goal of complexity theory is to understand how simple fundamental processes can combine to produce complex holistic systems (Gell-Mann 1994). Nonequilibrium systems with feedbacks can lead to nonlinearity and may evolve into systems that exhibit criticality. Research framed within complexity theory can address the rates and patterns of LCLU dynamics and possible nonlinear feedbacks between the processes of change and existing patterns. Changes depend on the existing patterns of LCLU, which may involve critical points where a small amount of LCLU change significantly alters feedback processes and leads to a new pattern or equilibrium.

Reconciling these distinct theoretical arguments involves (a) different response variables—economic (e.g., land intensification and extensification), demographic (e.g., fertility responses), and "economic-demographic" (e.g., out-migration); (b) LCLU dynamics and spatial/temporal patterns (e.g., structure of ecological systems viewed through their spatial and temporal organization), and (c) different response drivers (e.g., social, biophysical, geographic variables operating through scale-dependent relationships). Modeling LCLU dynamics using these theoretical approaches also draws upon the earlier work of Malthus (1798: exponential population growth combined with arithmetic increases in agricultural production lead to pressures to bring more land into production and hence land extensification); Boserup (1965, 1981: population pressures lead to increased inputs of labor per unit of land or labor intensity and hence to increased food production and land intensification); Bilsborrow (1987, n.d.: intensification and extensification are based on economic, demographic, and "economic-demographic" responses to the environment); Blakie and Brookfield (1987: household decisions are linked to broader socioeconomic processes); and McCracken et al. (1999) and Perz (2000: stages of household settlement serve as proximate determinants of LCLU dynamics).

These theoretical arguments and approaches are integrated in our studies to consider the scale-dependent drivers of LCLU dynamics that reflect social, biophysical, and geographical domains, exogenous and endogenous factors, and multilevel effects. Here we focus on linking mechanisms and their rationales to integrate data collected through socioeconomic surveys and spatial digital technologies.

2.2 Project Research Questions

The questions that guide our research evolve as we learn more about the study area; connections between people, place, and the environment; and new literature. This leads us to articulate new hypotheses and develop new ways of linking socioeconomic and demographic survey data to data acquired or enhanced through spatial digital technologies. Nevertheless, our research questions continue to center on land clearing (deforestation), agricultural extensification, land management, and methodological issues associated with spatial simulations and landscape characterization and modeling of LCLU dynamics. Some of the questions that we are currently addressing are as follows:

- What are the rates of different types of land conversions at the *finca,* sector, and regional levels?

- What are the social, biophysical, and geographical factors and processes that drive land-use/land-cover conversion? Are there clear and consistent patterns in the way these factors and processes are manifested in the landscape? Are there consistent patterns across *fincas* in household-level land management decisions that govern the structure of land use within *fincas?*
- Does the lag time between sector settlement and land titling affect land management? Does *finca* structure converge on a steady state over time? If so, how long does this process take?
- How do *finca* land management practices vary in response to social, biophysical, and geographical factors and processes?
- Is there evidence of increasing intensification in land use over time, and if so, where and why? Is there evidence of variability in agricultural sustainability in the study area, and how does it relate to LCLU practices? Do biophysical, social, and geographical factors explain this variability?
- Is there evidence of land abandonment or increasing long-term fallow on *fincas* in the study area? If so, where and why—that is, what is the nature of secondary plant succession on abandoned or fallow lands?
- Can we develop a remote sensing model of land-use/land-cover change (LCLUC) for the entire region by linking ground-based data to a time series of remotely sensed data? Can cellular automata methodologies be used to realistically simulate LCLU dynamics at the farm and landscape levels? Are there scale dependencies and multilevel links between social, biophysical, and geographical factors and observed land conversion rates and patterns at the *finca,* sector, and regional levels?

3. THE STUDY AREA

The study area is the so-called Oriente, which is taken here to refer to the northern Amazonian frontier of northeast Ecuador (Figure 1). The Oriente serves as a particularly useful laboratory to investigate human-environment interactions and the relationships to LCLU because of the absence of major confounding factors. Thus (a) Ecuador's Amazon region has no large urban areas; (b) settlers are predominantly small, poor farmers, rather than large ranchers, who in-migrated almost entirely as spontaneous colonists; (c) there have been no government subsidies or other major policies in Ecuador encouraging cattle ranching; and (d) there is no large-scale timber extraction by logging companies (though it occurs on a small, mainly illegal scale throughout the region). In addition, there is no season without rain, resulting in little slash-and-burn agriculture, in contrast to Brazil. Finally, the level of biodiversity is very high, so the ongoing loss of habitat is a serious concern.

Figure 1 shows a 90,000-ha Intensive Study Area (ISA) centered on the region's largest town, Lago Agrio (also called Nueva Loja), overlaid with roads and household farms or *fincas* on a 1999 Landsat TM image.

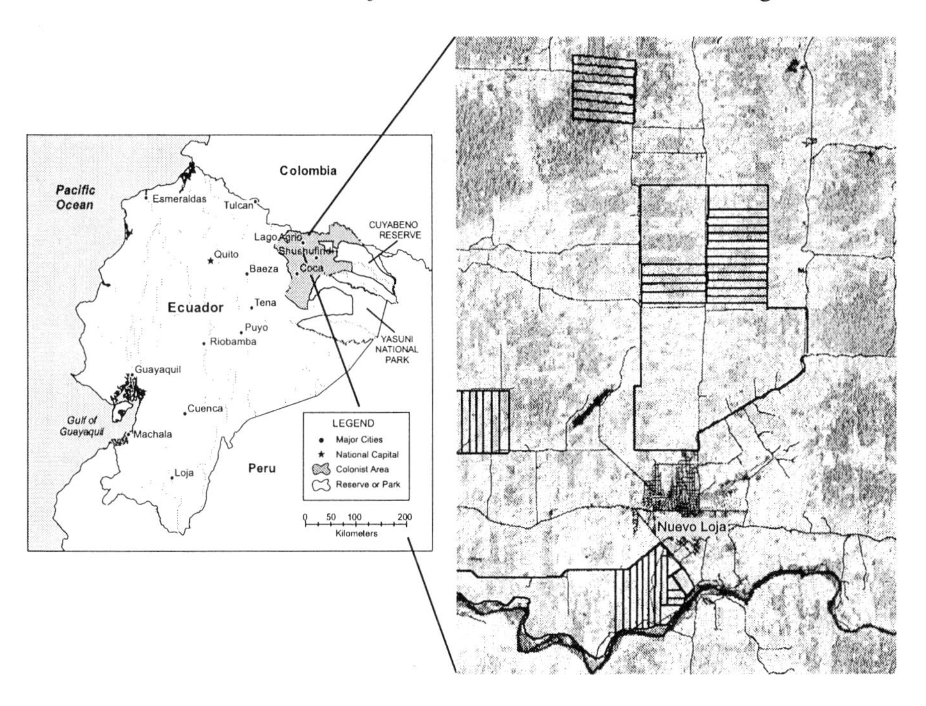

Figure 1. The study area location (northern Oriente, the Ecuadorian Amazon)

The Ecuadorian Amazon comprises the provinces of Sucumbios, Orellana, Napo, Pastaza, Morona Santiago, and Zamora Chinchipe. Our project focuses on Napo and Sucumbios. The 1990 census population of the region as a whole was 371,000, of which 273,000 or 74 percent was rural. The early results of the 2001 census, unadjusted for the likely significant undercount, show a population of 547,000, amounting to 4.5 percent of the total (of 12.1 million) in Ecuador. The population of the Amazon region has grown at over double the national rates over the last two intercensal periods—at 8 percent per year in 1974–82 and 5 percent per year in 1982–90—and it grew at 3.5 percent in 1990–2001 compared to about 2.2 percent at the national level, as fertility continues to fall in Ecuador. Almost half of the Amazon population was born outside the region, with two-thirds of the rural population in our study migrating from the Sierra, most since the mid-1970s. Government policies have encouraged migration to the Amazon region, as it has been perceived as an area with almost infinite space and resources and thus an "escape valve" to relieve socioeconomic imbalances in other regions. Access to the northern Amazon was initially made possible by the petroleum boom, which led to road construction (see Figure 2: a 1990 panchromatic aerial photograph is presented, with *finca*s readily apparent against the background of the residual forest). Petroleum has since consistently provided over half both of Ecuador's export earnings and federal government revenues and hence has been vital to its development. Clusters of dwellings evolved around major road intersections and petroleum encampments, sometimes growing into market towns. Lago Agrio is the largest city in the region, yet it had only 34,000 people in 2001.

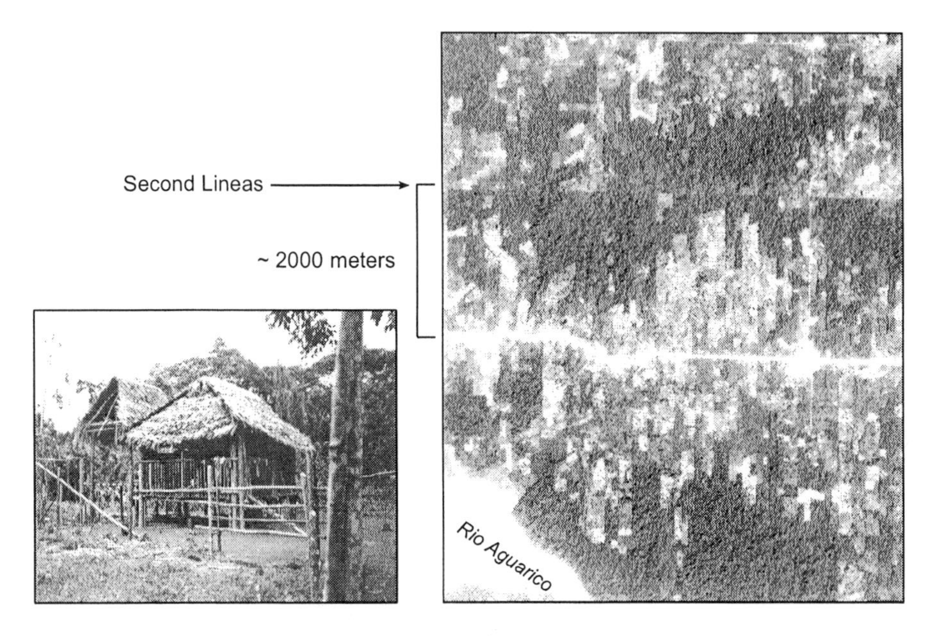

Figure 2. Settlement patterns in the Oriente associated with *lineas* and road position and type

4. METHODS

We describe below a number of data collection and data processing techniques used, with emphasis on methods, protocols, and operating procedures involved in linking people, place, and environment within a spatially and temporally explicit context. Because we are examining deforestation and agricultural extensification and the associated drivers for 1990–2000, it was important to maintain a connection between people and the environment not only across space but also across time. The household surveys involve interviews in 1990 and 1999 but collect data for times prior to and during those years, as well as plans for the future. Landsat satellite data cover the earlier period and aerial photography is available for 1990. Also, using relationships derived from the household surveys, image change-detections, and pixel histories, we integrate space-time effects across socioeconomic, biophysical, and geographic domains using multilevel statistical models and cellular automata (CA) methods that simulate LCLU dynamics for periods antecedent to and subsequent to the study period.

4.1 The Survey Instrument: Household and Community Surveys

A unique feature of the research is the implementation of a 1999 socioeconomic household survey in precisely the same geographic sites (farms) covered in the prior 1990 survey. This has produced a longitudinal household-level data set for a statistically representative sample of farm plots covering the entire major area of colonization in Ecuador's Amazon. A two-stage sampling design was used to select migrant farm households. In the first stage, sectors were sampled with probabilities of selection proportional to estimated size—that is, proportional to the number of plots in each sector. Thus, 64 sectors were selected from the total of 275 sectors settled by 1990. In the second stage, farm plots were selected from each sector in proportion to the size of the sector; a cluster of six to ten contiguous plots was thereby selected randomly from each sector. The final sample comprised 418 settler plots from 405 *fincas* selected from the 64 sectors that were visited, representing about 6 percent of all colonist plots in the main lowland settlement areas of the two northernmost Amazonian provinces of Sucumbios and Orellana.

In 1999, the same plots were revisited (except for one sector containing ten *fincas* on the Colombia border that was excluded for security reasons), and all farms and subdivisions within the same geographic space were revisited (and all new households and subdivisions separately interviewed), yielding a sample of 767 farms, plus 111 additional house lots or *solares* created by "parcelization" near several major towns, reflecting incipient urbanization. Detailed questionnaires, as in 1990, were again administered to both the head of the household (almost always male in this frontier region) and the spouse separately.

The questionnaire of the head included information on plot acquisition, land size, and tenure; agricultural production and inputs; livestock; contact with government and nongovernment agencies and credit and extension services received; use of hired labor and work off the farm; migration history and future intentions to migrate; and aspirations for children's education and migration. The spouse was asked to provide information on household composition, including out-migration of members from the household since the family arrived in the Amazon; housing conditions, water and fuel sources, and household possessions; health, fertility, and mortality; and gender roles and environmental problems. In addition, the spatial location of crops and other forms of land use and changes in LCLU since 1990 were noted on sketch maps for each farm, including the location of all subdivisions and houses. The locations of dwellings, plots, roads, and so on were spatially referenced using hand-held GPS units. Community questionnaires were administered in 2000 to community leaders, farmers, teachers, and groups of women and health workers in sixty places, ranging from tiny communities consisting of no more than a small *tienda* or store and a one-room primary school to the largest city, Lago Agrio.

Here we present two tables (Bilsborrow n.d.; Walsh et al. 2002) that compare some socioeconomic and LCLU characteristics of farm plots visited in 1990 and 1999. An unexpected finding in the 1999 survey was the widespread subdivision of *fincas* through inheritance and land sales. Informal and formal land transactions occurred throughout the sample sectors, even in remote areas, but the creation of *solares* or house lots was common near towns—especially Lago Agrio and Joya de los Sachas.

Table 1. Comparison of 1990 vs. 1999 household characteristics

Characteristic	**1990 (N = 418 HHs)**	**1999 (N = 763 HHs)**
Average people per HH	6.6	5.7
Population (farms only)*	2,761	3,813
Average distance, nearest market	28.2 km	20.4 km
Road distance to town	24.2 km (n=394)	19.1 km (n=682)
Walking distance to road	5.3 km (n=214)	2.6 km (n=333)
Canoe distance to town	34.1 km (n=35)	18.7 km (n=30)
Head born in Oriente	4.6 percent	8.0 percent
Electricity in dwelling	14 percent	32 percent
Own chainsaw	30 percent	20 percent
Have legal land title	50 percent	34 percent
Gini coeffs., landholdings	0.26	0.50

*Excludes population living in *solares*

Table 1 indicates some demographic characteristics and differences in households surveyed in 1990 and 1999. Two of the most important are the population and distance to market variables. The farm population increased by 38 percent, but if the population of *solares* on the same *finca madres* is included, population grew by 57 percent. The average distance from farm plots to market decreased from 28.2 km to 20.4 km. Although this partly reflects additional road construction, the substantial decrease indicates that properties closer to market towns tend to be subdivided more and that smaller communities also now have markets. This provides better market access for people working in Lago Agrio and other central market towns associated with the expansion of the service and nonfarm employment sectors. It is also evident that electricity is starting to reach more farms, presumably along the main roads. The fall in the proportion having a legal title to their plot is a consequence of subdivision, combined with the replacement of IERAC by an inactive government land agency in 1993 called INDA.

Table 2 below compares land use in 1999 of long-term owners (those there in 1990 who were still there in 1999) and new owners, showing the dramatic difference in mean farm size resulting from the creation of new farms through land subdivision. This in turn has implications for land use, as with smaller plots the new owners have less land in forest and pasture and more in the more intensive forms of land use—perennials (mostly coffee) and annual crops—as would be expected from Boserup's hypothesis of land intensification.

Table 2. Land use by long-term (>9 yrs.) vs. new owners, 1999

Land Size and Use	Long-Term Owner	New Owner
Mean plot size	34.2 ha	18.4 ha
Percent forest	42 percent	36 percent
Percent pasture	30 percent	23 percent
Percent perennial crops	21 percent	27 percent
Percent annual crops	7 percent	14 percent

4.2 SIIM Maps and the Survey

Survey Instrument Image Maps (SIIMs) are graphical products developed to support the 1999 survey of households. The SIIMs show the location of each sector or cluster of farm households and the location of each sample plot *(finca)* in that sector from the 1990 survey. A sector is the name of an area opened up for settlement by the Instituto Ecuatoriano de Reforma Agraria y Colonización (IERAC) during the period up to 1993. A sector—also called a cooperative or pre-cooperative—is an area comprised of ten to seventy contiguous *fincas* or farms of 40–50 hectares each. Farmers settling in an area form temporary pre-cooperatives to collectively solicit surveys and demarcations of their property borders, which is the first step toward individually acquiring actual land titles. The goal of the 1999 data collection was to revisit each of the farm plots or *fincas* surveyed in the 1990 sample of settlement sectors.

Three types of SIIM products were produced for the 1999 field survey (Frizzelle and McGregor 1999). First, small-scale regional SIIMs (1:250,000 scale) were created to help the 1999 survey teams navigate to the same survey sectors that were visited in 1990. Second, SIIM products at a scale of 1:25,000 were composed using aerial photographs from 1990 as the base layer, with corresponding annotation and shape files overlaid. These maps were used by survey teams to help the male head of the household recall what was growing on his property in 1990 and whether new areas had been deforested or forms of land use converted to other uses since that period. The third type of SIIM product was a large-scale map based on 1996 Landsat TM imagery to help the household head recall more recent LCLU on his farm plot in the 1999 household survey.

Once the aerial photography was scan-digitized and georeferenced, sample farm plot and settlement sector boundaries were estimated using manual interpretation and on-screen digitizing techniques. Sector and sample *finca* boundaries were annotated on the photographs as part of the SIIM products. Vector road files and place names from topographic maps included in the GIS were added to the SIIM products to assist survey team supervisors and interviewers in navigation and locating sector and farm lots. SIIM products were printed at several map scales to facilitate fieldwork planning and actual fieldwork. The general shape of each sector was determined from IERAC cadastral maps, while the relative locations of sample *fincas* within the sectors were based on sketch maps drawn by the 1990 sampling team under Bilsborrow. A UTM (Universal Transverse Mercator) coordinate system 1-km grid was overlaid on the 1990 SIIMs and used for comparison to GPS coordinates obtained in the field for local navigation and verification of site features.

4.3 Sketch Maps

The SIIM products described above were used to facilitate the preparation of two sketch maps for each sample *finca madre*. These maps, prepared together by the household head and interviewer, illustrate *finca* boundaries (and any subdivisions) and the location and area of each current form of land use in 1999 (along with special features, such as a road, creek, area of swamp, oil pipeline, etc.) (Figure 3). The 1990 SIIM airphoto map was then used to assist the farmer in recalling the same information for 1990; the 1996 Landsat TM image was used to determine land use at the *finca* level in between 1990 and 1999. The 1990 (airphoto-based) and 1996 (satellite-based) SIIM products were also used to assist recall whenever the farmer took possession of his plot after 1990.

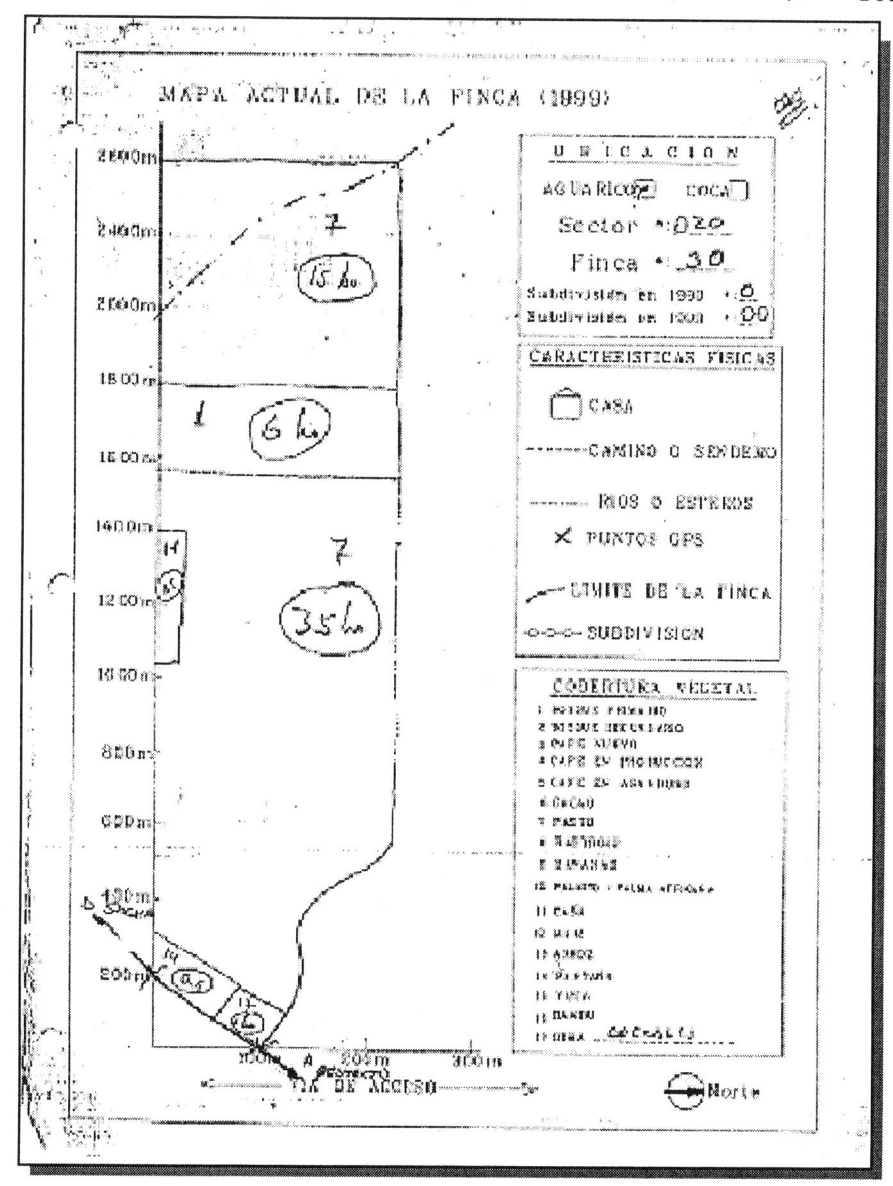

Figure 3. An example of a sketch map for a *finca*

These sketch maps were also used to assist in the classification of satellite imagery and to assess the accuracy and reliability of LCLU information derived from the 1990 and 1999 Landsat satellite data. Finally, recently acquired multispectral and panchromatic IKONOS data (1-m panchromatic and 4-m multispectral) for October 2000 are being related to the sketch maps. Subsequent discussions (1) emphasize the difference between compositional reliance versus spatial reliance in the use of the sketch maps for calibration and/or validation of image classification and/or image change detections, and (2) sketch maps will be related to the high-resolution IKONOS data as an initial test for scaling to Landsat TM data, which is more widely available throughout the study area. Details will be provided in the discussion section to follow.

4.4 GPS Technology

GPS coordinates were obtained by interviewers in the course of the 1999 household survey for the two "front" points of the property (on the road, or closest to a road) and for the house; thus at least three points—both averaged waypoints and rover points—were collected if there were subdivisions. In some cases, the owner of the subdivision lived somewhere other than on the plot of land being surveyed. In those cases, the GPS point was taken at the owner's dwelling unit off the property. The use of GPS receivers by interviewers to determine their location in the field, together with accurate SIIM products to navigate and locate specific farm lots in sample sectors, was an invaluable time-saver in fieldwork.

The different land-use patterns associated with the colonists versus the indigenous groups (not involved in the 1999 household survey but being analyzed as part of a subsequent NIH grant to Bilsborrow in 2000) required different approaches to linking spatial and nonspatial data. In the case of colonists, we created polygons for each 1990 household farm or *finca madre* and linked them to the spatial and nonspatial data corresponding to the *fincas*. However, the indigenous groups tend to live in clusters of households and often have small plots of land in agricultural use, known as *chacras,* which may be near their homes or several kilometers away. Thus the house is not located on or adjacent to the farm plot. Also, *chacras* are usually used only for a few years and then abandoned as soil fertility declines, but they remain linked to the indigenous household. So a file linking *chacras* to households over time was created.

To capture the locations of these areas, at least five GPS points were collected for each *chacra* included in the study: a point is collected at the center and at each corner of the plot, and an identification number is assigned to the *chacra* to link it to its household. If possible, dates of establishment or clearing of plots and dates of abandonment were obtained. For abandoned sites, the date was often only estimated by the respondent, which led us to focus on plots abandoned within the previous five years. The corner points for each *chacra* were converted to a polygon with an assigned identification number. This allows us to analyze all *chacras* belonging to a particular household in relation to the survey responses of that household.

4.5 Survey Linkages through GIS Attributes

Linkages between the household survey data and the GIS and remote sensing data were developed using Arc/Info software. This allows linking farm and household attributes from the 1990 and 1999 surveys with LCLU dynamics measured based on the satellite time series. Linkages between locational and nonlocational information within the database for each *finca* for 1990 and 1999 also facilitated the development, calibration, and verification of the LCLU simulations for antecedent and subsequent time periods, which will allow us to establish diffusion rules from the statistical relationships obtained from multivariate models.

In the first survey conducted in 1990, almost all *fincas* (94 percent) were intact plots of 40–50 hectares owned by one farmer. We refer to each of these plots as a *finca madre;* these are the *fincas* described earlier. But by 1999, this ownership pattern had changed, with 40 percent of all *finca madres* subdivided one or more times since 1990. Reasons for subdividing included giving part of one's plot to family members and selling portions of it to nonfamily members to obtain cash.

The predominance of *finca* subdivisions in the study area created a problem in linking the spatial and nonspatial data. Thus we collected survey data for each subdivision of every *finca*, which in some cases included as many as ten subdivided parcels. However, we obtained the spatial location only of the *finca madre*, not of each of the subdivisions. This leads to the question: How can one link multiple responses to a single location? We view this as a "many-to-one" matching problem that requires the development of database linkages that track subdivisions from an attribute perspective rather than a spatial perspective.

GIS software has the capability to deal with these "many-to-one" relationships, as well as the more common "one-to-one" relationships. Once a linking field, known as the *primary key,* has been set up with some type of identifier, linkages can be made. In our case, the primary key is the unique household ID number. To prepare the survey data for this linkage, all subdivisions were given an additional ID number that was the number of the *finca madre* in which the subdivisions were associated. Using the "many-to-one" linkage capabilities of the GIS, all subdivisions were linked to their *finca madre* in the spatial domain, allowing for all the data to be viewed interactively by selecting a *finca madre* polygon.

The next question is: How does one then analyze this linked data? The initial solution was to aggregate the responses (attribute data) for all subdivisions in a *finca madre* so that there was only one response for each variable for that location. However, this was only possible for continuous data, not for nominal or ordinal data. Thus it was not a solution for some of the variables in the survey data. Solutions for nominal and ordinal data were dependent on the particular type of data and its measure, which were decided on a case-by-case basis. Another option we investigated was to maintain the subdivision information as separate records so that descriptive statistics, for example, could be derived for each *finca madre* relative to household attributes for "all" households in the *finca madre*. This is analogous to maintaining subpixel resolution information for a group of pixels specified within a kernel window prior to aggregation to a coarser cell. Maintained at the previous aggregation level, information might include the mean value, minimum value, maximum value, coefficient of variation, and standard deviation. In short, the *fincas* created through land subdivision can be aggregated to serve as contextual information (at the *finca madre* level) for studying the behavior of the individual (subdivided) farm household units and, hence, as layered information within a hierarchy.

4.6 Integrating Remote Sensing Images and Resolutions

A satellite time series was assembled that extends from 1973 to 2000. While some synthetic aperture radar (SAR) data sets have been obtained for cloud penetration, soil moisture mapping, and subcanopy assessments, optical systems have been primarily emphasized. A host of Landsat TM images data has been acquired, preprocessed, and classified for Paths 8 and 9/Rows 60 and 61. The currently assembled Landsat TM time series includes images for 16 December 1984, 23 August 1986, 23 January 1987, 11 September 1987, 23 April 1988, 7 August 1989, 22 December 1989, 28 April 1990, 19 February 1991, 14 July 1992, 11 August 1996, 3 September 1996, 21 October 1996, 30 August 1997, 6 September 1997, 24 June 1999, 10 July 1999, 15 November 1999, 30 August 2000, 9 November 2000, 21 August 2001, 9 September 2001, and 8 January 2002. In addition, a number of Landsat MSS scenes from the 1970s and IKONOS panchromatic and multispectral data sets from 2000, 2001, and 2002 have been acquired (Figure 4). The IKONOS data sets are areally constrained to small test areas, but a number of sample *fincas* and sectors are included.

The Landsat TM and MSS time series has been used to characterize LCLU dynamics with special emphasis on the time periods of the longitudinal household surveys and the immediately preceding periods so that a LCLU context might be developed relative to the questionnaire responses. While defining the state (e.g., LCLU type) and condition (e.g., landscape greenness and/or plant biomass) of the landscape for target dates was critical, the power of the images contained within the time series was used to ascertain the nature of landscape change—that is, the change in composition, spatial organization, and direction of change for pairwise combinations of image dates, as well as the trajectory of change across the landscape by considering "pixel histories" as a panel data set where stability and/or dynamics at the pixel level might be examined.

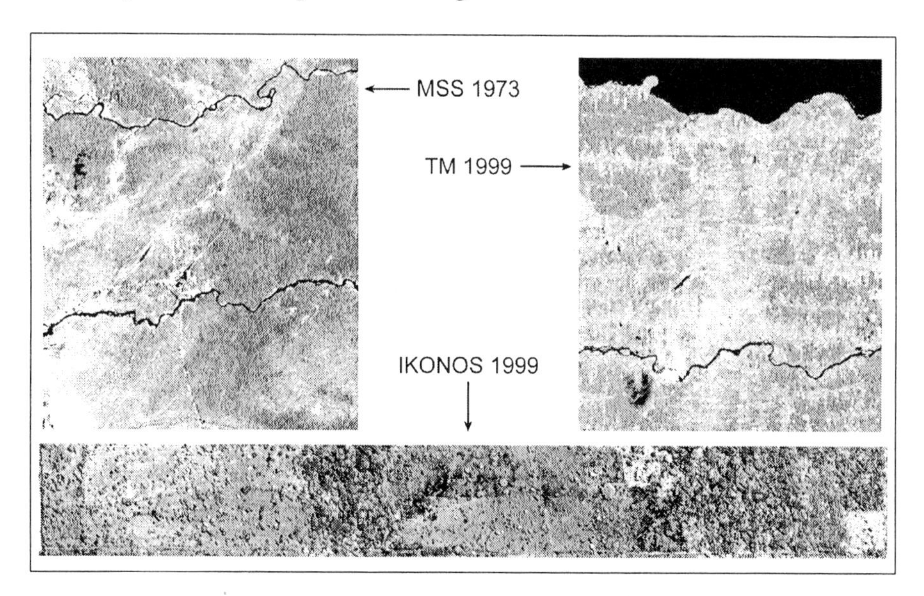

Figure 4. Digital image data used in this research

One of the keys to understanding how socioeconomic and demographic factors are related to LCLU change lies in the linking of the household survey data to the spatially and temporally explicit satellite data. Given that we are primarily relying on an optical Landsat MSS/TM time series to understand LCLU dynamics over a twenty-seven-year period, it is crucial to identify the locations of each of the sample *fincas*, or farms, on the satellite imagery and track their LCLU patterns over space and time. In the Oriente of Ecuador, as stated, most *fincas* tend to be rectilinear in shape and average 50 ha in size. The most common dimensions of these *fincas* are 250 x 2000 m. These *fincas* normally lie adjacent to one another in groups known as sectors. When viewed from a vantage point in space, these *fincas* are distributed in the classic "fishbone" settlement pattern common to the Amazon and are often associated with the location of roads. Figure 5 shows the general placement of *finca* boundaries on a Landsat TM image.

The first step in locating the sample *fincas* was to collect differentially corrected GPS data at the front corners of each property. In most cases, this resulted in two corner points. In a few situations, however, the *fincas* were irregularly shaped and required the collection of more than two points to satisfactorily identify the *finca's* shape. Differentially corrected GPS data were also collected at each of the farmhouses in which the survey was administered. In all cases, differentially corrected GPS processing was performed using base-station files collected at a base station in Quito and processed at UNC following the field data collection.

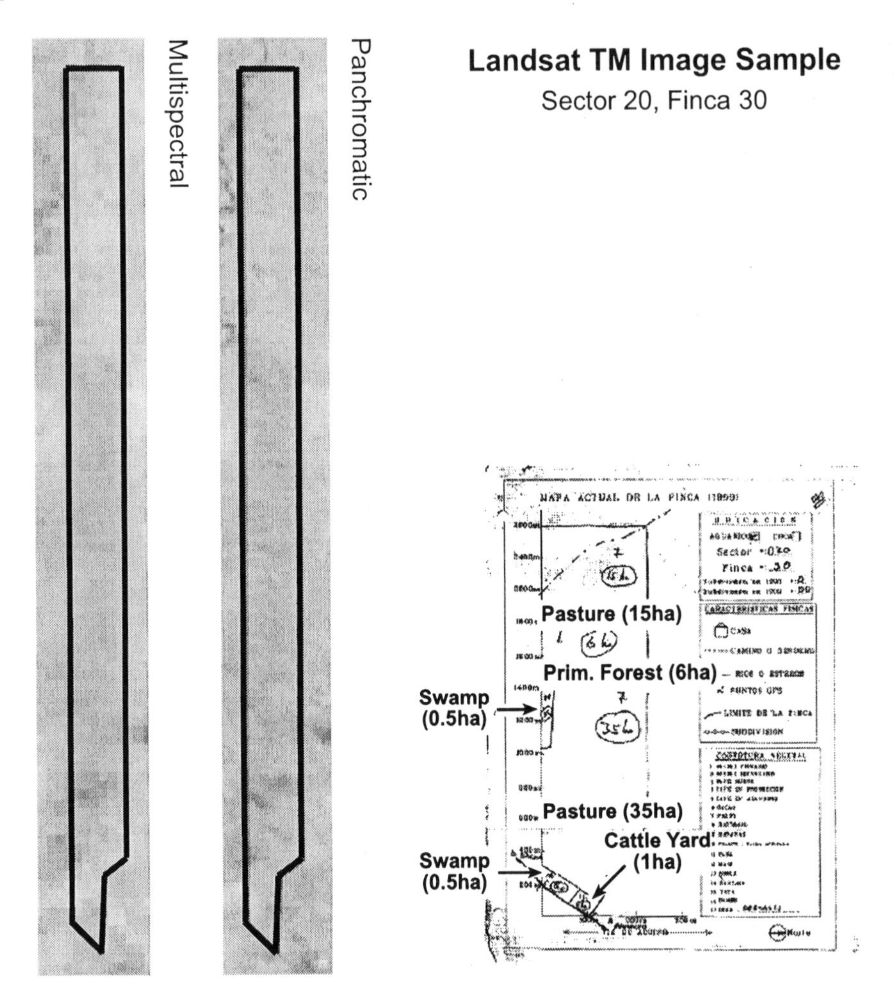

Figure 5. Landsat TM image with a sector and *finca* boundaries superimposed

The Landsat time series was rectified using the relative correction method, after a reference or master image was corrected using the absolute registration method. Geodetic control points were gathered in the field to aid in georeferencing, given the paucity of stable control points recognizable on the imagery as well as on the base maps. The master image, dated October

1996, was rectified to a large number of differentially corrected GPS points that were collected for geodetic control. The October 1996 image was selected as the master because it was the most recent cloud-free image that we had acquired at the time. A location was chosen for geodetic control if it was easily located on the image and if it was a static feature such as a bridge or intersection. The remaining images in the time series were then rectified to the 1996 master image. All images were rectified to within 0.5 pixels. The creation of a geodetic control network throughout the study area was crucial in establishing the horizontal and vertical control needed in a host of spatial operations.

Within the GIS, the *finca* corner points defined using a GPS were overlaid on the satellite imagery to spatially locate the *fincas*. The corner points and a variety of other data sources (e.g., sketch maps, ground photo, *finca* size reports from the 1990 survey, Landsat TM imagery, and aerial photography) were used to delineate the boundary of each sample *finca* on the satellite imagery and to create a GIS database of all sample *fincas* that contained spatial and attribute information. The result was a data set in which sectors and *fincas* were outlined by a cadastral polygon that contained a linked, unique household identification number as the key field.

After the satellite images were classified into LCLU classes, the *finca* polygons were used to clip the data for each individual *finca* so that intra- and inter-*finca* LCLU dynamics could be assessed as a stand-alone remote sensing analysis or as part of a study that compared LCLU as reported in the household surveys and represented on the sketch maps to derived remote sensing characterizations. Two analyses have been completed thus far: (1) comparing differences in reported versus calculated size of the *finca madres,* based on the cadastral data created using GPS points and Landsat TM images, and (2) comparing the percent of land cleared in 1999.

First, we determined how closely the area of each *finca madre* in the spatial data agreed with the area reported by the farmer in the 1999 survey. The calculated area of each *finca* was derived from the polygon that represented its border. These cadastral polygons were created using the methods described above. The areas of all subdivisions reported by household heads for each farm in the same *finca madre* were then summed to derive the total reported area based on the household survey. There are two identified sources of error in the reported areas for the *finca madres*. First, in many cases the farmer could not accurately locate one or more borders of his property, and second, in some cases—including in virtually all involving subdivisions—the property borders had never been surveyed, so the reported areas are estimates.

The comparison between the survey and the remote sensing calculations was performed on 133 sample *fincas* within three ISAs. Two of the *fincas* are outliers for this analysis, as the difference between the reported and calculated areas were extreme and were due to known errors. With these outliers removed, the calculated and reported areas for the remaining 131 *fincas* were compared. The differences in the areas ranged from 15.53 ha overreported to 26.35 ha underreported. The standard deviation of the differences was 7.02 ha. Of the 131 *finca madres* analyzed, 100 were within 1 standard deviation of the area calculated from the satellite images and 120 were within 2 standard deviations. Specifically, 30 *finca madres* are within 1 ha, 51 within 2 ha, 63 within 3 ha, 72 within 4 ha, and 83 within 5 ha. Thus 48—or about one-third—differ by more than 5 hectares or by over 20 percent.

Second, the analysis of the percent of forest cleared on a *finca madre* was based on a comparison of household survey data and spatial data. The comparison relied upon the 1999 survey and a LCLU classification from Landsat TM data. On the survey side, the amount of forested land in 1999 was calculated by summing primary or secondary forest and was subtracted from the total area of the *finca madre,* yielding percent cleared. On the spatial side, the amount of land with LCLU types other than forest was calculated for each *finca madre* using the cadastral data as spatial boundaries. This was divided by the total area to determine the percent cleared from the spatial data. For each *finca madre,* the percent cleared (survey) and percent cleared (spatial) were then compared. A correlation coefficient was then calculated between the two sets of figures. This analysis was performed based on data for the *finca madres* in the four ISAs (n = 191). The highest correlation was found for the northern ISA (Intensive Study Area) that contained Lago Agrio (r = 0.758), while the lowest was in the southwest ISA (r = 0.510), which happens to be a more heterogeneous area involving both established and newer *finca madres.* The overall correlation was 0.666 for the four ISAs.

Factors affecting the level and variation in the correlations include the nature of the land classification schemes used during the survey and the remote sensing image processing. The field surveys are more specific in describing LCLU types that occur on the farm, whereas the remote sensing image processing is, at present, mapping LCLU only at level-1 (e.g., agriculture) and some level-2 (e.g., agriculture-rice) categories. In addition, the distinctions between land use and land cover continue to be confounding at times. For instance, mature coffee trees may be viewed by farmers as either a functioning coffee plantation or as pastureland—or both—depending upon whether the land parcel was removed from production and relegated to unimproved pastureland for grazing of cattle and/or goats. From a remote sensing perspective, as a consequence of the closing of the plant canopy and increasing crown densities, mature coffee trees are often indistinguishable and hence are viewed as secondary forest succession. This alters the spectral response pattern and suggests forest and not coffee as the type of land use. The age of the farm, transitional nature of forms of land use, differences in the survey teams, and natural variations in site conditions on the farm itself all may lead to some variation in the strength of correlations. As our image analyses improve and we attain more detailed characterizations of LCLU types at the farm level from further fieldwork and lab analyses, the correlations should improve, but differences will no doubt continue to exist between estimates of the survey data and the satellite data.

In December 1999, the Complex Systems Research Center at the University of New Hampshire's Institute for the Study of Earth, Oceans, and Space submitted a request to the NASA Scientific Data Purchase Program for the acquisition of high-resolution IKONOS imagery for LBA projects. The IKONOS satellite is a commercial satellite owned and operated by Space Imaging. It provides both multispectral and panchromatic images with very high spatial resolution. The multispectral imagery has a 4 x 4 m spatial resolution, while the panchromatic images have a 1 x 1 m spatial resolution (Figure 6). Data were acquired (panchromatic and multispectral) for one of our test sites—La Joya de Los Sachas—and *finca* boundaries were overlaid for a sample sector.

In the summer of 2000, we performed a detailed, field-based LCLU analysis of one *finca* contained in the La Joya IKONOS scene. The entire 50-ha *finca* was walked with a GPS unit, and differentially corrected GPS data were collected for all LCLU types occurring within the *finca*. Using the GPS and IKONOS data, a highly accurate and detailed data set was created in which each plot on the *finca* was delineated as a polygon, spatially represented, and attributed with current and historical LCLU information, as well as future cultivation plans for the plot (Figure 7). Subsequent analyses will use this "reference" *finca* for linking between remote sensing, sketch maps, and household data. Additional IKONOS overflights were authorized for 2001 so that a time series can be constructed for selected survey *fincas*. Also, IKONOS data has been authorized for our study of indigenous communities where *chacras* are much smaller than the *fincas* of colonists and households and lands of the indigenous groups are generally areally distributed and spatially disjointed.

The LCLU polygons for the "reference" *finca* and subsequent *fincas* identified from future IKONOS images will be overlaid on Landsat TM imagery to develop an archive of LCLU spectral signatures. These spectral signatures will be used for future classifications, ground control activities, and validation and calibration efforts. Spectral signatures can also be created from the multispectral channels of the IKONOS data, allowing for classification of these high-resolution images.

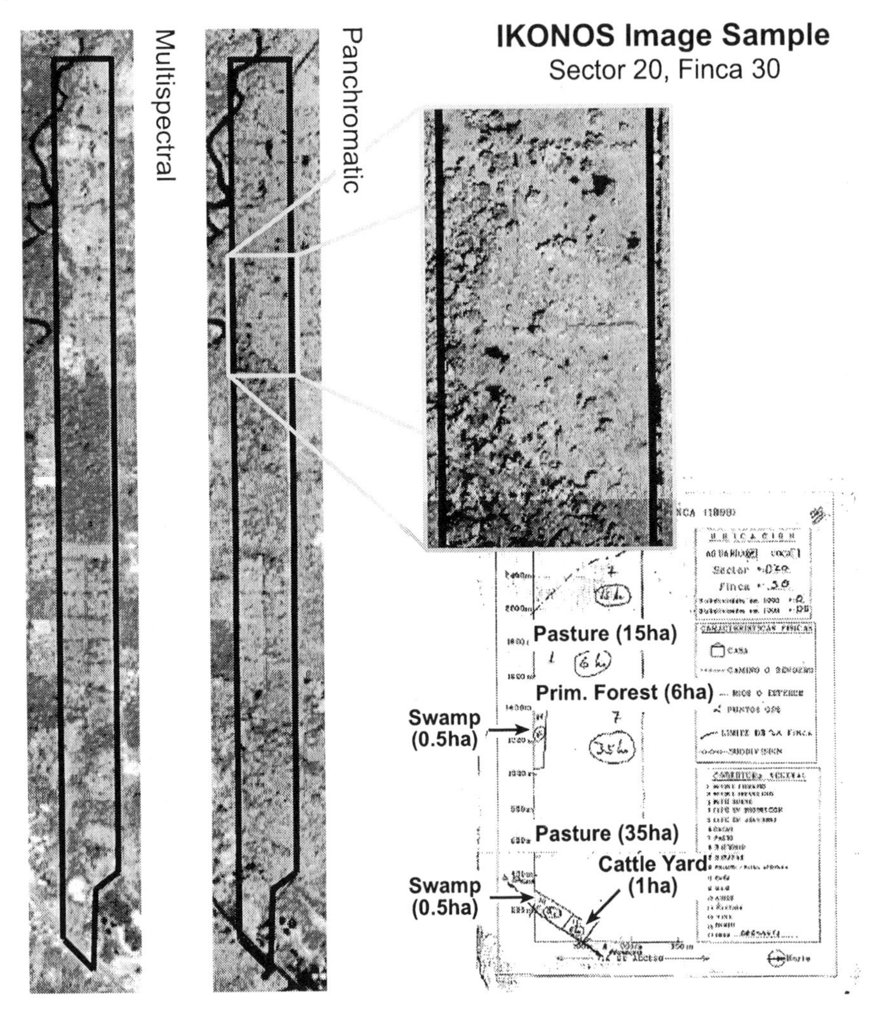

Figure 6. IKONOS multispectral and panchromatic images, with a sketch map for a corresponding *finca*

To extract more information from the satellite imagery, we anticipate performing spatial and spectral merges on concurrent TM and IKONOS images. The 1999 IKONOS and TM collection dates are only two weeks apart, making them ideal for this application. There are several options for spatial/spectral merging: (1) Merge the 30 x 30 m multispectral TM with the 1 x 1 m panchromatic IKONOS; (2) merge the 15 x 15 m panchromatic TM (Enhanced TM Data from Landsat-7) with the 4 x 4 m multispectral IKONOS; or (3) merge the 30 x 30 m multispectral TM with the 4 x 4 m multispectral IKONOS. A useful and relatively simple approach for fusing the two data sets is to alter the data model from the traditional RGB model

(Red-Green-Blue) to the IHS model (Intensity-Hue-Saturation) or a Principal Components-based transform for image representation. The IHS model integrates the spatial resolution of the sensor having the higher spatial resolution with the spectral resolution of the sensor having the higher spectral resolution, thereby generating an integrated image that has the best of both worlds: high spatial and high spectral resolution.

Intensive Finca LULC Surveys

IKONOS Multispectral

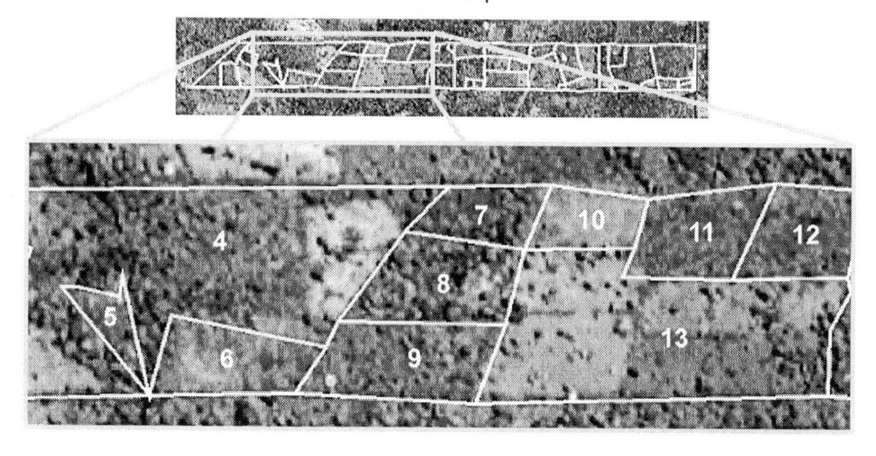

Poly ID	t - 2	t - 1	t	t + 1
4	Sec. Forest	Sec. Forest	Sec. Forest	Pasture
5	Sec. Forest	Pasture	Pasture	Pasture
6	Pasture	Coffee	Coffee	Fallow
7	Pasture	Pasture	Coffee	Fallow
8	Sec. Forest	Pasture	Palm	Palm
...	...	...	...	...

Figure 7. Referencing LCLU on a *finca* over time using an IKONOS image and a GIS database

Finally, the IKONOS data is being used to provide subpixel information in relation to Landsat TM data. The goal is to determine subpixel proportions of LCLU types for a Landsat TM pixel by considering the corresponding group of IKONOS pixels. Two tracks are being developed: (1) linear mixture modeling of Landsat TM data through the characterization of LCLU end members (Adams et al. 1995), and (2) defining the relationship between information represented on the survey sketch maps of LCLU (both composition and spatial organization) and Landsat TM data characterizations, using the IKONOS data as the intermediate data source for verification and/or calibration (see Figure 6). Figure 8 shows how the assigned household identification number reports the location (i.e., region, sector, and *finca*) of the survey unit and the number of farm subdivisions in 1990 and 1999. It also shows the database structure used to link two hypothetical variables (VAR1 and VAR2), and how the 1999 household identification numbers are linked to the *fincas madre* identification number through a GIS "join-item" operation, which is designed to link attribute data through the use of a key field.

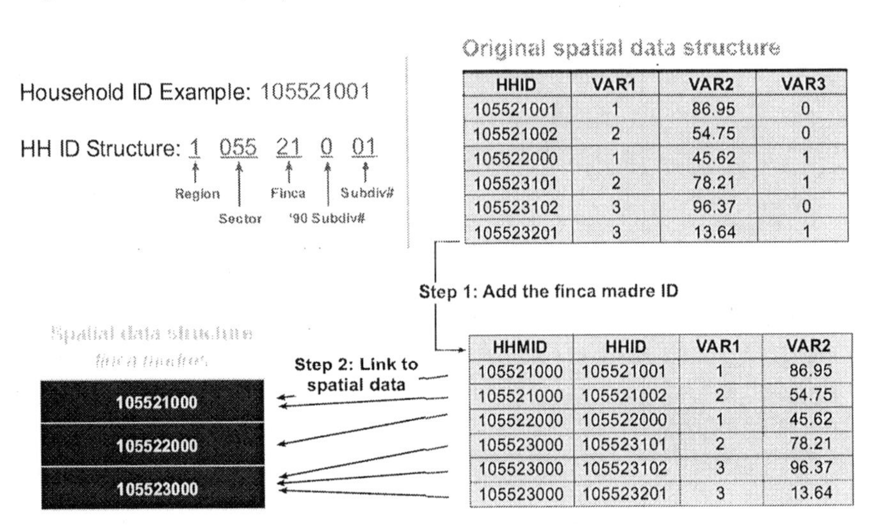

Figure 8. Tracking the links between subdivided *fincas* and *finca madres* through a GIS database

4.7 Pattern Metrics

Pattern metrics are commonly used to assess the nature of landscape composition and spatial organization, often characterized through remote sensing systems and image classifications. The metrics are used to generate spatial/temporal signatures of landscape patterns as part of the scale-pattern process paradigm articulated through Hierarchy Theory. The use of pattern metrics recognizes that the surrounding neighborhood and the regional spatial context in which local sites are juxtaposed influences landscape properties. The assumption is that landscape form is related to landscape function.

Computer software, such as "Fragstats," is available to assess landscape structure at the landscape, class, and patch levels. The landscape level represents patterns within a preset boundary such as a *finca* or a development sector. Class levels are set by the thematic nature of landscape types; for example, a forest class occurring throughout some areal unit, such as a *finca* or a sector. The patch level represents patterns associated with a single contiguous landscape type—for instance, a patch of forest. Metrics can be computed to assess landscape characteristics such as fragmentation, isolation, juxtaposition, edge, mean patch size, shape, fractal dimension, and so on. When metrics are calculated from a single-date LCLU classification, the results provide quantitative information on landscape composition and spatial organization; but when pattern metrics are calculated for an image time series, it is possible to define how the landscape structure has changed over space and time for various ecological scales and landscape strata.

Thus far, we have computed pattern metrics for three 90,000-ha ISAs (Intensive Study Areas) that have been strategically positioned within the Oriente so that we might examine the social, biophysical, and geographical drivers of LCLU dynamics related to patterns of LCLU organization. Each of these ISAs has a different age of colonization, topographic makeup, geological structure, relationship to market and service centers, and road and river networks. We hypothesize that each ISA is representative of a different stage of LCLU dynamics occurring within the broader region.

A basic question in the application of pattern metrics is: At what spatial level do social units (e.g., households or communities) or environmental units (e.g., hill slopes or watersheds) influence the composition and spatial structure of the landscape through LCLUC? Conversely, what are the feedbacks between LCLUC, the environment, and human behavior? Perhaps each farmer uses land in unique ways, regardless of the action of neighbors. Or perhaps all farmers within a particular sector use their land in a similar fashion because they talk to each other or are constrained by similar environmental conditions or socioeconomic forces that operate at coarser scales. Thus land use may be influenced by proximity or access to a market town or the connection of a *finca* to other *fincas*. Spatial and social networks refer to different types of connections across the landscape. But how does one define proximity, access, and connection? One method is distance from the nearest market town. For example, radial buffers can be used to identify Euclidean distance from Lago Agrio to all survey *fincas* and sectors in the northern ISA with network distance based on road types and travel speeds (McDaniel 2000). Spatial buffers can also be used to define spatial strata hypothesized to affect human behavior.

To answer these questions and others, both landscape and class metrics were calculated for selected areal units and ISAs. First, the metrics were run for each sample *finca* focusing on the *finca madre*. Second, each sample sector was defined as the landscape, and metrics were again computed. Third, proximity to Lago Agrio of each surveyed *finca madre* was estimated using 1-km radial buffers, and the metrics were run for each buffered territory. Fourth, proximity to "all-weather" roads was computed using spatial buffers, and the metrics were run for those defined distance strata. Finally, the metrics were run for each of the ISAs to assess the influence of remoteness versus connectedness on landscape structure.

4.8 Remote Sensing Image Change Detections

LCLU composition and spatial organization are critical indicators of landscape form and function. But so too is the nature of change characterized between selected image dates, where the spatial-temporal and compositional-organizational elements of change can be assessed. A number of image change-detection approaches were used in the Ecuador project to characterize landscape dynamics—some might be classified as traditional, others as experimental. Strategically, these change images correspond in space and time to the longitudinal survey data, with appropriate space and time lags accounted for. Here, we briefly describe a number of image-change approaches we used to characterize the Oriente through time:

- Channel/scene integration: near-infrared channels (e.g., channel 4 of Landsat TM) from scenes of different periods composited as a qualitative method of assessing regional change.
- Multidate composite: multiscene data stack representing different time periods used as the input or feature set for an unsupervised classification to identify changing and nonchanging spectral clusters.
- Principal components analysis: change is assessed through the derivation of eigenvectors that relate spectral channels from scenes collected at different time periods to generate components and eigenvalues that indicate the percent variance associated with each of the defined components.
- Image algebra: the ratio is found for two channels of the same spectral region and wavelength for two time periods, and image differencing is achieved by subtracting the spectral responses of one date from that of the other.
- Binary mask: uses a multidate image composite recoded into a binary mask consisting of areas, noting areas that have changed and not changed between two dates.
- Post-classification: classifies scenes from two different dates on a pixel-by-pixel basis and reports through a change matrix.
- Minimum and maximum change: records the year of lowest and highest changes in NDVI across the image time series.
- Change vector analysis: characterizes change magnitude and direction in spectral space. An extension of this technique measures absolute angular

changes and total magnitude of Tasselled Cap indices (brightness, greenness, and wetness); polar plots and spherical statistics summarize change vectors to quantify and visualize both magnitude and direction of change (Allen and Kupfer 2001).

- Transition probabilities: generates from-to change classes across the image time series that are based upon probabilities of change (Allen and Walsh 1993).

The above approaches are designed to characterize LCLU dynamics. They are by and large sensor independent, although certain spectral regions are necessary to compute the NDVI, a measure of surface greenness. It is customary to apply these change-detection approaches on two-image sets, but multiple images and periods can be assessed through the more sophisticated techniques, such as change vector analysis and principal components analysis. Linking people and environment to change images is potentially complex because of the high number of possible change classes when a number of images are contained within the assembled time series. Selecting certain trajectories of change and formulating hypotheses about how the human agent has shaped the structure and function of the landscape with time is generally not well studied. But characterizing the nature of change is among the initial challenges and operations.

4.9 Remote Sensing Image Change Trajectories

Change trajectories of LCLU are considered analogous to panel data analysis, in which survey respondents are mapped over time through multiple observations. Instead, here the given unit of observation is the pixel, and its "life history" is constituted by the values derived from the component images of a satellite time series. For each pixel, values derived from satellite images captured at multiple points in time are combined into a "from-to" change detection in a panel data or pixel history approach. To represent pixel histories or trajectories, consider, for example, the classification of five Landsat TM images into forest (F), secondary forest (SF), agriculture (A), pasture (P), and other (O). Once classified, the per-image LCLU categorizations can be grouped relative to time-specific LCLU pixel histories or trajectories [e.g., $image_1$ (F), $image_2$ (F), $image_3$ (F), $image_4$ (A), and $image_5$ (A) *versus* $image_1$ (F), $image_2$ (SF), $image_3$ (SF), $image_4$ (P), and $image_5$ (P)] mapped at the pixel level to indicate possible patterns or trajectories of LCLUC (Crews-Meyer 2000, 2001). Our hypothesis is that the nature of the trajectory is associated with the function of the land at that pixel and its neighborhood of similarly related pixels. Hence, different trajectories of LCLU may suggest, for example, differences in the stability or dynamics of LCLU over time and space, which is further suggestive of land sustainability or resilience. These differences in space-time patterns are related to landscape form and function, important factors in understanding the interactions between population and the environment. Figure 9 shows the trajectories of LCLU change in the northern ISA, showing agriculture, new forest, and new agriculture for 1973–99, using four classified Landsat TM images for 1973, 1986, 1989, and 1999. Landscape form and function are related, so the past offers insights for the future, where space and time concepts are integrated through LCLU change trajectories and pixel histories.

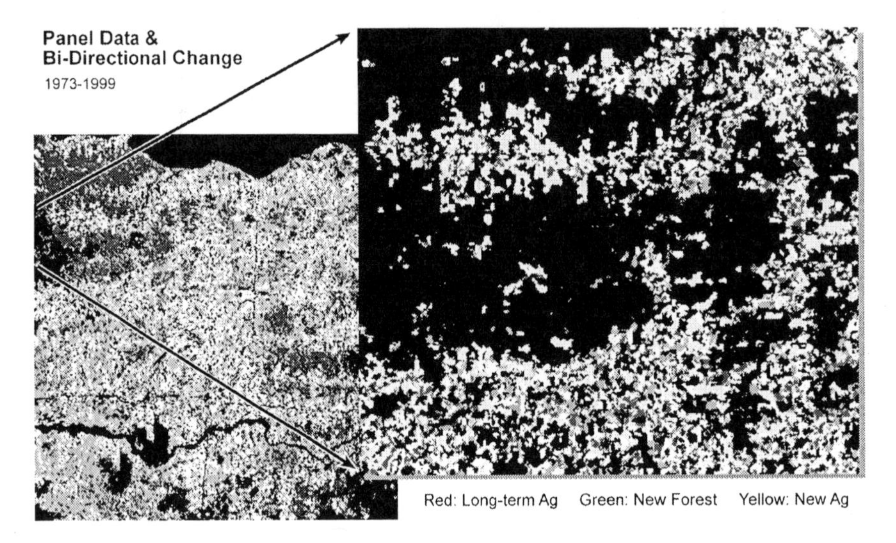

Figure 9. Pixel change trajectories for 1973, 1986, 1989, and 1999 Landsat TM images

4.10 Empirical Models of Household Decision Making: Linking People, Place, Environment, and Time

Each part of the amalgam of land-use decisions a migrant farmer makes at a given point in time has implications for the other parts; a full analysis must take this into account. Given the interrelated and zero-sum nature of land-allocation decisions, a statistical model that simultaneously takes into account the influences of spatial factors, individual and household characteristics, and community factors is being developed that integrates LCLU, LCLU dynamics, resource endowments, geographic accessibility, and socioeconomic characteristics of households on the survey *fincas*. Thus, satellite imagery, GPS technology, GIS tools and techniques, and data from household survey instruments will be analyzed using statistical methods to determine how many categories of LCLU can be distinguished and used in the multivariate models. The ability to link observations spatially and temporally across socioeconomic, biophysical, and geographical domains is critical. Multilevel models will be used to examine variations in household decisions viewed in the larger context in which those decisions are made. Some models will require linking methods previously discussed.

An important estimation issue in the multivariate models is combining areal (community) and individual/household-level independent variables in the same equation, while still satisfying the required statistical assumptions regarding error terms. To do this, a generalization of the Laird and Ware (1982) Generalized Linear Mixed Model (also referred to as a Multilevel Model) will be used with a two-component error structure for each of the land-use share equations. Each equation will then take the following general form (the dependent variables being the proportions of the plot in each of the LCLU communities):

$$Y_{ij} = b_{ij} X_{ij} + g_j Z_j + d_{ij} + v_{ij} \tag{1}$$

where Y_{ij} is the outcome for farm plot i in community j; X_{ij} are values of the independent variables for the ith household in the jth community; Z_j is a predictor for a measure in the jth community; b_{ij} and g_j are corresponding regression coefficients; d_{ij} and v_{ij} are the respective community- and household-level error terms for Y_{ij}, with $d_j \sim N(0,\sigma_0^2)$ and $v_{ij} \sim N(0,\sigma^2)$.

The two error components in equation (1) allow for the specification of a spatial and/or temporal correlation structure across the error terms for households within a given sector or community. Such a correlation is likely since households within a sector or linked to a particular community are often similar to each other, since they may confront similar environmental conditions and market access and may also have certain similar socioeconomic characteristics (such as coming from the same region of origin), which may lead to similarities in patterns of LCLU.

4.11 Cellular Automata and LCLU Simulations

Space-time simulations of LCLU patterns have been developed for the northern ISA (containing Lago Agrio) through the use of cellular automata (CA) models (Messina and Walsh 2001; Messina 2001). The CA models integrate data contained within the GIS on LCLU dynamics from the image time series and postulate landscape conversion based on the transportation network.

A CA system consists of a regular grid of cells, each of which can be in one of a finite number of k possible states, updated synchronously in discrete time steps according to a local interaction rule. The state of a cell is determined by the previous states of a surrounding neighborhood of cells (Wolfram 1984). The rule contained in each cell is essentially a finite state machine, usually specified in the form of a transition function or growth rule that addresses every possible neighborhood configuration of states. The neighborhood of a cell consists of the surrounding (adjacent) cells.

Within CA and the broader theoretical framework of Complexity Theory, it is expected that there will be both nonlinear and hierarchical relationships among the biophysical, spatial, and socioeconomic factors that govern LCLU change within the Oriente. Part of the modeling requirement, then, is the focus on modeling the highest level of abstraction necessary to capture the details of the system (Dale et al. 1993). Within the Ecuadorian Amazon, local and regional characteristics are driven by decision making at the household level and captured in the longitudinal socioeconomic surveys. These decisions manifest as changes upon the landscape in both the composition and spatial organization of LCLU types. The integration of this household decision and the landscape change is through the assumption of neighborhood interaction. The functional neighborhood defines both the knowledge-base diffusion constraints as well as a biophysical control on landscape elements and associated change conditions or possibilities.

Under the Ecuadorian Amazon scenario, a CA model was used as the LCLU change-based predictive, dynamic, spatial simulation model. As a CA theory-based simulation, model building focused on small localities or functional neighborhoods surrounding sampled and/or surveyed locations where high-quality data were available for model development and validation (Messina et al. 1999). The development of the model relied upon relationships between the social survey, infrastructure, and biophysical data, as well as remotely sensed time series data that were processed to characterize LCLU dynamics (Figure 10). Back simulations were run for calibration and validation that involved a comparison to images in the time-series not used for rule development (Figure 11); forward simulations were run for selected time periods, where derived LCLU patterns were interpreted within a policy context.

5. DISCUSSION

The broader context in which households make land use decisions is reflected in a number of community-level or contextual factors hypothesized to affect land use. Among the potentially most relevant in the Amazon region are:

- Characterization as an urban area or proximity to the nearest large town, since this indicates access to off-farm employment providing an alternative source of income that may alleviate pressures on settlers to clear more of their plots.
- Population size of local towns, since a larger local market may facilitate market sales of farm produce and/or better prices.
- Distance by road from the local produce market to the provincial capital or other major market town in the Amazon region, for similar reasons.
- Availability of credit or distance to the nearest source—Banco Nacional de Fomento—are positively related to land use, including the area in pasture, since cattle are often accepted as collateral.
- Proximity to an agricultural extension center may reduce land clearing by enhancing agricultural intensification through greater contact with extension agents.
- Land subdivisions are more likely the older the head of household (and therefore the older his children) and the larger the number of children resulting from past fertility, through subdivision among heirs; they are also more likely if the *finca madre* is near major roads and towns.
- The resource endowments of a site, particularly soil fertility and topographic features, affect drainage and geomorphic conditions and land productivity and use.

The spatial organization of the landscape may influence the relationships between the environment and human behavior, such as LCLU conversion practices, resource flows along connected patches, and possibly migration patterns.

In the following section, we discuss how broad characteristics of household forces can influence LCLU patterns and how we have linked data across thematic domains and space-time dimensions to reflect hypothesized drivers of LCLU dynamics.

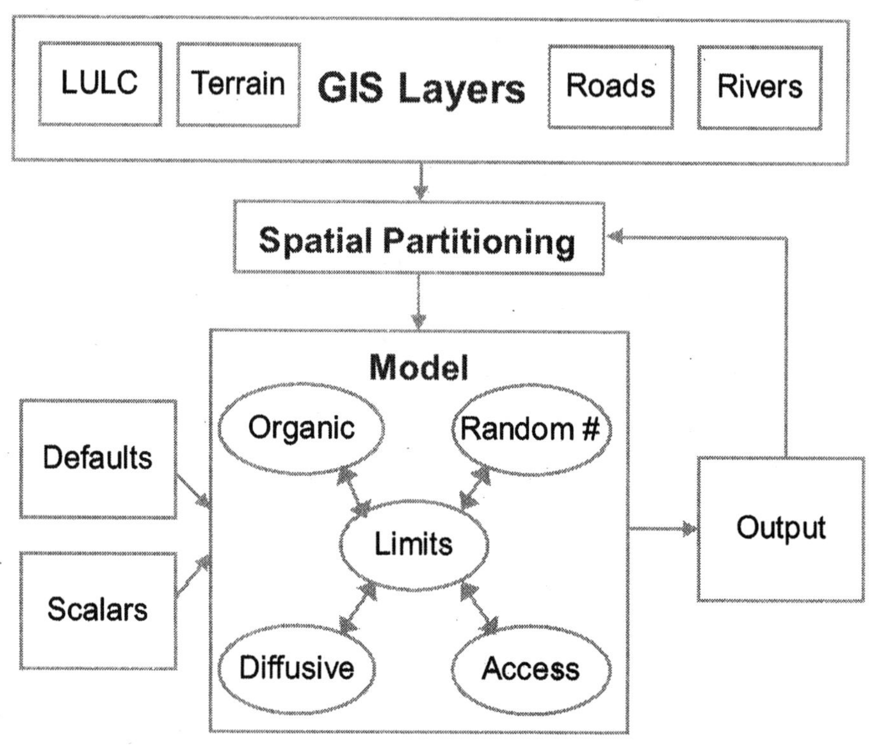

Figure 10. General structure of the CA model used for LCLU simulation

5.1 Linking People to the Environment

Regarding the household surveys, we included retrospective and prospective questions in the questionnaire and used event seeding as temporal markers to aid in event reconstruction and LCLU chronologies. We asked questions about "what," "where," and "when," and can identify intra- and inter-*finca* connectivities through the survey design. We developed spatial and temporal hypotheses about LCLU dynamics and have begun to explore drivers of LCLUC taking into account socioeconomic, biophysical, and geographical factors in the data collection and research design. Some of these scale-dependent drivers we have hypothesized to operate with time lags.

During the data collection, for budgetary reasons, we did not implement full household surveys of households living on *solares* or obtain GPS coordinates of *solares*, the small, subdivided plots. The *finca* or agricultural farm was our spatial unit of observation, whereas the household was our social unit of observation. They are not always the same. This dichotomy in the definition of the study units produced differences and introduced the need to develop special linking protocols. We did, however, obtain GPS coordinates for the *finca madre,* and hence our spatial unit of observation was well represented. We also had the dwelling of the household linked to each farm (subdivision) of the *finca madre* spatially referenced, so our socioeconomic units of observation are well specified. But we did not obtain the spatial location of the *solares*, and hence we did not maintain consistency in the characterization of the spatial dimension. While we anticipated the possibility of subdivisions prior to the fieldwork, we could not anticipate the large number of subdivisions or—especially—the many *solares*. The latter reflected some incipient aspects of urbanization as it occurred on *fincas* on main roads and near larger communities. With most heads of household of *solares* work on nearby farms or in nearby towns, interviewing these new landowners and taking their GPS points was too difficult, time consuming, and expensive to undertake. Instead, we developed GIS database management procedures that spatially linked *solares* (as well as the agricultural subdivisions) to the *finca madre.*

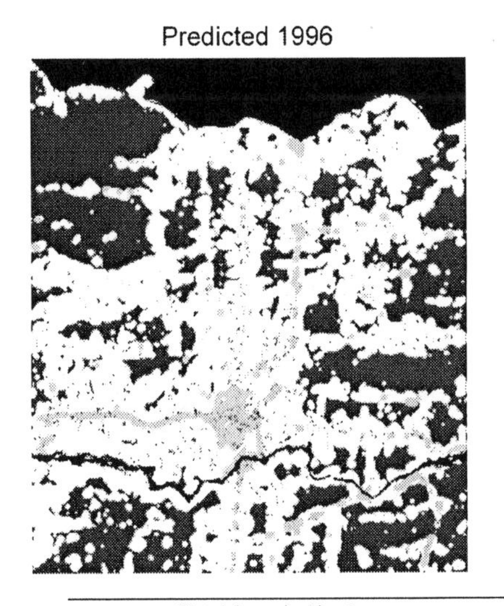

Total Area in Hectares

	1986	1996 Predicted	1996 Actual
Urbanized	48036	30015	31853
Forested	1050	3591	3774
Agriculture	27158	45072	43029

Summary Correlations

Forested	Urbanized	Agriculture
70.95%	7.80%	23.08%
2.15%	54.50%	3.94%
26.90%	37.70%	72.98%

Figure 11. Comparison of "observed" and "expected" LCLU patterns

From statistical analyses of the surveys and the remote sensing images, we know that the landscape is becoming more connected because of increases in the population base, land subdivisions, road network, and off-farm employment. It is also clear that farm size is decreasing and perennial and annual crops are increasing at the expense of forest and pasture. The spatial connectivity of households to the road network and to the expanding labor market and consumption opportunities being provided in Lago Agrio and other larger communities in the region is probably altering the trajectories of LCLU dynamics—increasing the geographic reach of households as push factors on development and expanding the geographic reach of communities on *fincas* as pull factors in the LCLUC calculus.

Sketch maps produced as part of the household surveys generated a detailed description of LCLU through a list of attributes, but less consistently useful data were obtained regarding the spatial position of cover-type boundaries or edge definitions and the planimetric position of features on the landscape. As the sketch maps were related to Landsat TM and IKONOS multispectral and panchromatic data, problems with spatial ambiguity became apparent, but so too did the high quality of the compositional information. It may be that the new IKONOS data have a spatial resolution that is too detailed to compare with the data collected in the survey; that is, it may be too much to expect household heads and interviewers to represent LCLU patterns in *fincas* through sketch maps that can be compared with IKONOS imagery. Although field plots are normally small in the Oriente, they are generally less than the 30 x 30 m pixel of Landsat TM, closer to the Enhanced Landsat TM of 15 x 15 m, and coarser than the resolutions of IKONOS. While spatially imprecise, the sketch maps, along with ground photographs taken from the front (or road) edge of the *finca* and oriented toward the house, provide insights not otherwise available. But more needs to be done to ascertain the appropriate use of the sketch maps and whether additional postprocessing of them would add to their value. While scanned, indexed, and read to a CD-ROM, cartographic approaches that relax feature outlines on sketch maps but maintain the position of features through a centroid or core representation may reduce the need to fit the sketches to planimetric standards without minimizing their value in characterizing LCLU within *fincas*.

GPS technology continues to be a powerful approach for spatially referencing landscape features. Survey teams equipped with relatively low cost GPS units and trained in their use were able to collect accurate locational readings in the field, though this added to the expense of data collection.

GIS software systems and database management schemes offer a number of approaches for linking data across thematic domains and spatial and temporal dimensions. Using ID numbers assigned to land parcels and households and clever record management schemes, "one-to-one," "one-to-many," and "many-to-one" relationships can be examined. When these relationships are considered for multiple time periods and multiple spatial resolutions, the complexity increases, but so does the power of the analyses.

Multiresolution systems are being integrated into more and more LCLU studies. While optical systems are primarily being used, nonoptical synthetic aperture radar systems are being examined—particularly for regions plagued by persistent cloud cover and dense plant canopies. The integration of NOAA AVHRR, MODIS, Landsat, SPOT, and IKONOS data is a powerful approach that affords numerous opportunities. Certainly, the issue of decomposing complex scale-dependent relationships is important, as well as the compatibility of spatial, spectral, temporal, and radiometric resolutions of the various remote sensing systems relative to the spatial and temporal signatures that characterize the socioeconomic and biophysical variables and systems under study. In the Oriente, household *fincas* continue to be central to the storyline of deforestation and agricultural extensification, but because of limitations in labor, capital, and technology, *fincas* appear to be larger than required for a single household to manage as an integrated unit. Subdivision through sale and kinship ties is reducing the mean size of household properties and altering the population density on lands adjacent to improved roads and service and market towns. The diffusion of LCLU changes from Lago Argio outward in a radial pattern is readily apparent in satellite images. Using higher resolution satellite data for targeted test areas to scale up to coarser-grain satellite systems offers an approach for "training" at fine scales but implementation at coarser grains. The temporal dimension has been much harder to anticipate in Ecuador. Cloud cover continues to severely limit our satellite time series, which leads to serious gaps in our temporal coverage. Systems with reduced spatial coverages, such as IKONOS, also are susceptible to the vagaries of clouds, offering severely degraded images with a high percentage of cloud cover for small test areas. With Landsat's larger areal extent, we generally have obtained usable data only for portions of the study area, thereby necessitating the concept of Intensive Study Areas to serve to consolidate portions of cloud-free imagery into a useful time series. The temporal depth of new systems is problematic unless compatibility with longer-run systems can be achieved.

Pattern metrics continue to offer a mechanism for introducing principles of landscape ecology into human dimensions research. Deriving measures of landscape structure and relating them to social, biophysical, and geographical drivers at distinct space and time scales offers a powerful approach to population and environmental studies. Using image classification for a single satellite image, for multiple satellite images, and for change images at decadal, interannual, and intra-annual scales, pattern metrics offer an approach for characterizing landscape states and conditions across space and time as an indication of landscape form and function.

Interpreting the meaning of landscape form offers a challenge to the research community.

Developing pixel histories or trajectories across the assembled time series indicates the importance of the duration of LCLU at a place in determining its function and the likely directionality of future changes. The stability of LCLU at a site may suggest forest conservation, inaccessibility of place, or production consistency, whereas LCLU variation might suggest a declining resource base, changes in crop prices, or swidden agriculture. Inductive and deductive approaches can be used to define the temporal signature of LCLU dynamics, and spatial tools can be used to assess the temporal imprint of such patterns.

A conceptual model of LCLU dynamics in the Oriente has been developed that emphasizes household decision making. Statistical analyses have been carried out that integrate data from satellite time series, GIS coverages, and the household longitudinal survey. It is these relationships (or rules) that will also be used to improve the cellular automata modeling. Thus, while linkages between people and the environment have been primarily carried out at the data collection stage, linkage considerations and protocols have also occurred in analyses and modeling to date as well. Still, much more can be done.

CA models are being used to integrate across the socioeconomic, biophysical, and geographical domains through simulations of LCLU. The ability of a system to grow and then alter its rate of growth and possibly reverse itself or "die" is a fundamental attribute in biological or human system CA modeling. The systems modeled by Clarke et al. (1996, 1997) and the modeling scheme followed here both attempt to emulate biological patterns. Transition probabilities for the cellular automata models depend on the state of a cell, the state of its surrounding cells, the physical characteristics of the cell (e.g., terrain, soil quality, vegetation, hydrology), and the weights associated with the infrastructure context of the cell (e.g., proximity to roads, markets, schools, and medical facilities and time since settlement). These weights are determined from models of land use based on the household survey data and the GIS database. Transition probabilities are determined empirically using field data and the time series of satellite data, in combination with parametric and nonparametric models. In essence, the CA models give us the ability to explore LCLU patterns for future time periods and to consider them relative to perturbations in the base variables and stochastic elements that yield a set of LCLU patterns that can be interpreted relative to environmental policy and a range of outcomes.

6. CONCLUSIONS

This chapter is written somewhat more from a natural science–spatial science perspective than a social science perspective and likely would have different foci if the latter had been the main perspective. The "how" and "why" of links to connect people and the environment through LCLU and LCLUC signatures might also have been different. But this is one of the many major challenges we face in this field—to have broad representation across the social, natural, and spatial sciences through assembling multidisciplinary research teams to create a dialogue that allows one to be engaged in a scientific discourse from various complementary perspectives. For us, thinking about the importance of space and time leads us to emphasize techniques for linking across thematic domains through approaches that maintain a temporally and spatially explicit context. When working with longitudinal survey data as well as a satellite time series, a space-time context seems appropriate. But what are the challenges and opportunities that a space-time context provides to establishing links between people and the environment?

From a spatial perspective, the discrete nature of survey data versus the continuous nature of remotely sensed data can be problematic. (In our case, however, remotely sensed data are also available only intermittently due to extreme problems with cloud cover.) Data transformations that involve approaches to distribute people across the landscape from discrete to continuous surfaces—for example, through the concept of population potential—offer one such solution. Calculating the areal percent of LCLU occurring within a farm or community territory or deriving an aspatial measure of spatial organization of LCLU within that spatial domain are approaches for moving from the continuous to the discrete as a way of reconciling between socioeconomic and environmental data.

We have found that it is much easier to create a research design that includes links between thematic domains and space-time dimensions for new research than for retrofitting already collected data. The existence and content of longitudinal data and an image time series sets the template for future data collection and analyses, so defining mechanisms to link between previously collected data is necessary. With the improving resolutions of new remote sensing systems, the ability to link detailed information on the environment with detailed information on population has empowered the research community with new opportunities based on spatial imagery. But defining the connections and compatibilities between differing data types, theoretical traditions, and varying methodological approaches continues to challenge—and excite—those engaged in population-environment interactions research. And the existence of better and less expensive spatial data does not solve the problems of getting more and better data on the human decision makers involved in land-use transformations of the earth's surface or on the context within which they function. Yet the prospects are ever more exciting for collecting better data and undertaking better studies based on linked data sets.

ACKNOWLEDGEMENTS

This work is based on a collaborative project of the Carolina Population Center (CPC), Department of Geography, and the Department of Biostatistics at the University of North Carolina, in conjunction with EcoCiencia, a leading nonprofit ecological research organization in Quito, Ecuador. Professors Richard E. Bilsborrow and Stephen J. Walsh are the Principal Investigators. We are grateful to the U.S. National Aeronautics and Space Administration (NASA) (NCC5-295) and the Mellon Foundation for supporting all aspects of the project and to the Summit and Compton Foundations and PROFORS (Proyecto Forestal, in Lago Agrio, Ecuador, funded by GTZ, the German foreign aid program) for supporting the household survey data collection in the Oriente region of the Amazon in 1999. In addition, NASA provided assistance with access to Landsat satellite imagery, the CPC Spatial Analysis Unit, and the Department of Geography's Landscape Characterization and Spatial Analysis Lab supported spatial and biophysical data collection and processing of remote sensing and GIS data, and the Centro de Estudios de Poblacion y Desarrollo Social (CEPAR) in Quito entered and processed the 1999 household survey data and the 2000 community survey data. These data sets have been further processed and analyzed at CPC with the assistance of graduate students supported by traineeships funded by the National Institute for Child Health and Human Development (HD07168) and other sources. Data from the earlier 1990 Amazon household survey were collected and processed with funding provided by the U.S. National Science Foundation, the World Wildlife Foundation, and the CPC, and logistical support was received from several Ecuadorian government agencies, including CONADE (the former National Planning Agency), the Ministry of Agriculture, and INIAP (Agricultural Research Center in Francisco de Orrellana).

REFERENCES

Adams, J. B., D. E. Sabol, V. Kapos, R. A. Filho, D. A. Roberts, M. O. Smith, and A. R. Gillespie. 1995. "Classification of Multispectral Images based on Fractions of Endmembers: Application to Land-Cover Change in the Brazilian Amazon." *Remote Sensing of Environment* 52: 137–154.

Allen, T. R., and J. A. Kupfer. 2001. "Application of Spherical Statistics to Change Vector Analysis of Landsat Data: Southern Appalachian Spruce-Fir Forests." *Remote Sensing of Environment* 74(3): 482–493.

Allen, T. R., and S. J. Walsh. 1993. "Characterizing Multitemporal Alpine Snowmelt Patterns for Ecological Inferences." *Photogrammetric Engineering and Remote Sensing* 59(10): 1521–1529.

Bilsborrow, R. E. 1987. "Population Pressure and Agricultural Development in Developing Countries: A Conceptual Framework and Recent Evidence." *World Development* 15(2): 183–203.

———. n.d. (forthcoming) *Population and Land Use in the Ecuadorian Amazon: Proximate and Underlying Linkages.* Rome: UN Food and Agricultural Organization, in press.

Blaikie, P., and H. Brookfield. 1987. *Land Degradation and Society.* London: Methuen.

Boserup, E. 1965. *The Conditions of Agricultural Growth: The Economics of Agrarian Change under Population Pressure.* New York: Aldine.

———. 1981. *Population and Technological Change: A Study of Long Term Trends.* Chicago: University of Chicago Press.

Bromley, D. 1989. *Economic Interests and Institutions: The Conceptual Foundations of Public Policy.* Oxford: Blackwell.

Clarke, K. C., L. Gaydos, and S. Hoppen. 1997. "A Self-Modifying Cellular Automaton Model of Historical Urbanization in the San Francisco Bay Area." *Environment and Planning B* 23: 247–261.

Clarke, K. C., S. Hoppen, and L. Gaydos. 1996. "Methods and Techniques for Rigorous Calibration of a Cellular Automaton Model of Urban Growth." Paper presented at Third International Conference/Workshop on Integrating GIS and Environmental Modeling. Santa Barbara: National Center for Geographic Information and Analysis.

Crews-Meyer, K. A., 2001. "Assessing Landscape Change and Population-Environment Interactions via Panel Analysis." *Geocarto International* 16(4): 69–79.

———. 2000. "Integrated Landscape Characterization via Landscape Ecology and GIScience: A Policy Ecology of Northeast Thailand." Ph.D. dissertation, Department of Geography, University of North Carolina–Chapel Hill.

Dale, V. H., R. V. O'Neil, M. Pedlowski, and F. Southworth. 1993. "Causes and Effects of Land-Use Change in Central Rondonia, Brazil." *Photogrammetric Engineering and Remote Sensing* 59(6): 997–1005.

Forman, R. T. T., and M. Godron. 1986. *Landscape Ecology.* New York: John Wiley & Sons.

Frizzelle, B. G., and S. J. McGregor. 1999. "Integrating Geographic Information Science (GISc) Techniques in the Data Collection Phase of Population-Environment Research." *Papers and Proceedings of the Applied Geography Conference* 22: 199–207.

Gell-Mann, M. 1994. *The Quark and the Jaguar.* New York: Freeman.

Jolly, C., ed. 1994. *Population and Land Use in Developing Countries.* Washington, D.C.: National Academy Press.

Laird, N. M., and J. H. Ware. 1982. "Random-Effects Models for Longitudinal Data." *Biometrics* 38: 963–974.

Luhman, N. 1985. *A Sociological Theory of Law.* London: Routledge and Kegan-Paul.

Malthus, T. R. 1798. *On Population.* Edited (1960) and introduced by G. Himmelfarb. New York: Modern Library (Random House).

Marquette, C. and Bilsborrow, R. E. 1994. *Population and the Environment in Developing Countries: Literature Survey and Research Bibliography.* New York: UN Population Division.

McCracken, S. D., A. Siqueira, E. Moran, and E. Brondizio. 1999. "Domestic Life Course and Land Use Patterns in an Agricultural Frontier in Brazil." Paper presented at Conference on Patterns and Processes of Land Use and Forest Change, March 23–26, University of Florida.

McDaniel, P. M. 2000. "Household and Community Drivers of Land Clearing: Human-Environment Interactions in the Ecuadorian Amazon." Master's thesis, Department of Geography, University of North Carolina–Chapel Hill.

Messina, J. P. 2001. "A Complex Systems Approach to Dynamic Spatial Simulation Modeling: Landuse and Landcover Change in the Ecuadorian Amazon." Ph.D. dissertation, Department of Geography, University of North Carolina–Chapel Hill.

Messina, J. P., and S. J. Walsh. 2001. "2.5D Morphogenesis: Modeling Landuse and Landcover Dynamics in the Ecuadorian Amazon." *Plant Ecology* 156(1): 75–88.

Messina, J. P., S. J. Walsh, G. N. Taff, and G. Valdivia. 1999. "The Application of Cellular Automata Modeling for Enhanced Landcover Classification in the Ecuadorian Amazon." Proceedings, 4th International Conference on GeoComputation (on CD-ROM), Mary Washington College, Fredericksburg, VA, pp. gc–081.

Moran, E. F., and E. S. Brondizio. 1998. "Land Use Change after Deforestation in Amazonia." In D. Liverman, E. F. Moran, R. R. Rindfuss, and P. Stern, eds., *People and Pixels: Linking Remote Sensing and Social Science* (Washington, D.C.: National Academy Press): 94–120.

Perz, S. 2000. *Household Demographic Factors as Life Cycle Determinants of Land Use in the Amazon.* Miami: Latin American Studies Association.

Pichón, F., and R. E. Bilsborrow. 1999. "Land Use Systems, Deforestation, and Demographic Factors in the Humid Tropics: Farm-Level Evidence from Ecuador." In R. E. Bilsborrow and D. Hogan, eds., *Population and Deforestation in the Humid Tropics* (Liège, Belgium: International Union for the Scientific Study of Population), 175–207.

Rudel, T., and B. Horowitz. 1993. *Tropical Deforestation: Small Farmers and Forest Clearing in the Ecuadorian Amazon.* New York: Columbia University Press.

Southgate, D., and M. Whitaker. 1994. *Economic Progress and the Environment: One Developing Country's Policy Crisis.* Oxford: Oxford University Press.

Walsh, S. J., J. P. Messina, K. A. Crews-Meyer, R. E. Bilsborrow, and W. K. Y. Pan. 2002. "Characterizing and Modeling Patterns of Deforestation and Agricultural Extensification in the Ecuadorian Amazon." In S. J. Walsh and K. A. Crews-Meyer, eds., *Linking People, Place, and Policy: A GIScience Approach* (Boston: Kluwer Academic Publishers), 187–214.

Wolfram, S. 1984. "Cellular Automata as Models of Complexity." *Nature* 311: 419–424.

Wood, C. H., and D. Skole. 1998. "Linking Satellite, Census, and Survey Data to Study Deforestation in the Brazilian Amazon." In D. Liverman, E. F. Moran, R. R. Rindfuss, and P. Stern, eds., *People and Pixels: Linking Remote Sensing and Social Science* (Washington, D.C.: National Academy Press), 70–93.

Chapter 5

HOUSEHOLD-PARCEL LINKAGES IN NANG RONG, THAILAND

Challenges of Large Samples

Ronald R. Rindfuss
Department of Sociology and Carolina Population Center, University of North Carolina
ron_rindfuss@unc.edu
Pramote Prasartkul
Institute for Population and Social Research, Mahidol University
Stephen J. Walsh
Department of Geography and Carolina Population Center, University of North Carolina
Barbara Entwisle
Department of Sociology and Carolina Population Center, University of North Carolina
Yothin Sawangdee
Institute for Population and Social Research, Mahidol University
John B. Vogler
East-West Center

Abstract An understanding of the dynamic connections between human behavior and the biophysical environment requires that people and land be spatially linked, conceptually and operationally. This paper describes the design and execution of a plan for spatially linking at a fine grain level: households and land parcels in Nang Rong, Thailand. The overall goal was to relate household dynamics to land use. There were several challenges that had to be surmounted: the large number of links to be determined given the sample size; a residential pattern with clustered dwelling units located away from the land farmed); a complex pattern of ownership and use; and the absence of a clear one-to-one relationship between households and parcels (a household using several plots; several households using the same plot). The paper reviews decisions about the design of the data collection, including the decision to start with households and then link to plots, to focus on use rather than ownership, to rely on a village headman informant to collect GPS data on the location of dwelling units, to collect information from households about each plot used and its proximate neighbors, to collect locational information about the plots used by each household from a group interview, and to manually match the last two. The paper describes in detail the options available at each point in the design process, and the reasons for choosing as we did; the map products that were prepared, their cost, and mode of use; the interviewing that took place; problems that arose; and the quality of the links, as far as we are able to evaluate at this point.

Keywords: household-land-use links, theoretical and applied considerations, social-spatial data collection, northeast Thailand

1. INTRODUCTION

Scientists from a variety of fields are increasingly acknowledging the need to understand the nature and causes of change in land cover. In turn, this has led to recognition of the necessity to comprehend the role of human behavior in changes in land cover and the accompanying distinction between "land cover" and "land use."[1] It has also led to the examination of the nature of feedbacks between population and the environment. If we are to understand the causes in the change in land cover—and especially if we are to move toward prediction—we need to be able to know initial conditions, how people are using the land, the factors that lead to trajectories of land-use change, whether land-use change also involves land-cover change, and how people relate to composition and change.

Identifying land use and the factors leading to land-use change requires data on the social entities that influence or make land-use decisions. We also need to know how initial land-cover conditions affect these decisions. There is a range of such social entities, including international and governmental bodies, corporations, religious and other organizations, households, and individuals. This range can be thought of as the social organization of people and behavior, and such social organization is itself subject to change over time. Further, just as social organization affects the nature of decision making influencing land use, there is also a biophysical and geographical organization of the land that affects land use and land cover. Examples here include elevation, slope, soil fertility, and water quality, as well as land fragmentation and geographic accessibility.

For linguistic ease, consider the components of social organization and the spatial organization of land as being scaled. On the social side, the smallest element is an individual and then a household. On the spatial side, the smallest element is the finest areal unit that can be measured with available measuring instruments. The minimum mapping unit—the pixel—and the instantaneous field of view of imaging devises are applicable concepts for remote sensing characterization and ground-based locational referencing. Finally, on both the social and spatial side, there is a temporal scale.

This essay recounts our experience with linking fine-grain elements on both the spatial and social side in Nang Rong, Thailand, and our specific rationale for doing so. As such, the chapter is about a seemingly narrow topic: the "how" and "why" of data collection linkage in a specific study. It is set, however, within a much broader scientific context. Just as there are organizational aspects of human behavior and the spatial organization of the biophysical landscape, there are also organizational aspects of science, and right now numerous components of the scientific community are clamoring for better understanding of human impacts on land-cover/land-use change (LCLUC).

From a public policy perspective, perhaps the most visible example involves estimates of global warming and their implications. President Bush has rejected the Kyoto protocols, suggesting that his decision was based on a lack of scientific consensus. Ignoring the question of the wisdom of this decision, it is clear that there is a great deal of uncertainty and disagreement among those projecting future global warming. The Third Assessment Report of the Intergovernmental Panel on Climate Change had a predictive range of 1.4 to 5.8 degrees C (Celsius) for global warming between 1990 and 2100 (Houghton et al. 2001). One of the key factors leading to such a wide range is uncertainty about human behavior (Allen, Raper, and Mitchell 2001), with greenhouse gas emissions as a principal component of this uncertainty. LCLUC, with its implications for emissions and carbon sinks (Wofsy 2001), is, in turn, a potentially important component of human factors affecting future global warming.

Other examples come from a range of scientific and policy applications. In recent years, a wide variety of issues have arisen under such labels as land degradation, biodiversity, resilience, and sustainability. While frequently vaguely conceptualized, the basic idea is that there is a threshold or tipping point beyond which the land can no longer support the plant or animal activity of which it had previously been capable. Desertification is perhaps the most extreme example. Almost invariably, reaching such thresholds is attributed to some form of human land use: gathering of firewood, overuse of fertilizer or pesticide, overgrazing by livestock, to name a few. This issue of thresholds is one that is receiving increased attention, especially in the land-use modeling community. It is also important from a policy perspective to the extent that reaching such thresholds is considered problematic and that effective policy measures are available.

Related is the issue of biodiversity and the way in which LCLUC affects it. A recent National Academy of Sciences report suggests that recreated wetlands do a relatively poor job of recreating the biodiversity found in naturally existing wetlands (Committee on Mitigating Wetland Losses 2001). Similar issues arise with respect to reforestation and other mitigation processes and point to the importance.of understanding *change* in land cover and land use—its characterization, trajectories of change, and the multidimensional, scale-dependent dimensions of change.

So far the examples have come from the biophysical side of LCLUC dynamics, but there are also important theoretical and policy issues related to human behavior and LCLUC. Migration and urbanization are perhaps the most obvious. If, for whatever reason, LCLUC processes lead to a situation in which the agricultural population in a given area is land constrained, out-migration is one likely outcome, and this out-migration typically is toward urban industrial centers. Once rural to urban migration streams have started, they not only affect the life course of individuals and households participating in the stream, but they also affect land use in areas surrounding the destination cities (e.g., the orange groves in southern California or the fertile rice fields surrounding Bangkok).

2. WHY LINKAGE AT THE HOUSEHOLD LEVEL?

We now address two nested questions. Why is data linkage a potentially troublesome problem for students of LCLUC, and, given that we require linkage, why do we want it at the household level? The linkage problem arises because data on humans and data on land cover come from quite different data-gathering approaches, which in turn reflect the very nature of humans versus land. Land is stationary, continuous, and numerous aspects of land cover are measurable by remote sensing (whether instruments are mounted on satellites or aircraft). The fact that land is stationary in space provides analysts an anchor to link assorted remotely sensed images and their landscape characterization, especially if an adequate number of geodetic control points are available for vertical and horizontal control. It also permits linkage to map products and integrated GIS coverages.

Humans, at various levels of aggregation, do not possess these appealing (from a data perspective) qualities of land. Humans are mobile, discrete, and not easily captured by remote sensing methods.[2] Within the social sciences, censuses (complete coverage) and surveys (sampled coverage) have been the primary methods for collecting data on human behavior. Both tend to be cross-sectional and require considerable prior thought and proper procedures if they are to be longitudinal, which in turn facilitate change analyses.

To link data on human behavior and land cover requires defining the boundaries or territory that human behavior affects, with the provision of sufficiently specific geographic coordinates. At some scales and for some purposes, this is relatively straightforward. For example, if the human units under consideration are administrative/political (such as countries, states or provinces, counties or districts), they are likely to have known geographic boundaries that can be linked to various types of land-cover data. But for most scales and purposes, humans and their behavior are unlikely to have easily known and delimited territories. At a more macro scale, consider corporations and/or religious institutions. They frequently occupy as well as have control over land and its use. This land can be widely dispersed geographically, as in the case of multinational corporations, and there is unlikely to be an easily accessible data source that provides geographic coordinates for the land associated with a given class of corporations and/or religious institutions.

For an example at a more micro scale, consider households. All households have a residence and affect land cover/land use in that locale. Some might have more than one place where they live—consider nomads and members of the middle class who own a vacation home. Households might own or control land, thereby affecting the use of that land. The land owned by a household might be one continuous parcel or multiple parcels widely dispersed. Or possibly a given parcel of land could be owned by multiple households. Household members through their work and leisure activities can affect LCLUC. Certain occupations (e.g., logger, land developer) can have quite direct and profound implications for land-cover change; for others, such as an accountant or a factory operative, the implications could be indirect. Similarly, numerous leisure-time activities have the potential to affect land cover; examples might include gardening, camping, and wetland preservation. To the best of our knowledge, there is no simple, straightforward way to link households to the land they affect, even in countries such as Norway or the Netherlands where many kinds of records are administratively centralized.

Given that effort—and sometimes *considerable* effort—is needed to link human behavior to LCLUC, at what level or scale should the linkage occur? Clearly the answer to this question depends on the particular research issue being addressed and specifics about the place where the study is being conducted. In our case, we were working in northeast Thailand, where most of the population is rural and agricultural, where much of the land is held and used by individual households (as opposed to communities, corporations, or other entities), where households are free to make their own decisions about how to use their land, and where households are responding to local, national, and international markets, as well as their own household circumstances (e.g., household size, availability of household members to work on household land parcels, and so forth). Hence, from the sheer force of the locus of much land-use decision making in the Nang Rong District, households are essential to understanding the forces leading to land-use change and, concomitantly, land-cover change. Further, as discussed below, many of the theoretical perspectives guiding our approach to the development of the Nang Rong studies are at the individual or household level, hence demanding data linkage at least at the household level.

Why not link individuals to the land? The answer involves theoretical, ethical, and practical issues of studying land use in a rural setting such as Nang Rong. Most of the social science literature with which we are conversant presumes that decisions on land use in rural areas are made at the household level rather than the individual level (e.g., Becker 1991; Chayanov 1966; Kriedte, Medick, and Schlumbohm 1981; Rosenfeld, 1985; Stark 1991; Wilk and Netting 1984). While specific individuals within a household might hold land-use decision preferences that differ from the positions of other household members, these differences are thrashed out in the household's decision-making style and then a decision emerges. Our ethical concerns revolve around not wanting to stir up differences within the household that can potentially ignite conflict that had been smoldering. It is our judgment that the research benefits do not outweigh the risks involved. The practical reasons for not linking individuals to the land revolve around the difficulty of conducting private interviews with a single individual in Nang Rong villages. Interviews are typically conducted outside (in the shade of a porch or tree), and usually family members, neighbors, and/or friends will gather to listen (and perhaps offer advice on the proffered answer).

It should be noted that focusing on households rather than individuals raises thorny issues with respect to longitudinal data. Unlike individuals, who remain unique from birth to death, there is no agreed-upon definition of what constitutes the continuation of a household over time (Citro and Watts 1986; Duncan and Hill 1985; Keilman and Keyfitz 1988; McMillan and Herriot 1985). As individuals leave a household and join or form another household, which is the successor household to the earlier one? In our case, we first have a set of rules that guided the fieldwork to define successor households.[3]

Why not infer household-level relationships from village or district-level relationships? Doing so would require considerably less data collection effort and permit the use of data that was collected for other purposes. Part of the answer is that, as we have discussed elsewhere (Rindfuss et al. this volume), to obtain village-level boundaries we need household-level linkage. The more crucial answer, however, involves what has been known in sociology for more than half a century as the *ecological correlation fallacy.* In a classic article on the topic, W. A. Robinson distinguishes between a correlation in which the units are individuals and an ecological correlation in which the units are groups of individuals. He describes the arithmetical relationship between individual and ecological correlations, and concludes: "The relation between ecological and individual correlations which is discussed in this paper provides a definite answer as to whether ecological correlations can validly be used as substitutes for individual correlations. They cannot. While it is theoretically possible for the two to be equal, the conditions under which this can happen are far removed from those ordinarily encountered in data. From a practical standpoint, therefore, the only reasonable assumption is that an ecological correlation is almost certainly not equal to its corresponding individual correlation" (Robinson 1950: 357). In short, if we want to understand how household behavior affects or is affected by LCLUC, then we need to be able to link LCLUC data to human behavioral data at the household level and not at the village or district level.

As a final note, we should add that even though we were interested in relationships at the household level, we are also interested in relationships at the village level and have collected the necessary data to examine such village-level relationships. The approaches and methods are discussed elsewhere (Rindfuss et al. this volume). Thus we will be able to vary the scale at which we conduct our analyses and see if it matters—and how.

3. GUIDING THEORETICAL PERSPECTIVES

Before discussing the theoretical perspectives guiding our work, it is important to emphasize that our efforts to link households with the land parcels they own/use is set within a larger project—indeed a set of projects—that seeks to understand the extensive nature of social change occurring in Nang Rong, Thailand. Nang Rong has been undergoing extensive demographic, technological, economic, agricultural, transportation, and land-use change. These changes are interrelated, driven partially by factors inside the district and partially by factors outside the district. We have been gathering social and biophysical data to put ourselves in a position to document, explain, and understand the profound changes that have been occurring. We view the assembled data sets, described below, as a "laboratory" that we can repeatedly enter and use as we examine new analytical questions (Entwisle et al. 1998). Hence, no single theoretical perspective has directed our data-gathering approach. Rather, there are a number of theoretical perspectives that have provided guidance. But we have also been opportunistic, incorporating data that becomes available even if we do not have an immediate theoretical need. The field is changing rapidly, and frequently these auxiliary data sets have moved to center stage as we recognize new research questions and opportunities.

In keeping with this multipurpose laboratory approach, perhaps the most influential theoretical orientation has been Kingsley Davis' (1963) Multiphasic Theory of Demographic Change and Response, which emphasizes multiple demographic reactions to economic strain and population pressure. Davis argues that in the face of high rates of natural increase and rising material expectations, rural residents would react in a variety of ways, including changes in the timing and likelihood of marriage, reduction of childbearing, and out-migration. At their core, these are arguments about the behavior of individuals and households responding to a larger social and economic context and the consequences of their adjustments for that context. While Davis did not address the issue of land-use change, extension to land-use change is straightforward. One would expect individuals and households facing rising material expectations and increased population pressure to alter their land-use strategies in a variety of ways: move toward cash crops; migrate to urban areas; bring additional land under cultivation; take nonagricultural jobs and rent their land; and intensify the use of their lands through the use of fertilizer, herbicides, and mechanical assistance. A set of research questions develops with respect to which households follow which strategies, as well as where and when on the landscape.

The life-course framework has also provided general guidance. Rooted in sociocultural theories of age and social relations (Ryder 1965; Elder 1974, 1998; Featherman 1983), interconnections among multiple role domains and demographic processes, intertemporal dependencies, links among individuals (in marital unions, households, kinship groups, and social networks), and the importance of the larger historical context are all fundamental to this framework (Clausen 1972; Elder 1985; Hogan 1981). A central premise is that the timing and sequence of occupying these roles, in and of itself, can impact on later life-course development. While most life-course work has been done at the individual level, there are possible extensions to the household level. Households go through a series of stages or phases where the role of the household is changing and where the stage itself has strong implications for the activities of the household. Just compare a household with young children to one that consists of a retired married couple. Further, links among households (in kinship groups, neighborhoods, and other social networks) are crucial.

Additionally, extensions of the life-course framework to land parcels are also straightforward. First and foremost life-course thinking applied to the land forces us to consider the history of specific land-use parcels. In the northeast Thailand setting, certain crops such as cassava tend to deplete the soil; hence we expect that as the number of years increases for a given land parcel involved in a crop use that depletes soil fertility, the tendency for land cover/use to change as a consequence also increases. In addition, the emphasis on links among individuals within the life-course framework ties naturally to the concept of spatial autocorrelation within ecology and might provide the basis for a social understanding of the processes leading to spatial autocorrelation and other forms of scale dependence. The links or ties between neighboring farmers provide the opportunity to exchange information and help, which in turn, other things being equal, would lead to an increased degree of spatial autocorrelation. In other words, a well-functioning social network might increase spatial autocorrelation. For a final example, consider Thailand's recent economic crisis, marked by the dramatic decline in the value of the baht (the Thai currency) in 1997. The life-course framework, in conjunction with various theories about migration, would suggest that young adults would be less likely to migrate out of Nang Rong and that some of those who had already migrated to urban destinations would return. This in turn would change the labor supply within households, making it easier to plant labor-intensive crops. Further, the change in the value of the baht might reduce the value of certain farm commodities, affecting decisions about what to plant.

Context—social context and by extension biophysical context—is also of critical theoretical importance. Indeed, it is at the heart of much contemporary sociology and can trace its roots to such classical theorists as Durkheim and Weber. Similarly, within geography, hierarchy theory addresses the importance of considering multiple scales versus a single scale of analysis. Often the scale above and the scale below the "characteristic" scale are considered as context and mechanisms, respectively. Walsh et al. (1999b) integrated social, biophysical, and geographical domains in northeast Thailand through a scale-dependent analysis of NDVI (Normalized Difference Vegetation Index) variation in the Nang Rong District. Findings suggested the importance of social variables at fine scales and biophysical variables at coarse scales. Within social demography, this concern with context has led to the development of several multilevel models that blend micro- and macro-level variables into a single analysis. The implication with respect to our data collection is the importance of collecting data at a variety of levels or scales, both time and space, and at the appropriate scales.

Individually and together, the guiding theoretical frameworks emphasize multiplicity on both the land and human sides: multiple responses to social change, multiple levels of analysis, multiple aspects of the life course of individuals, households, and land parcels, multiple connections in social and geographical space, and multiple ties between people and land in rural agricultural areas. As a result, there is a push for data collection strategies that incorporate the myriad aspects of individuals and land parcels, including their biographies, the households within which individuals live, the neighborhoods within which parcels of land are located, and the broader context within which households and neighborhoods are located. These contexts include time as well as space parameters of the social, biophysical, and geographical domains. Put differently, our guiding theoretical issues are data hungry and demanding of innovative data collection strategies that go beyond traditional surveys and the application of digital spatial technologies with emphasis on remote sensing. Creating such data sets "from scratch" is admittedly difficult. Our strategy has been to capitalize on an existing database for Nang Rong. This has the advantage of feasibility but the disadvantage typically associated with longitudinal data sets: early design decisions constrain subsequent design decisions. For example, the extreme southwest corner of our study area has become quite dynamic from a land-cover perspective, but unfortunately we have no sample villages there because the area was very remote and sparsely settled in 1984 when the social side of the longitudinal study began. Hence, our linked data cannot inform us about this dynamic region.[4] This is a function of our decision to continue with the villages and households originally included in the study in 1984—thus pointing out that a longitudinal design may not be best suited to capturing land-cover hot spots.

Finally, it should be noted that longitudinal designs by their very nature mean that the researcher is interviewing the same people repeatedly. Because of the repeated need to contact respondents, it is critical to maintain good relationships with them. We do this in a variety of ways, ranging from making the interview experience as pleasant as possible to working with the local administrative/political leadership to hiring interviewers from a nearby teaching university to providing a small gift at the end of the interview. Our response rates have remained exceptionally high and cooperation during the interview has been excellent.

4. STUDY SITE

Nang Rong District is located in Buriram Province in northeast Thailand (see Figure 1). The district is relatively small at approximately 1,300 km^2. It lies on the Korat Plateau, characterized by relatively infertile soils, poor drainage, and inconsistent precipitation levels caused by a highly variable monsoonal rainfall pattern (Parnwell 1988; Rigg 1987; Kaida and Surarerks 1984). Over 80 percent of the average annual precipitation occurs between April and November, with soil moisture deficits common at other times (Rigg 1991).

Nang Rong is also relatively small from a demographic perspective, containing 183,000 people in 1990 (National Statistics Office 1990). There were 356 administratively defined rural villages in 2000, plus several market centers and administrative towns. There is size variation over time and across villages, but typically they average about 100 households. Villages were the smallest unit of local government. Based on naturally occurring clusters of dwelling units arranged in a nuclear fashion, the administrative intent was for villages to contain approximately 100 households. If a cluster of dwelling units was not large enough to constitute a village, for administrative purposes it would be joined to another nearby cluster to form a village. Should a cluster of dwelling units grow much larger than 100 to 150 households, there was a tendency to administratively subdivide that village into two or more coterminous villages.

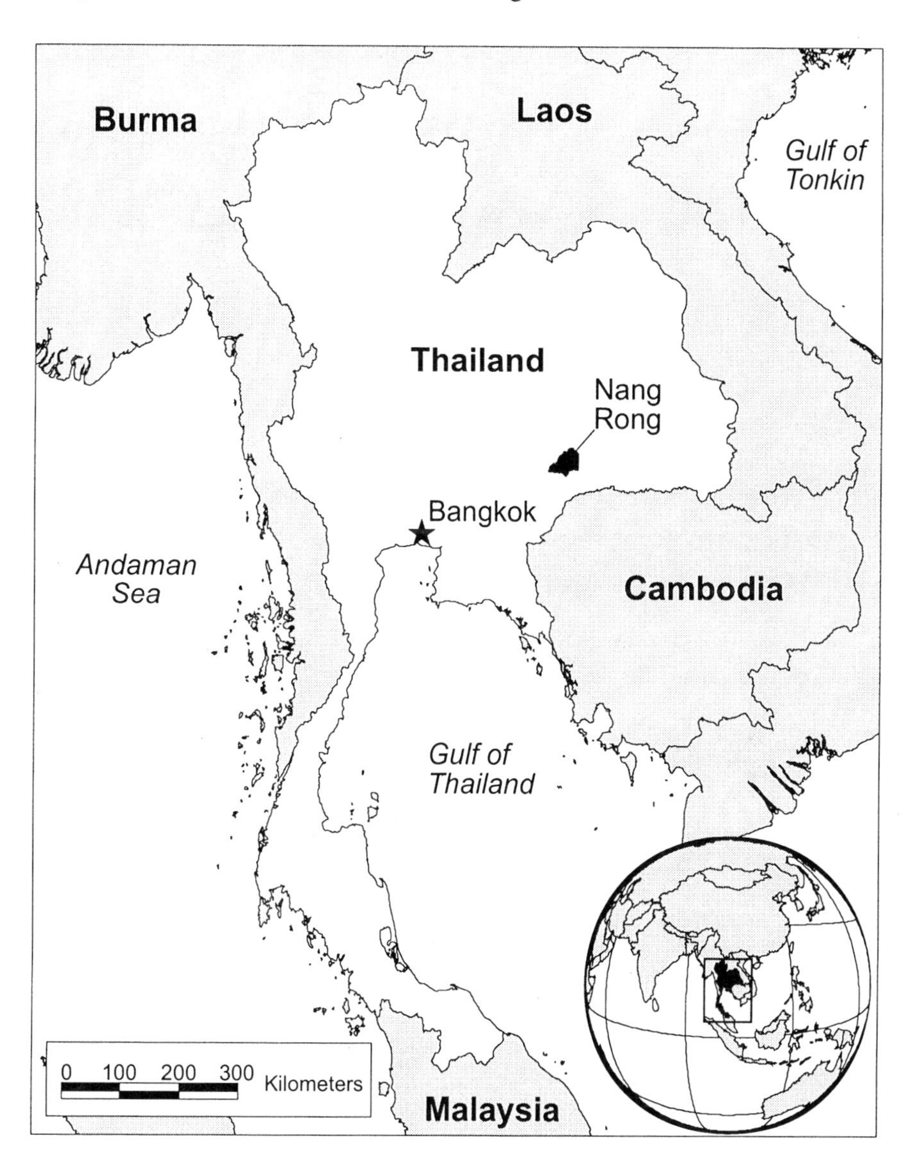

Figure 1. Study area location, Nang Rong District, northeast Thailand

Throughout our data collection efforts, we have kept track of the village divisions that have occurred for administrative purposes, and we can, for data analysis purposes, recombine them to their prior administrative organization. Our original 1984 sample of fifty-one villages—for which, as described below, we collected household data—now contains ninety-two administrative villages after subdivision.

Within the villages, dwelling units are generally organized in a cluster, surrounded by agricultural land. The typical household uses two or three parcels of land, and these parcels tend not to be contiguous. See Figure 2 (B) for an illustration. This pattern contrasts markedly with patterns found in the Amazon and other parts of the world, as illustrated by Figure 2 (A). Further complicating matters, some parcels are farmed by multiple households (presumably these are households related to one another through kinship ties). The parcels are fairly small, and most agriculture in the district is rain fed rather than irrigated.

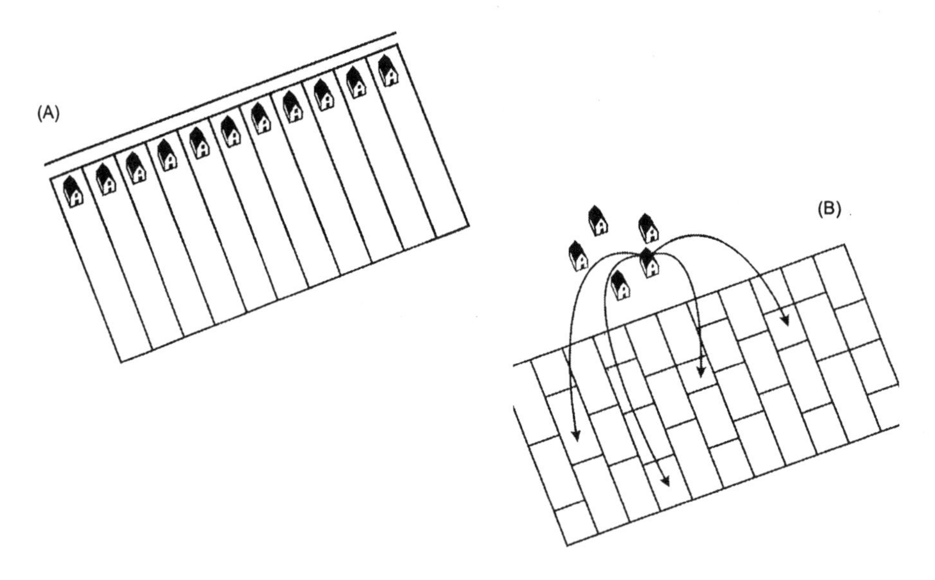

Figure 2. Illustration of households in nonclustered (A) and clustered (B) villages

Households engaged in farming activities are dependent on the timing of the annual monsoon. They are likely to be engaged in agricultural activities at the same time as other households and have limited agricultural work to do in the "off-season." The monsoonal rains are especially critical for paddy rice, which is the dominant crop, grown for both subsistence and commercial purposes. In more recent years, upland agriculture—chiefly cassava and sugarcane—has developed. Cassava is primarily sold to the

European market as a calorie-rich livestock feed, and sugarcane is sold on the national and international markets. Upland crops are not tied to the same monsoonal rhythms as paddy rice.

Until relatively recently, Nang Rong was an isolated area—a frontier with respect to patterns of land settlement.[5] The main highway, the "Choke Chai-Dech-Udom Highway," linking Nang Rong to Korat (the second-largest city in Thailand and a regional node of the northeast) and then to Bangkok, was built primarily for military purposes in the late 1960s. Many villages were established before that. One of the results is that villages tend to follow the main rivers rather than the main roads (see Figure 3). In Nang Rong, human settlement occurred in several waves, the largest beginning in the 1950s and extending through the 1970s. The frontier was largely closed by the 1970s, with some available land remaining until the 1990s. After the closing of the frontier, the dominant direction of migration shifted from in to out in response to the rapid growth in urban job opportunities in construction, manufacturing, and services.

There are many different types of land titles in Nang Rong, ranging from the equivalent of squatter's rights to fully titled land that the owner can sell, mortgage, or deed to one's heirs. Traditionally, if land were not titled, by default it was owned by the King; relatively little land was held under community ownership. In recent years, attempts have been made to clarify land titles and prepare cadastral maps. Figure 4 shows cadastral map coverage in Nang Rong in 1999—maps that we used in our fieldwork as described below. As can be seen, large portions of the district are not covered by these maps.

In Thailand, deforestation has been underway for a century or more (Feeney 1988), much of it associated with agricultural expansion. In Nang Rong, prior to World War II, agricultural expansion involved increased production of paddy rice, largely for subsistence. Beginning in the 1960s, there has also been extensive deforestation in the upland areas of Nang Rong associated with the cultivation of cash crops such as cassava and sugarcane. Figure 5 shows Landsat Thematic Mapper (TM) land-cover classifications for Nang Rong District and a 10-km buffer around the district for 1972–73 and 1997. Over time, the lowlands have been transformed into a landscape matrix dominated by rice paddies with isolated trees in and around the paddies, riparian forests, and forests retained near villages (Walsh 1999). In the uplands, forests are still significant in terms of area covered, but cash crops (cassava and sugarcane) now also comprise a substantial area. The natural vegetation of the district consists of a dry monsoon forest predominated by dwarf dipterocarp trees and containing areas of grassland, thorny shrubs, and bamboo thickets.

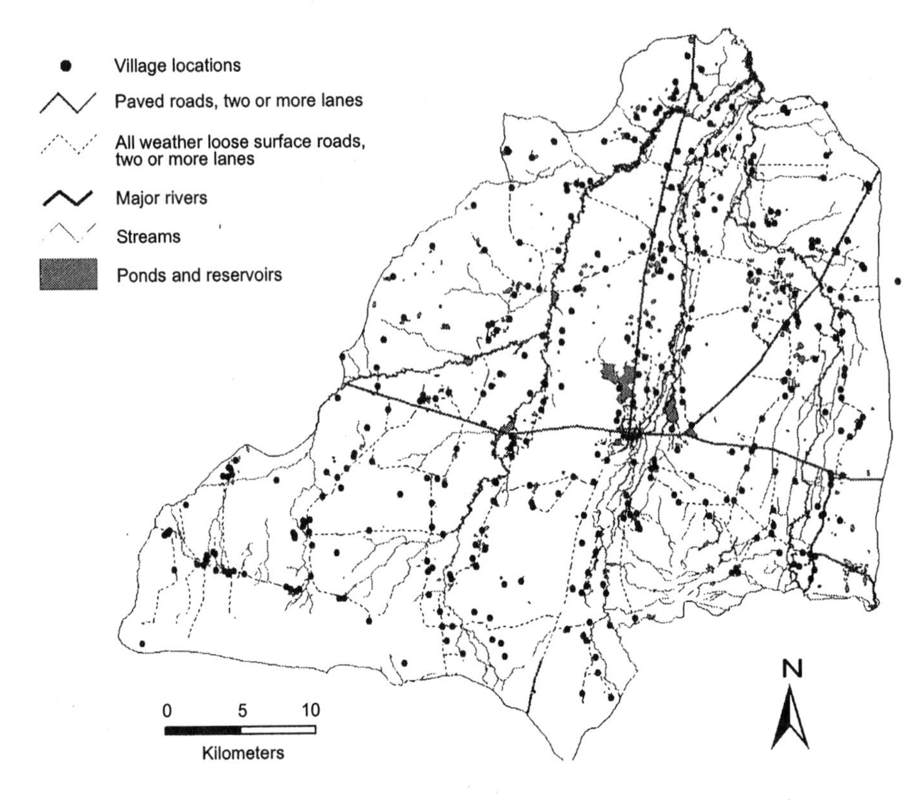

Figure 3. Nang Rong District with village locations, major roads, and hydrography

5. PRE-2000 DATA STRUCTURE

The 2000 data collection, which included our linking of households with the specific parcels they farmed, built upon previous data collection efforts in Nang Rong. To understand what we did in 2000, it is important to be familiar with the basic contours of the earlier data on which the 2000 data collection is built. Since this earlier data collection is described in detail elsewhere (Chamratrithirong and Sethaput 1997; Entwisle et al. 1998; Rindfuss et al. 2000a, n.d., and this volume; Walsh et al. 1999a and b), here we sketch the bare essentials.

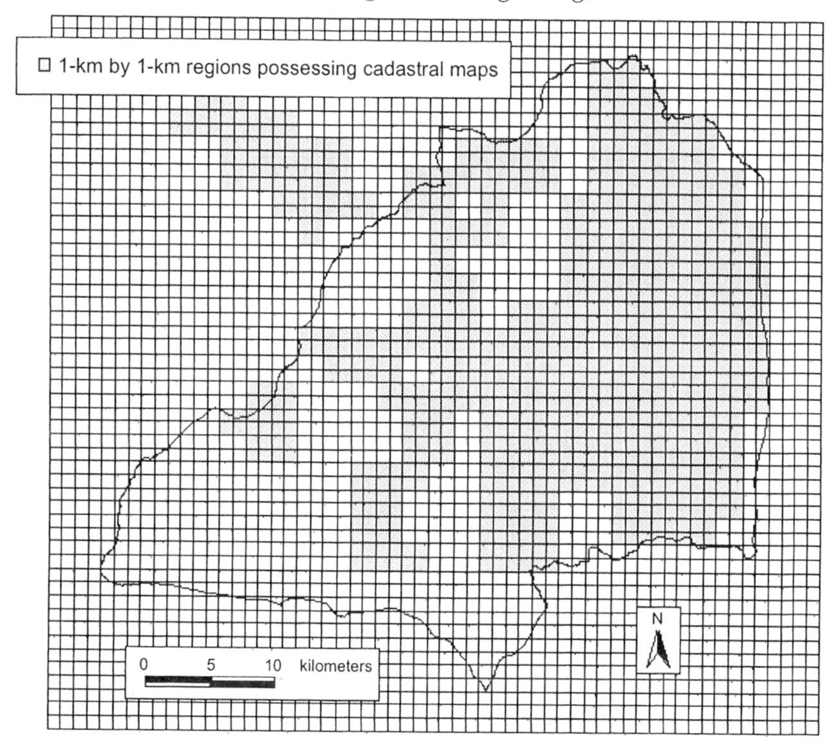

Figure 4. Cadastral map coverage of Nang Rong District in 1999

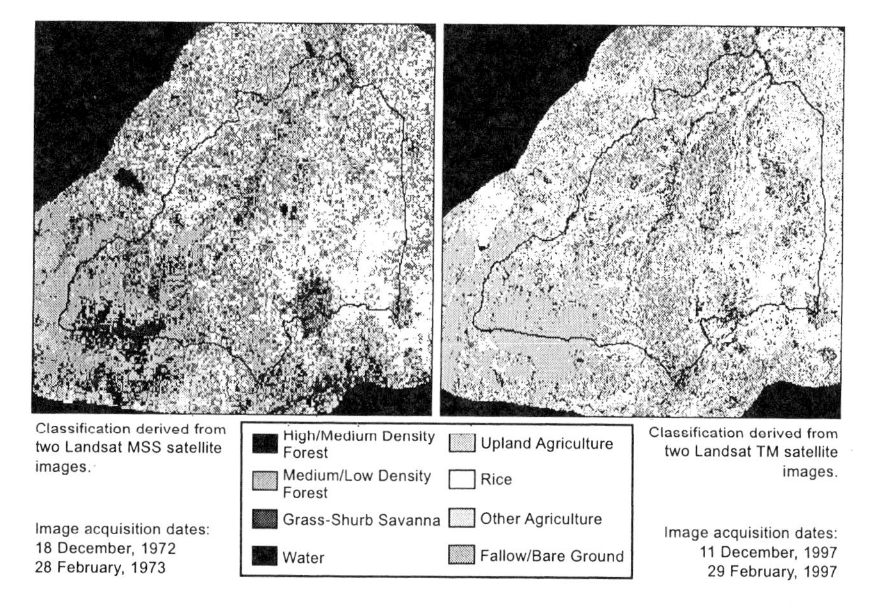

Figure 5. Landsat TM classifications of the study area for 1972–73 and 1997

Figure 6 provides an overview of the various components of the Nang Rong data, including the 2000 component. On the social side, data collection began in 1984, when fifty-one villages were selected to be included. Each household in the fifty-one villages was interviewed, with information obtained on every member of the household. In 1994, reinterviews were conducted with all the surviving 1984 households as well as with all new households. If the original village had split administratively during the intervening ten years, all households in the successor villages were included in the follow-up.

Community data were collected in each of the fifty-one villages in 1984. In 1994, the community data collection was expanded to include all 310 villages within the 1984 boundaries of Nang Rong. Finally, in 1994–95, for a sample of twenty-two of the original fifty-one villages we followed out-migrants to the most common destinations. The questionnaires used in 1984, 1994, and 2000 can be found at www.cpc.unc.edu/project/nangrong.

We have already alluded to the complications posed by shifting administrative boundaries for a longitudinal study. Our approach is to track all of the original component units, making it possible to use whichever one is best for analytic purposes. Households can be reconstructed; villages can be reconstituted.

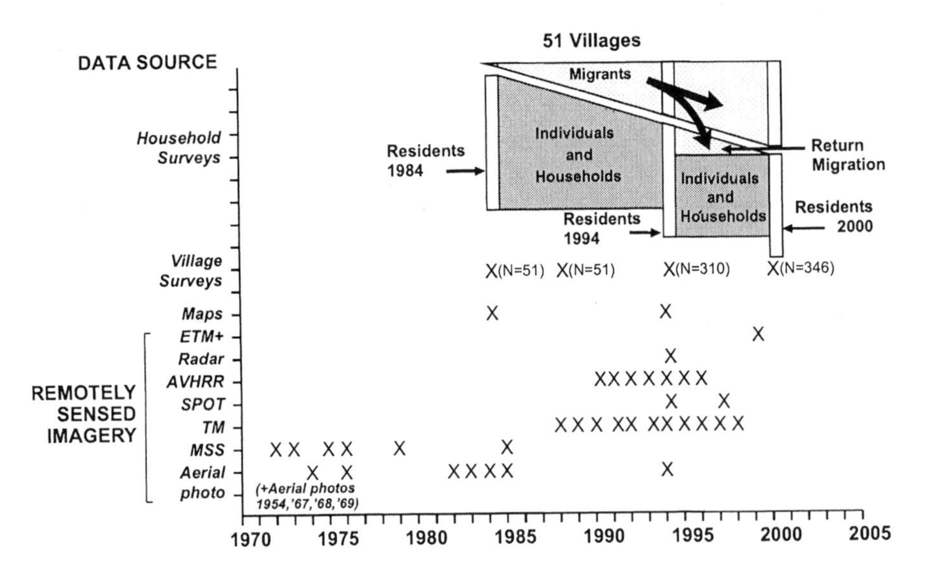

Figure 6. Nang Rong data

As Figure 6 indicates, we also have assembled detailed maps and a collection of remotely sensed images extending back into the 1950s. The maps include a series of 1:50,000-scale topographic base maps compiled in

1984 and a series of administrative planning maps dated 1993–94. The former maps depict road types ranging from major paved roads to foot and cart paths, detailed permanent and intermittent hydrographic features, village locations, and district administrative boundaries. The 1993–94 administrative planning maps contain more detailed administrative boundaries as well as village locations.

The satellite time series includes a relatively recent scene acquired from the Enhanced Thematic Mapper (12/99); multiple scenes acquired from the Landsat Multispectral Scanner (12/72, 2/73, 11/75, 1/76, 10/79, and 12/85); from the Landsat Thematic Mapper (12/88, 2/89, 12/89, 3/90, 12/90, 3/91, 3/92, 11/92, 12/92, 2/93, 3/94, 11/94, 1/95, 2/95, 12/95, 1/96, 12/96, 3/97, 9/97, 10/97, 11/97, 12/97, 10/98), from the SPOT Multispectral (4/94, 11/97) and Panchromatic (3/94, 4/94, 12/97), and from the NOAA Advanced Very High Resolution Radiometer (134 frames systematically sampled—approximately every two weeks—from 6/90 to 9/96), and from the JERS-1 Synthetic Aperture Radar (11/94). The satellite time series is capable of representing land cover at the plot and landscape levels through analog and digital techniques. In addition to the satellite data, we have acquired vertical panchromatic aerial photography for the following years: 1954, 1967, 1968, 1969, 1974, 1976, 1982, 1983–84, 1985, and 1994. Scales range from 1:6,000 to 1:50,000.

We also have a field database of differentially corrected GPS coordinates for point locations taken throughout Nang Rong District during 1997–2000. Attributes connected to point locations include: locational coordinates, video streams, photographs, digital LCLUC class information for specific satellite dates, and actual LCLUC–type information at the point location. Approximately 650 points are contained in this database that provide for geodetic control as well as land-cover validation.

6. THE 2000 DATA COLLECTION: DESIGN AND IMPLEMENTATION CONSIDERATIONS

The 2000 round of data collection shared numerous features with the 1994 round, and we will not discuss those here. Rather, we will note the additional features, with special emphasis on linking households to the plots of land they use. Our new procedures built on several major design decisions about coverage, focus, and scale. We discuss those decisions here, leaving a description of our field procedures to the next section.

6.1 Land or People?

In linking land and people, a decision needs to be made about the starting point. Do you start with a specific land area and identify all the people using that land? Or do you start with a defined group of people and identify the land they use? Or do you do both? There is no one-to-one match. Some households living in the area under consideration are likely to use land outside this area; conversely, some land within the study area is likely used by households living outside the study area (see Figure 1 in Rindfuss et al. this volume).

Early on, we decided to start with households and link them to their land parcels. Although there were theoretical considerations, this decision built on strengths of preexisting data. The linking of households and parcels was set within a broader project that had a history. Part of that history was following households over time. We already had longitudinal data for most households, in many cases stretching back to 1984. Further, the broader project had additional goals that led us to contact and interview all households in the original fifty-one villages, no matter what decision was reached about direction of linkage. Once we were interviewing the household, we could include an interview module that inquired about the parcels of land that they farmed.

The decision to start with households and locate their land parcels meant that we would not have continuous coverage of the land around the village. Missing would be household information for land associated with households from other villages (unless it was land adjacent to land used by households from the village in question), land associated with some governmental, religious, or corporate entity, and land that no one claimed. Figure 7 shows the results for one of our villages. The land parcels in yellow are lands used by households from one of our fifty-one sample villages. For the other land parcels (shown in the background black and white), we have no information from our household survey. From other more qualitative information, we know that the vast majority of the remaining land in Figure 7 is used by households from neighboring villages. Furthermore, the decision to start with households leads naturally to a focus on household decision-making processes. More corporate kinds of decision making are not necessarily covered by this approach. We recognize that households are crucial stakeholders in Nang Rong, but not the only ones.

We considered strategies for contacting people who were associated with the land but not from our sample villages, but all those that seemed to have a reasonable chance for success had associated costs that far exceeded our budget. From an analytical perspective, we will still have a lot of information on land not associated with households from our sample villages. We will have all the land-cover information derived from remotely sensed data, as well as information derived from the various available maps formalized within our GIS. Derived measures of geographic site and situation will be generated for all land parcels, irrespective of whether they are linked to specific households, and the changes in LCLU will be assessed for all parts of Nang Rong. Further, for many of the parcels we will know the village of the user of the land parcel and the associated characteristics of that village. Hence we will be able to know the extent to which our linked land is similar to the land that was not linked to a survey household.

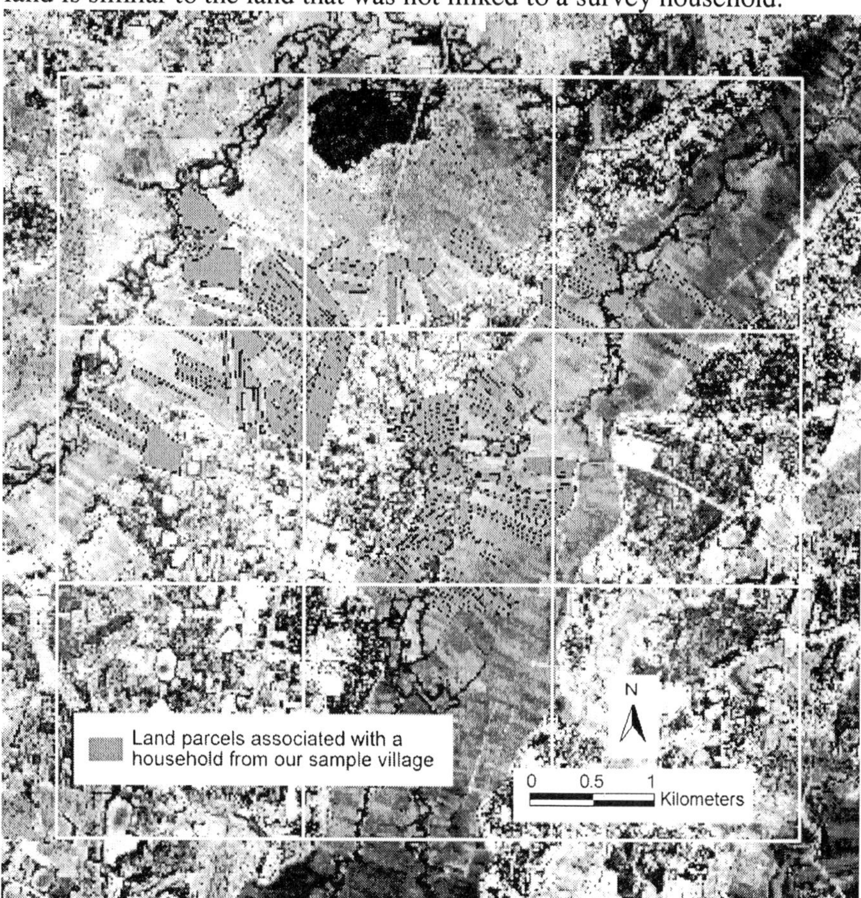

Figure 7. Map showing land parcels associated with our sample village

6.2 Ownership versus Use

The household to land linkage could be based on land ownership, land use, or some combination. From a theoretical perspective, both ownership and use are important. Indeed, we know relatively little about whether households treat their land differently if they own and use it, own and rent it to others, own and let others use it for free, rent it from someone else, or simply occupy it. Also, as mentioned earlier, there are many different types of land title. In Nang Rong, a process has begun to standardize, regularize, and regulate land tenure. This is a process that is being played out in many parts of the world. To what extent do changes in land tenure systems affect land cover and land use? To answer this and related questions, one would like linkage information based on ownership, where ownership is broadly defined to include the full range of land titles that might exist. On the other hand, stewardship of the land might depend on whether a household owned it or just used it, the length of time involved in both, and the site conditions for the parcel. To explore this, an analyst would like to know both use and ownership.

Ultimately, our decision on use versus ownership hinged on practical and ethical issues. Practical considerations meant that household interviews frequently were semipublic, with neighbors listening and sometimes participating. Some of our qualitative work and early pretests suggested that the question of who owned which parcel of land could be sensitive. Sometimes, households were using land to which they had no title. Sometimes, owners did not want others to know that they owned a particular parcel of land. Sometimes there was ambiguity within a kinship group as to exactly who owned a parcel of land. Indeed, for some parcels the land was jointly used and perhaps jointly owned. In such situations, pushing to find out exactly who owned a given parcel might generate conflict among relatives—typically siblings expecting to inherit a particular parcel. For ethical reasons, we wanted to avoid divulging to others information that the household might want to keep private, and we wanted to avoid creating any conflict as a result of our interviews. Since land use tends to be common knowledge within a village and ownership does not,[6] we decided that use of a parcel would be the linking mechanism rather than ownership. Within the household interview, we explicitly defined *use* for the respondents so that all respondents had the same definition in mind:

> By use, I mean the *plangs*[7] that your household uses to grow crops, livestock, fruit trees, fishponds, or anything else that benefits your household. This includes joint use. By joint use, I mean your household and one or more other households cultivate this land jointly or together.

Even though use is the primary linking mechanism, we also have two sources of information on ownership. First, within the household interview, once we established that the household was using a given parcel, we asked whether or not they owned that parcel of land. Hence we know the ownership status of land used. But if they owned a piece of land and did not use it, we do not know about this linkage from the household survey. The second source of information on ownership comes from cadastral maps and associated information available from the Land Office. For all land parcels that have been cadastrally mapped, we know the name of the land owner and can match this to the names listed on household rosters.[8] This will provide numerous examples of parcels owned by a household but not currently used by those households. In those cases where the user is some other household within the village, we will know the characteristics of both the owner and user households.

6.3 Scale and Production Considerations

As explained, the 2000 data collection built upon a relatively large longitudinal data set. The 1994 round of data collection interviewed 7,337 households containing 42,219 current and former members. Given the rate at which new households were forming, we anticipated that the original fifty-one villages would contain approximately 10,000 households in 2000. By most social science standards, this is a large sample. Further, based on our experience in 1994, we anticipated that the average interview would be one hour or longer. Any procedure we used had to be relatively simple, short, and capable of being included in a household interview designed for other purposes as well. Further, since we were hiring and training a relatively large number of interviewers, the tasks required of the interviewers had to be tasks that could be easily learned as well as standardized across interviewers and supervisors.

Given these requirements, certain approaches were eliminated from the beginning and as a result of early pretests (Rindfuss et al. forthcoming). For example, asking interviewers to accompany household members to their fields and then taking GPS readings to identify the fields was unthinkable at the scale of our operation—although we did pretest this to confirm that it was too time intensive. Because of the heterogeneity that exists within Nang Rong, both with respect to site conditions and household characteristics, we wanted to keep the analytical sample as large as possible, and hence we rejected the possibility of collecting the linkage data only on a subsample of households or villages. We considered an approach whereby households would place a marker in their fields and the interview team would find the markers, GPS the location, and record the information on a form. We actually did not even pretest this approach because villagers convinced us that it was totally unworkable. We also tried bringing detailed maps into the household and having household members point out on the maps the locations of their land parcels. This approach was abandoned for a number of reasons. First, the maps were large and unwieldy. The interviewers did not like carrying them around the village. They were also awkward to use in a household setting. Further, given the tropical conditions, they needed to be laminated, and our cost estimates suggested that this would be prohibitive. Further, it was quite common for household members not to be able to read the maps. While it is conceivable that we might have been able to train household members in the art of map reading, the time costs would have been too large for the scale of our operation. Sometimes only the elderly and children were at home, particularly during planting season when many members of the household were working in the fields. While the older folks may have known about the land, many could not see the map well, and the children lacked the knowledge of the land to provide any input.

7. COLLECTION OF LINKED DATA IN 2000

One might think of our approach as a process of acquiring pieces of a jigsaw puzzle from several sources, spreading them all out on a table, and then assembling them into a plausible picture. One key, and perhaps the most indispensable key to assembling a jigsaw puzzle, is the shape of a puzzle piece relative to the shape of other pieces. In our household-to-land linking case, the most fundamental key is social networks. Our approach works because people know neighboring relationships among users of land parcels, and we can match these relationships across different providers of information of neighboring relationships. Some of the providers of neighboring relationships know details about various characteristics of the household but would not be able to provide specifics on the location of the parcels the household uses. Other providers of neighboring relationships know the location of parcels but not the detailed characteristics of the household. We "solve the puzzle" by matching neighboring relationships across assorted information providers. To continue the jigsaw analogy, the social network information on households using neighboring parcels provides the equivalent of the color of the adjacent puzzle pieces—but not the exact shape.

Figure 8 summarizes the steps in our data collection and matching process, described below. Figure 9 is an illustration to which we will refer in the course of describing our fieldwork. It shows three village clusters and their associated households (HH). It also shows a set of twenty *plang*s that lie in the space adjacent to the three villages, with each *plang* associated with a specific household in one of the three villages. Village 2 is one of our fifty-one sample villages; Villages 1 and 3 are not.

A Prior to data collection
- Obtain cadastral maps
- List of names from earlier rounds
- Digitize cadastral maps
- Create field maps

B Spatial team-phase I
- HH listing
- Dwelling unit GPSing
- GPS village centroid
- Community interviews

C Household interview team
- Locate and interview old HHs
- Interview new HHs

D Spatial team-phase II
- Group interviews
- Spot checking

E Matching and checks
- Matching household and group interview data
- Matching cadastral and household data

Figure 8. Steps in the Nang Rong household-land parcel linking, 2000

7.1 Activities Prior to Data Collection

Prior to the fieldwork, we went to two different offices to obtain existing hard-copy cadastral maps. The Land Office, which is under the Ministry of the Interior, had cadastral maps for areas that had the highest level of deeds (land that can be sold, mortgaged, and/or given to heirs). The Land Reform Office, under the Ministry of Agriculture and Cooperation, had cadastral-like maps for land in the land-reform areas. The deeds or documents for these lands are referred to as: "sor por kor." This is land that can be cultivated only by the person on the sor por kor document. The land cannot be sold, mortgaged, rented to someone else, or deeded to an heir. We digitized these maps (about 1,100 altogether) and used them in the production of our field maps. (They will also be used repeatedly in our analyses.) Each of the parcels on the maps had a unique identifying number. The Land and Land Reform Offices also provided us with a list of all the parcel ID numbers and the names of the owners of each parcel.

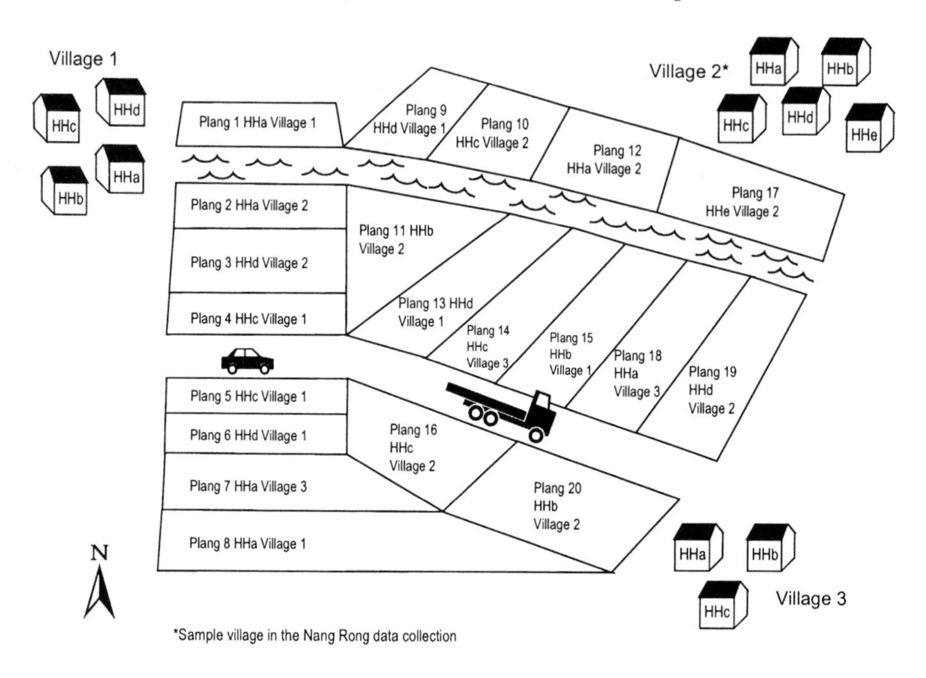

Figure 9. Illustration of households in clustered villages and parcels of land they use

In addition, from our previous rounds of data collection, we had the names of all individuals who lived in (or had lived but were temporarily absent from) our fifty-one villages in 1984 or 1994. We entered these names (over 40,000) into a computerized data file so that they could later be sorted and used for matching purposes. (We also used names from the 2000 round of household data collection, which were entered into the file during the course of the fieldwork.)

7.2 Initial Activities of the Spatial Data Collection Team

Our field-workers were divided into two different types of data gathers or two different teams: the spatial team and the household interview team. The spatial data collection team received specialized training in administering the community questionnaire, use of GPS devices, map reading, and related spatial skills. The household interview team received training in administering a very complex household questionnaire that included detailed questions about each household member, social network ties to others in and outside the village, life-history data, and various characteristics of the household.

The spatial team conducted the interviews for the community questionnaire. Participants in these interviews included the village headman, his or her assistant, and other knowledgeable village informants. The interview was a group interview, with discussion on various questions allowed to proceed until a consensus was reached.[9] They also took a GPS reading at the point indicated by the village headman as the social center of the village. (The social center varied by village but was typically the headman's house, village temple, demonstration market center, school, or other public building—in general, a gathering place of some sort.) This GPS reading became, in essence, an ID that can be used to locate the village in any of the spatial data layers that are part of our GIS.

During this initial phase of the data collection, the spatial team was also responsible for obtaining a complete listing of all households, the name of the household head, and the household's *Ban Lek Ti* number.[10] They then assigned a unique ID number to each household. This list of all households in the village was later used during the household interviews and during the matching process.

The final component of the initial phase of spatial data collection was to take GPS readings for all dwelling units. This involved going by foot with the village headman or an assistant headman to each dwelling unit in the village, taking a GPS reading of the location of the dwelling unit, confirming that the household occupying this dwelling unit was on the list of all households in the village, and recording identifying information.[11] In a typical village, there might be one or two households not on the original list provided by the village headman. These would be added to the list and given a unique ID number. The dwelling unit GPS information would be recorded on a form that also included the name of the household head and the unique ID number of the household. This form links households to the land parcel that contains their dwelling unit. The household questionnaire asked a variety of questions about use of the land parcel that contains their dwelling unit, and this could in turn be linked to the various remotely sensed images and derived GIS coverages available through the collected GPS readings. It is also possible to link to the information on the cadastral maps.

7.3 Household Data Collection

After the first phase of the spatial data collection was completed, a team of household interviewers would move into the village and begin conducting the household interviews. They would first interview all the "old" households—that is, households determined to be the successors to the 1994 households that were part of our earlier data collection. The team would then interview the remainder of the households, termed "new households." New households might contain persons who were part of the earlier data collection who moved out of their old household, or they might contain members entirely new to the village.

In both old and new household interviews, in addition to questions about the land parcel containing their dwelling unit, households were asked about any other land parcels they used during the previous growing season and the growing season prior to that.[12] After all these parcels were listed, households were asked a series of questions about their use of each parcel, such as crops planted, fertilizer used, and so forth. In addition, for each *plang* or parcel of land, the household was asked who used the neighboring parcels in each of the four cardinal directions (N, S, E, and W). Interviewers recorded the first and last name of the user, the household ID number, and the name and ID number of the village where the neighboring user lived. The interviewers had the list of all households in the study village with them during the interview, and these lists were used to clear up any ambiguities. Quite frequently the household did not know all the pieces of information, in which case the interviewer would record what the household knew. If there were two or more neighbors in one direction, then information for all would be recorded.

Figure 9 above provides an illustration. Imagine interviewing HHb in Village 2. They would report on *plang* 11. To the west, there are three neighboring *plang*s (numbers 2, 3, and 4), and, if they could, they would report on all three. Exactly who their neighbors are in the other three directions is somewhat ambiguous. For the south they might report on *plang* 13. Since this *plang* is used by a household from another village, they might know only the user's first name[13] and village. For the north they might report on *plang* 9. Since it is across a major stream, they might not know the user. The information on neighboring land users becomes a set of attributes of the *plang* of interest (in this case, *plang* 11) that can be used to match with information collected in the next step. The information on neighboring users provides location in what might be thought of as "social network space," but it does not yet provide a geographic location. For geographic location, we need to match with information in the next step.

7.4 Second Phase of Spatial Data Collection

The principal aim of the second phase of the spatial data collection was to obtain spatially explicit information on who used neighboring *plangs* that could be linked or matched with the information that was collected in the household interviews. This was accomplished by conducting a series of group discussions in each sample village.[14] The participants in these group discussions included the village headman and other selected village residents who were both knowledgeable about which households used which parcels of land around the village and who had the ability to read map products. They tended to be longtime village residents. Sometimes they were individuals who had occupations that required them to traverse land used by others from the village or adjacent villages. For example, a trapper who needed permission to trap on land used by others was one of the most knowledgeable individuals in one village.

Our goal was to collect spatially explicit information on use (not ownership) of lands by village households. In most cases, we expected that the lands used by village households were in reasonably close proximity to the village, but we would not know exactly how close until we finished the fieldwork. Hence we wanted to collect information as far out from the village as we could. At the same time, we wanted the scale of the map to be as fine as possible so that our informants could pick out recognizable landmarks. Finally, we wanted these map products to be of such a physical size that they could be easily managed and maneuvered during the process of the group interview.

Figures 10a and 10b show our solution for the map products. We produced nine map tiles for each village. Each tile represented 4 km^2. The nine tiles were centered on the geographic center of the village, and together they represented 36 km^2 surrounding the village. Physically, the maps were 76 cm^2, scaled at 1:2,900. The background for the maps was composed of a series of air photos flown in November-December 1994. Cadastral lines were overlaid on this base.[15] Figure 11 is an illustration from one of the maps for one village.

The group discussion had three main tasks. First, if there were areas on the maps that did not have cadastral lines, members of the group were asked to indicate how this noncadastral land was subdivided for use. These lines would then be drawn on the map—and redrawn if group discussion so necessitated. The lines would be digitized later. The new land parcels would be given unique ID numbers. Areas on the maps with existing cadastral lines were updated if necessary. For example, an existing parcel may have been subdivided by the time of the group discussion. In this case, the new lines were drawn and unique ID numbers assigned to the resulting parcels.

Second, the groups were asked to indicate which parcels were used by which households in the village. This information was recorded. If a household from the village used a parcel that was bounded by parcels used by households from some other village, this information was also collected, for parcels immediately adjacent to the one in question and for parcels one parcel removed. Finally, using a map for the entire Nang Rong District (containing village locations, village names and ID numbers, a UTM grid, roads, and hydrogaphy), the group was asked if some village households used land outside the nine map tiles—that is, the 36 km2 surrounding the village center. If so, they were asked which households and which village was closest to their land.

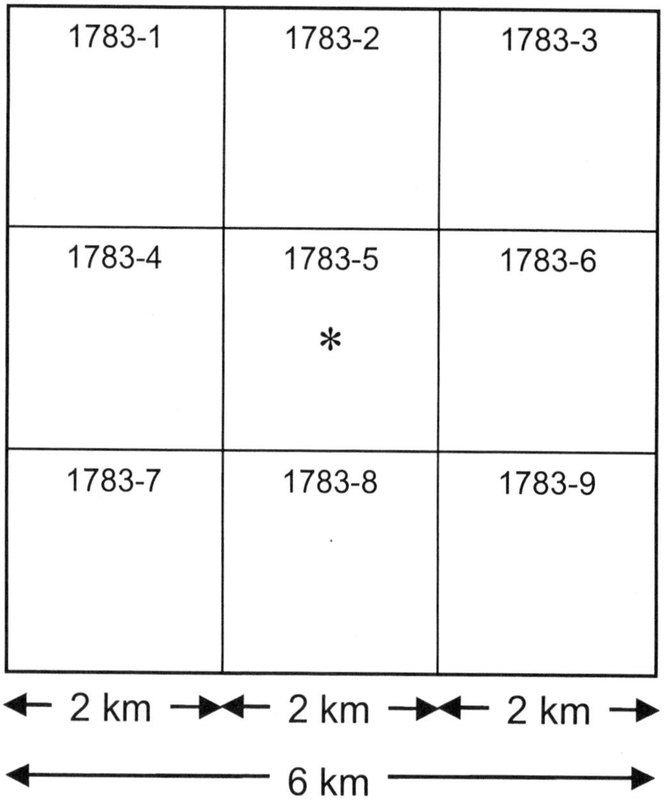

1783-1	1783-2	1783-3
1783-4	1783-5 *	1783-6
1783-7	1783-8	1783-9

← 2 km →← 2 km →← 2 km →

← 6 km →

*The center of the map set (*, tile 5) is the GPS location of the study village.

Figure 10a. Villages and map sets

Figure 10b. Group discussion and map sets

For these group interviews, the spatial team consisted of three members: the group moderator, the list manager, and the data recorder. The group moderator was responsible for introducing the informants to the project, orienting the informants to the maps and the tasks of the group discussion, and leading/facilitating the group discussion. The moderator also drew additional cadastral-type lines on the maps or aided the informants in such drawing, as well as assigning IDs to each newly drawn parcel. The list manager was responsible for maintaining tallies and answering queries about various documents, such as a list of all individual residents of the village and associated split villages, a list of all households, and a tally sheet of the number of plangs reported to have been used by each household in the village. Finally, the data recorder was responsible for recording and checking all information supplied by the group discussion and suggesting when further discussion was needed to fill in missing information on the various lists or the maps.

The group discussion began by assembling all nine map tiles as shown in Figure 10a and in the photograph in Figure 10b. The group moderator then asked the informants to stand around the periphery of the maps and imagine they were flying like a bird or in an airplane over the village or viewing the village from a high vantage point. The group moderator would point out significant orienting features on the nine map tiles and allow discussion from group members as they were getting oriented to the maps. Then the group was asked to identify on the maps all major areas of cultivation used by villagers and to consider who would be knowledgeable to invite for subsequent group discussions. After this, cadastral type lines were drawn, as necessary, and users of specific parcels identified.

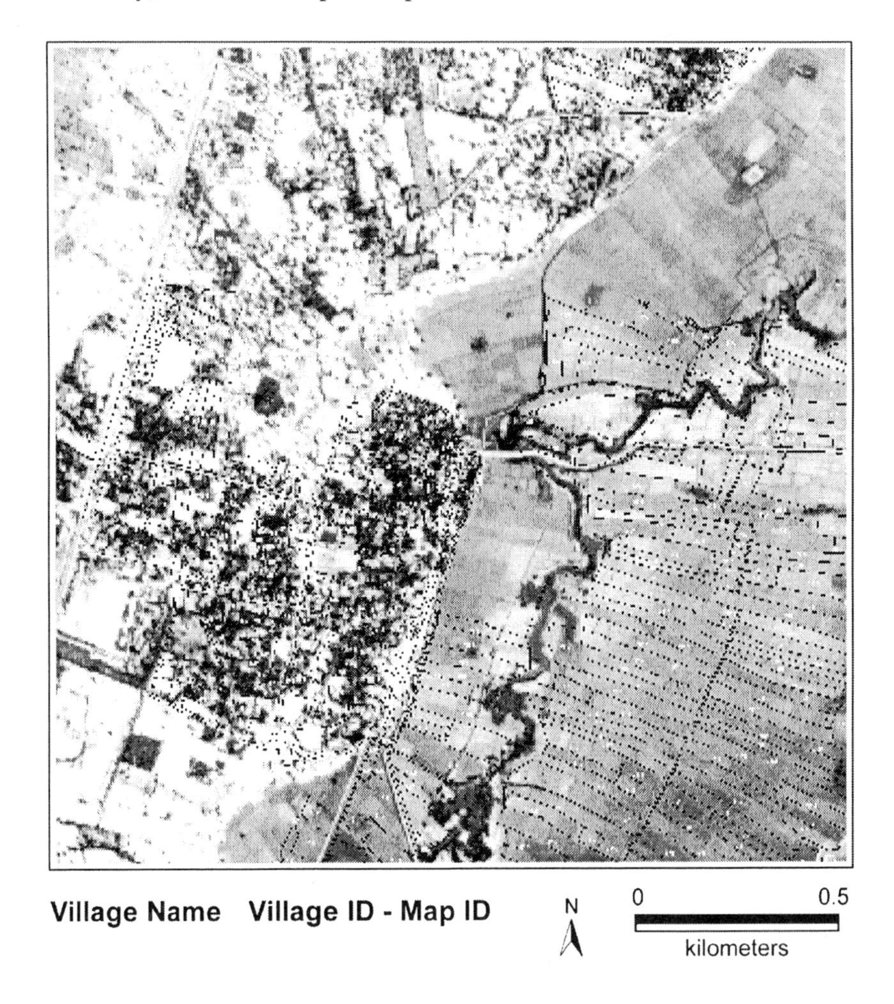

Figure 11. Field maps used by the spatial team for year 2000 survey

In addition to the group discussion, the second phase of the spatial data collection included some spot-checking for accuracy. In each of the ninety-two villages, six parcels used by village households were randomly selected, using a 1 baht coin that was tossed on the nine map tiles. They were selected in two clusters of *plangs*, each cluster representing at least three different households. For each of these six parcels, arrangements were made to have a member of that household accompany the spatial team to that field to obtain

GPS coordinates and check that we were obtaining accurate user as well as location linkages.

7.5 Matching Household and Group Interview Data

Following the logic explained earlier, the goal of the linking/matching process was to associate each *plang* listed on the household questionnaires with the *plang*s identified during the second phase of spatial data gathering. The matching process was painstaking and labor intensive.[16] The procedure was to simply take a *plang* listed on a household questionnaire and search for it in the group discussion data. This was done one adjacency at a time. If the adjacency name matched but the direction did not, we considered this to be a match. (That is, if the household questionnaire indicated that Person X used the *plang* to the south and the group discussion said that Person X used the *plang* to the west, this was considered a match.) This was to allow for the ambiguity of direction if the *plang*s were not lined up perfectly north-south (see Figure 9 for an illustration) and to recognize that some respondents might not know north from west. We looked for adjacency name matches for *plang*s that were immediately contiguous to the *plang* in question, as well as a match one *plang* removed from the *plang* in question, and we recorded whether the match was contiguous or one step removed. Finally, for each *plang,* the matching team provided their subjective estimate of match quality, with the estimate ranging from 1 (best) to 5 (worst).

8. REMAINING ISSUES, CONCERNS, REGRETS, AND FEELING OF GRATIFICATION

As the foregoing has made clear, collecting information necessary to link individual households to the land parcels they used was a time-consuming and costly process. We have not attempted to calculate the marginal or actual costs of collecting the linking information, but surely it was in the hundreds of thousands of dollars; this places it beyond the budgets frequently available to the human dimensions research community—but less than hardware budgets available for remote sensing. To give some idea of costs, our expenses just for laminating the maps used in the fieldwork were approximately $10,000. During the fieldwork, the spatial team, which was primarily devoted to collecting linking information, consisted on average of eleven persons working from February to November, or a little over nine person-years. In addition, there were supervisors from both the Institute for Population and Social Research and the Carolina Population Center who spent time in the field. We are now in the process of digitizing the additional information drawn on the field maps and expect the labor costs for this digitization to be approximately $30,000.

Was it worth it? A complete answer will come in several years, after we have had a chance to analyze the data and know what we have learned that we could not have learned if we had restricted our linkage to the village level as we did in 1994. (See Rindfuss et al. n.d. for a discussion of the drawbacks to and assumptions in doing so.) We think, however, that the answer will be yes. We are now in a position to address a wide range of analytical issues with a sample size sufficient to provide us with serious statistical power. Our analytical models can be multilevel and spatially and temporally explicit.

We now have considerable temporal detail on a range of critical social and economic variables for every individual and household in fifty-one original sample villages. With our social network data (see Rindfuss et al. n.d.) we can link these individuals and households to other households and individuals. We can link these individuals to the location of their dwelling units and the land parcels they use.

All this specificity, however, creates a problem regarding the maintenance of confidentiality while sharing our data with other researchers. We promised—and ethical considerations demand—that we not divulge the identities of our responding households. Yet with all the linkages in the data set, identifying respondents would be straightforward and easy, and this is particularly the case for anyone who can speak Thai. Hence we are not in a position to release our data set to other researchers.[17] In particular, we cannot release the spatially explicit data. Yet science would clearly be better served if we could. Emerging scientific norms, with which we agree, argue that data should be shared with the broader scientific community. Scientific advancement will clearly be more rapid to the extent that multiple researchers, with diverse research questions and talents, are able to use common data sets. We do not yet have an answer for this conflict between protecting confidentiality and making fine-grain, spatially explicit social science data available to the research community—but it is one with which we are grappling.

To what extent is the strategy we have used transferable to other settings? To best answer this question, we address the features of Nang Rong that made it easier and more difficult to collect the linked data we collected. Then the reader can decide relative to his or her study site. Perhaps the aspect most favorable to collecting the linkage data is the cooperation of residents of Nang Rong. This is partly a function of rural (as opposed to urban) populations in general, partly a function of the specifics of Thai culture, and partly a function of the longstanding positive relationships that we (and more specifically, researchers at the Institute for Population and

Social Research, Mahidol University) have maintained with governmental and nongovernmental organizations in Nang Rong, as well as the residents of our original fifty-one villages. In the 2000 fieldwork, only six of 9,422 households refused to cooperate with our interviewers, which is a remarkably high level of cooperation.

The interviewers were recruited from Buriram Rajchapbhat Institute in Buriram, the provincial town nearest to Nang Rong. They were already familiar with northeastern Thailand dialects and culture. In addition, the household interviewers lived in the village while they were conducting the household interviews. This allowed them an opportunity to get to know villagers in an informal manner and eased questions about who they were and why they were in the village. There were numerous stories about warm relationships between the interviewers and village residents. There was even a case where food treats were brought to an interview team by a household from a village where the interviewing had already been completed. In addition, having interviewers stay in the villages meant that they could work without the usual time constraints facing interviewers. They did not have to arrive at the village at a specific time in the morning or leave at dark. They could arrange interviews at the villagers' convenience.

During the peak of data collection, our staff in Nang Rong included over sixty-five individuals. The availability of experienced supervisors from previous rounds of data collection made the management of such a large field staff feasible. In addition, daily wage rates in Thailand are moderately low compared to the United States or Europe, allowing us to employ a large field staff and still stay within our budget.

In rural parts of Thailand there has long been a village headman system that was set up for administrative purposes. Villages are relatively small, and as a result, the average headman knows every household within the village. Further, village headmen report to district-level administrators and have periodic meetings that include other village headmen. As a result, we were able to attend a meeting of all village headmen, explain the purpose of the study, and answer any questions that arose. In addition, prior to doing any work in any given village, we would meet with the village headman, explain what we planned to do, and again answer any questions. Having access to and cooperation from such administrators greatly facilitated our work.

The locations of the plots that villagers used were both problematic and had some facilitating features. As Figure 2 above illustrates, a typical household had multiple plots that they used and that were not contiguous with the plot containing their dwelling unit. This made the fieldwork much more challenging than if we had to link them only to the plot containing their dwelling unit. On the other hand, most of the plots that they used were reasonably close to their village, thus allowing us to produce field maps at a sufficiently detailed scale to allow informants to "see" actual parcels. It is unlikely that our approach would have worked if household plots were widely scattered throughout Nang Rong District and neighboring districts.

Finally, the nature of typical farming activities in Nang Rong tended to facilitate and then hinder our fieldwork. The vast majority of agriculture in Nang Rong is rain fed and hence dependent on the timing of the monsoon. As a result, during the dry season there is limited work needed for those who are farmers, which is the occupation of most workers in Nang Rong. This in turn means that households have more time available for a variety of nonwork activities, including talking with our interviewers. Much of our fieldwork occurred during this off-agricultural season, and as a result it proved easier for our respondents to cooperate with our field-workers. Unfortunately, some of the fieldwork went past the end of the off-agricultural season and into the rainy season, when farmers were preparing their fields, planting, and transplanting (in the case of rice). At this point, most households were much busier, and this in turn slowed down our remaining fieldwork. We still were obtaining excellent cooperation from village residents, but scheduling interviews took more effort, and sometimes we had to settle for times that were less than optimal from the perspective of maximizing the efficiency of our fieldwork staff.

This is a natural segue into a discussion of those features of Nang Rong that might have made it more difficult for us to conduct the fieldwork linking households and their farm parcels. Perhaps the most obvious are aspects of the climate in Nang Rong. It is a tropical climate, where the local joke is that there are three seasons: hot, hotter, and hottest. One should not underestimate the toll this takes on fieldwork staff, as well its affect on the ability of respondents to concentrate throughout a long interview or group discussion. Similarly, flooding during the rainy season created transportation problems. These were not insurmountable, but they made the fieldwork more difficult and delayed its completion.

Fiscal problems were caused by the changing value of the baht (B), the Thai currency. For many years, the Thai currency was pegged to the U.S. dollar, and the exchange rate was approximately 25B = $1. This changed during the fiscal and economic crisis of 1997, when the baht was allowed to float. When we wrote the proposal, the exchange rate was approximately 42B = $1. By the time the proposal was reviewed and funded, the baht had strengthened relative to the U.S. dollar. During the peak of our fieldwork, the exchange rate was about 36B = $1. Given the scale of our fieldwork, this deterioration in the value of the U.S. dollar created a problem for us, which we solved in a variety of ways, including raising some additional funds, cutting back in some areas, and "borrowing" some from the coming years in the project. Ironically, as we write these words, the exchange rate now is 44B = $1.

A fluctuating exchange rate is something that, obviously, the researcher cannot control and funding agencies are unlikely to cover. Yet they seem to be happening with increasing frequency and could wreak havoc with a well-conceived research plan. We do not have a solution but would suggest that it ought to be up for discussion within the broader land-use research and funding community.

Paradoxically, our zeal to have fieldwork staff carefully check their work and go back to make further inquires if necessary proved problematic in the one case where we wanted errors to remain so that we could check error level and pattern. Recall that in the second phase of the spatial data collection, in each village, six parcels were selected for spot-checking. A member of the associated household would go with the spatial team to the parcel to insure that we had the right parcel in the correct location. Our original thinking was that this data would provide us an opportunity to check the accuracy of the larger endeavor and a chance to see whether there were any systematic error patterns (e.g., were errors more common in smaller *plang*s, isolated *plang*s, or *plang*s used by those who were only loosely connected within the social networks of the village). A preliminary look at these data suggested that the process was error free. We have since determined that when errors were discovered, they were corrected. As a result, we now have only impressionistic evidence suggesting that the error rate was low.

In Thailand, during the period of our longitudinal research, administrative boundaries have been changing continually. As villages grow, they are split into two or more administrative villages. Similarly, as the population of districts grows, portions of them are split off to create new districts. While this process has obvious administrative advantages (and perhaps some disadvantages), it has created problems for us. At numerous junctures in our data gathering process, we are asking respondents and informants about "this village" or "Nang Rong." When we do, we are referring to the village or Nang Rong in terms of its 1984 boundaries. We attempt to make this very clear to respondents and informants. We verbally remind them. For the group discussions, we have a map showing boundaries for Nang Rong District as they existed in 1984. At the village level, we give them the names of all the current villages that had split off from the 1984 village. Yet we have to assume that these changing boundaries cause some confusion in the minds of respondents and informants. Unfortunately, we have no straightforward way of knowing how much.

From an analytical perspective, adjusting for the changing boundaries is less problematic. We have collected our data in such a manner that we can go back to the 1984 boundaries (assuming our villagers have also done so) and thus have longitudinal comparability. For most comparisons we can also go back to the 1994 boundaries if there is an analytical need.

The distinction between use and ownership was an issue with which we continually struggled. Even in conversations among senior members of the research project, we found ourselves sometimes using the term *own* when we meant *use.* In the context of Nang Rong, it should be noted that this distinction is even more ambiguous than it might be in other settings. The recent frontierlike past, in conjunction with multiple land-tenure types, means that in some cases the distinction might be ambiguous. For the least desirable land title, ownership is actually ambiguous. Those who use these *plang*s have the right to continue to use them. But they do not have the right to sell or mortgage the *plang,* nor do they have the right to deed it to their heirs. We also have anecdotal evidence suggesting some confusion for those who rented out their land. During the interview, we had defined *use* as growing crops, raising animals, or generating income for the household. Some interviewers/interviewees interpreted this to include rent.

As explained above, we had good reasons to concentrate on use, but this means that we do not have information on any absentee owners. Until we begin our analyses, we will not know how common absentee ownership is. But we do expect that absentee owners make decisions using a different calculus than resident owners. We will be able to test this with our land-cover data, but if it is true, we will not have any information about these absentee owners to enable us to begin to speculate on what might be motivating their decision making.

The nature of our data will also permit checks on the quality of the data on both sides of the household-parcel link. For example, household reports of what they grew on each of their parcels could be used to cross-check against the classified satellite data for that year. When they do not match, we can explore reasons, considering such variables as topography, proximity to water, the spatial mosaic, the number of parcels upon which the household was reporting, and the number of years the household had been using that parcel.

We end this chapter by asking: How successful were we in collecting high-quality linkages between households and the land parcels they use? Because we are writing this narrative while we are still digitizing the maps used in the data collection process, completing the task of linking household interviews longitudinally, and in general just beginning to exercise the data, we can give only a preliminary answer. In light of this, the numbers we report here should be taken as preliminary; we have not been able to do all the necessary cross-checks.

First, approximately 10 percent of all the *plang*s used by households fall outside the 36 km^2 area covered by our nine map tiles. For these *plang*s, at best we know where they are relative to the village nearest them. While this gives us a clue to their location, we would not be in a position to locate them in any of the remotely sensed images available to us. Not obtaining more precise locational information on these *plangs* was a practical decision. In our estimation, had we attempted to obtain more precise information for them, the quality of the household-*plang* linkage for the other 90 percent of *plangs* within the 36 km^2 area centered on the village would have deteriorated.

For the remaining *plangs* inside the 36 km^2 area, we allowed matching of immediately adjacent *plangs* (that is, 0 jumps) and then matching with *plangs* that were one *plang* removed from the *plang* we were attempting to match (that is, 1 jump). Under the most generous definition of matching—that is, a match in at least one direction in 0 or 1 jump—we calculate a match rate of 76 percent. If we back off from this definition and insist that the matches be immediately adjacent to the *plang* (i.e., 0 jumps), then the percent matching in at least one direction is 75 percent—essentially the same.

We also know the number of directions in which matches were identified. Only 20 percent matched in all four directions with 0 jumps. While this might be considered the strictest definition of a match, we would argue that such a standard is unrealistic for a variety of reasons. First, many of the *plangs* border a road, stream, or some other feature that would make it unlikely that the user would know a neighbor. Second, many immediate farming neighbors are from a nearby village rather than from the household's village. Hence it is unlikely that the household or the group of village informants would know the surnames of those farmers. Third, asking about neighboring *plangs* in the household interview occurs toward the end of a very long interview. There might be a tendency on the part of the respondents, the interviewers, or both, to feel that they had accomplished their task if one or two neighbors had been named and recorded. Fourth, the *plang* might have a triangular shape and hence not have four sides. Finally, to record a match, in all but a few rare exceptions, all we need is one neighbor recorded on both the household interview and the group discussion.

To the best of our knowledge, this is the first time that the neighbor network approach has been used to link households to the farm plots that they use, which raises the question of how we evaluate our level of matches. What is our yardstick? In remote sensing work, checks to insure that one has correctly classified level-1 land-cover type commonly result in percentages above 90. On the other hand, in social science surveys—which are the workhorse of many social science disciplines—typical response rates in North America, Europe, and Japan are quite frequently in the 70 to 75 percent range and are considered acceptable.[18] Further, there is routine data error embedded in responses provided by respondents from more than 9,000 households and recorded by interviewers. Our household-to-*plang* linkage rates are well within the bounds set by survey response rates, and as such we are pleased with our efforts using the neighbor network approach to link households with the land parcels they use. However, we reserve the right to revise our opinion as more household-to-land parcel linkage attempts are reported in the literature and as we have a chance to examine our data quality in greater detail.

NOTES

Authors' note: Address Correspondence to: Ronald R. Rindfuss, Carolina Population Center, University Square, CB# 8120, 123 West Franklin Street, University of North Carolina at Chapel Hill, Chapel Hill, NC 27599-3997, phone (919) 966-7779.

[1] *Land cover* refers to the various types of coverages of the earth's surface, including vegetative, water, and such clearly man-made surfaces as roads and buildings. *Land use,* on the other hand, refers to the way in which humans use a particular land-cover type.

[2] Even if they can be "seen" by remote sensing procedures, they cannot be identified and thus linked into a longitudinal data file.

[3] The successor household was defined in terms of age, gender, and kinship relations. At time 2, the successor household was defined as the household that contained the oldest female member of that household from time 1. Should no household be found that contained her, then the successor household was defined in terms of her husband from the household at time 1. Should that person not be found in any household, then it would be the second-oldest female from time 1 and so forth until a successor household was found or it was determined that no successor household existed at time 2. The last persons to be used for searching and linking purposes would be servants and boarders.

We collected sufficient linkage data that we can change these rules to suit any definition of successor household. Hence the definition can be varied to suit the theoretical interests of the investigator using the data and to test the sensitivity of results to alternative specifications of successor households.

[4] But our satellite time series can define the land cover dynamics of this region as a context to what is happening elsewhere in Nang Rong.

[5] "Frontier" is used here in terms of the history of the past 500–700 years. At the time of the great Ankor civilization (roughly 802–1432), Nang Rong was part of the Ankor empire (Rooney 1999). Indeed, the ruins at Prasat Phnom Rung attest to the Khmer influence in the earlier history of Nang Rong.

[6] In areas where cadastral maps exist, ownership is a matter of public record. But even though it is a matter of public record, it need not be common knowledge within the village.

[7] A *plang* is a contiguous parcel of land, perhaps equivalent to the English term *land plot.* During the process of constructing the questionnaire, we did some qualitative, cognitive lab work (e.g., Tanur 1992) to determine the best term to use to refer to farm plots (Rindfuss et al. 2001). Basically, this involved talking with small groups of individuals from Nang Rong and asking them about the different Thai terms that could possibly be used to refer to a parcel of land. We went both forward and backward: asking them to tell us what certain terms meant to them and asking them what term would be best to describe our concept of farm plot. The Thai word *"plang"* emerged as being closest to our concept, but it also was clear that using this term still left some ambiguity. Thus in the questionnaire we used the term *plang* and then followed this with a definition of our concept of a farm plot as a single contiguous piece of land.

[8] Matching on names is feasible in Thailand because names tend to be unique. There are a number of reasons why Thai names are much more likely to be unique than American, European, Korean, or Chinese names. Put differently, in Thailand the John Smith problem is

less likely to arise. Surnames were not used in Thailand until King Rama VI decreed that they should be used. This decree came out about 1920, and thus family names have been used only for a few generations.

Further, when families were registering their surnames, there were checks to make sure that there were no duplicates. Thus in principle, in the 1920s each family had a unique surname. While the checking for duplicates at registration time was by hand and hence not perfect, the overall effect was to have the vast majority of surnames unique. Today, if you want to change your surname, you have to go to the registrar's office and they will check to see if your proposed new surname duplicates the surname of anyone else in the country. If it is a duplicate, you have to pick another one.

Unlike in many Western countries, there is no tendency for fathers to give their first names to their son or mothers to give their first name to their daughter. Instead, sometimes a syllable from the mother's name and a syllable from the father's name might be joined to create a name for the child. In short, frequently first names are made up and thus have a higher probability of being unique.

The Thai alphabet has 44 consonants, 21 vowels, and 5 intonations. Hence there is a greater possibility for creating a wide range of unique names than in the Western alphabet.
A computerized file exists containing the names of approximately 52,000 individuals who were listed in the household rosters of the 1994 data collection. We are still checking the number of duplicate names in the file. There are fewer than 100 cases where more that one person had the same first and last name. For some of these cases, however, it appears that the same person was found recorded as residing in more than one household, presumably in the transition of moving from one to the other. It is our expectation that when we finish resolving this issue, there will be fewer than twenty cases of duplicate names. For these reasons, matching by name should be less problematic in Thailand than in most other countries, although it should be noted that we have not yet attempted the matching.

[9] Given the nature of Thai culture, the homogeneity of our villages along ethnic, religious, and other lines, and given the nature of the questions being asked, reaching a consensus was not problematic.

[10] A *Ban Lek Ti* number is similar to a dwelling unit number (e.g., a street address). It is a number assigned to households and used for administrative purposes. It is a number that we sometimes used in the matching process.

[11] We should make clear that "dwelling unit" refers to a physical structure and "household" refers to a collection of individuals, typically related by blood or marriage, who share meals and a common kitchen. One household might have more than one structure that constitutes their dwelling unit; and one dwelling unit might contain more than one household.

[12] The interviews occurred between April and July 2000. The growing season for rice and most other crops in Nang Rong begins with the monsoons, which typically begin in June. Harvesting of rice usually occurs in December-January. So the "previous growing season" refers to the time period from July 1999 to January 2000. As part of the interview process, we unambiguously defined this term for respondents.

[13] In Thailand, surnames are rarely used in everyday conversation. It is possible to know someone fairly well and never learn their last name.

[14] By 2000, the original fifty-one villages had subdivided into ninety-two villages. This second phase of the spatial data collection was repeated in each of the ninety-two villages.

[15] During the middle to late 1990s, land-cover patterns were relatively stable within the district, particularly for the lowland rice-producing areas. Even when trees are removed, they are generally cut from within existing rice paddies, where they had been retained for shade for field-workers and water buffalo. In the upland areas, often more distant from the village center, some forest lands were cleared in the middle to late 1990s. These lands were typically converted to cassava and sugarcane. But most of the deforestation and agricultural extensification was completed by the late 1980s. As a result, the 1994 air photos were sufficient for our purposes, especially since more recent air photos were not available to us.

[16] We did not attempt to automate the matching process, but we might consider doing so in the next round of data collection. Automating the matching process would mean postponing the matching until all the data has been entered and cleaned. This in turn would mean that the matching is attempted well after the field staff had found other employment and after they had forgotten details that might not have been recorded. We decided it would be best to have the field staff do the matching while the data collection process was fresh in their minds and while they could still return to a village to collect additional data if necessary. The matching was done while the second phase of the spatial data gathering was also ongoing, and field staff rotated between matching work and data gathering.
[17] See www.cpc.unc.edu/project/nangrong for the data that have already been released.
[18] In rural areas of developing countries, response rates are typically higher. In the household portion of the Nang Rong data collection, only six households out of more than 9,000 refused to participate, which is an extraordinarily high response rate for a social science survey.

ACKNOWLEDGEMENTS

The design and research reported here were primarily supported by a grant from the National Institute of Child Health and Human Development (RO1-HD33570 and RO1-HD25482). Additional support was provided by the Mellon Foundation (through a grant to the Carolina Population Center), the National Aeronautics and Space Administration grant (NAG5-6002), the National Science Foundation (SBR 93-10366), the Evaluation Project (USAID Contract #DPE-3060-C-00-1054), the MacArthur Foundation (95-31576A-POP), and a centers grant to the Carolina Population Center from the National Institute of Child Health and Human Development, HD05798. We would also like to acknowledge, with gratitude, the help and cooperation of numerous individuals who assisted in the design and collection of the data described here. Numerous staff members and graduate students at the Institute for Population and Social Research, Mahidol University, and the Carolina Population Center, University of North Carolina participated in the design, pretest, and actual data collection. A large field staff of supervisors, interviewers, and spatial data collectors did an extraordinary job during the fieldwork and data entry phases of the project. And finally, and perhaps most importantly, we would like to acknowledge the cooperation of the people of Nang Rong. Our requests for information were sometimes taxing, and Nang Rong residents cooperated in a manner that was truly heartwarming; for that we are in their debt.

REFERENCES

Allen, M., S. Raper, and J. Mitchell. 2001. "Uncertainty in the IPPC's Third Assessment Report." *Science* 293: 430–433.

Becker, G. S. 1991. *A Treatise on the Family*. Cambridge: Harvard University Press.

Chamratrithirong, A., and C. Sethaput, eds. 1997. "Fieldwork Experiences Related to the Longitudinal Study of the Demographic Responses to a Changing Environment in Nang Rong, 1994." Mahidol University, Institute for Population and Social Research.

Chayanov, A. V. 1966. *The Theory of Peasant Economy.* Homewood, IL: R. D. Irwin.

Citro, C. F. and H. W. Watts. 1986. *Patterns of Household Composition and Family Status Change.* Washington, D.C.: U.S. Bureau of the Census.

Clausen, John A. 1972. "The Life Course of Individuals." In M. W. Riley, M. Johnson, and A. Foner, eds., *Aging and Society,* vol. 3: *A Sociology of Age Stratification* (New York: Russell Sage Foundation), 457–514.

Committee on Mitigating Wetland Losses. 2001. *Compensating for Wetland Losses under the Clean Water Act.* Washington: National Academy Press.

Duncan, G. J., and M. Hill. 1985. "Conceptions of Longitudinal Households: Fertile or Futile." *Journal of Economic and Social Measurement* 13(3-4): 361–375.

Elder, G. H., Jr. 1974. "Age Differentiation and the Life Course." *Annual Review of Sociology* 1: 165–190.

———. 1985. "Perspectives on the Life Course." In G. H. Elder Jr., ed., *Life Course Dynamics: Trajectories and Transitions, 1968–1980* (Ithaca, NY: Cornell University Press), 23–49.

———. 1998. "The Life Course and Human Development." In R. M. Lerner, ed., *Theoretical Models of Human Development,* vol. 1 (New York: Wiley), 931–991.

Entwisle, B., S. J. Walsh, R. R. Rindfuss, and A. Chamratrithirong. 1998. "Land Use/Land Cover and Population Dynamics, Nang Rong, Thailand." In D. Liverman, E. F. Moran, R. R. Rindfuss, and P. Stern, eds., *People and Pixels: Linking Remote Sensing and Social Science* (Washington, D.C.: National Academy Press), 121–144.

Featherman, D. L. 1983. "The Life-Span Perspective in Social Science Research." In P. B. Baltes and O. G. Brim Jr., eds., *Life Span Development and Behavior* (New York: Academic Press), 1–57.

Feeney, D. 1988. "Agricultural Expansion and Forest Depletion in Thailand, 1900–1975." In J. F. Richards and R. P. Tucker, eds., *World Deforestation in the Twentieth Century* (Durham: Duke University Press), 112–143.

Hogan, D. P. 1981. *Transitions and Social Change: The Early Lives of American Men.* New York: Academic Press.

Houghton, J. T., Y. Ding, D. J. Griggs, M. Noguer, P. J. van der Linden, X. Dai, K. Maskell, C. A. Johnson, eds. 2001. *Climate Change 2001: The Scientific Basis.* Cambridge: Cambridge University Press.

Kaida, Y., and V. Surarerks. 1984. "Climate and Agricultural Land Use in Thailand." In M. M. Yoshino, ed., *Climate and Agricultural Land Use in Monsoon Asia* (Tokyo: University of Tokyo Press), 231–254.

Keilman, N., and N. Keyfitz. 1988. "Recurrent Issues in Dynamic Household Modeling." In N. Keilman, A. C. Kuijsten, and A. Vossen, eds., *Modelling Household Formation and Dissolution* (Oxford: Clarendon Press), 254–286.

Kriedte, P., H. Medick, and J. Schlumbohm. 1981. *Industrialization before Industrialization: Rural Industry in the Genesis of Capitalism.* New York: Cambridge University Press.

McMillan, D. B., and R. Herriot. 1985. "Toward a Longitudinal Definition of Households." *Journal of Economic and Social Measurement* 13(3-4): 349–360.

National Statistics Office. 1990. *Population and Housing Census.* Changwat, Buriram, Thailand: Office of the Prime Minister.

Parnwell, M. J. G. 1988. "Rural Poverty, Development and the Environment: The Case of Northeast Thailand." *Journal of Biogeography* 15: 199–208.

Rigg, J. 1987. "Forces and Influences behind the Development of Upland Cash Cropping in Northeast Thailand." *Geographical Journal* 153(3): 370–382.

———. 1991. "Homogeneity and Heterogeneity: An Analysis of the Nature of Variation in Northeast Thailand." *Malaysian Journal of Tropical Geography* 22(1): 63–72.

Rindfuss, R. R., A. Chattopadhyay, T. Keneda, and C. Sethaput. 2000a. "Migration and Longitudinal Data Analysis: Implications of Individual and Family Processes." Paper presented at the Population Association of America Annual Meeting, Los Angeles.

Rindfuss, R. R., B. Entwisle, S. J. Walsh, P. Prasartkul, Y. Sawangdee, T. W. Crawford, and J. Reade. 2002. "Continuous and Discrete: Where They Have Met in Nang Rong, Thailand." In S. J. Walsh and K. A. Crews-Meyer, eds., *Linking People, Place, and Policy* (Boston: Kluwer Academic Publishers), 7–37.

Rindfuss, R. R., A. Jampaklay, B. Entwisle, Y. Sawangdee, K. Faust, and P. Prasartkul. n.d. (forthcoming) "The Collection and Analysis of Social Network Data in Nang Rong, Thailand." IUSSP.

Robinson, W. S. 1950. "Ecological Correlations and the Behavior of Individuals." *American Sociological Review* 15(3): 351–357.

Rooney, D. 1999. *Angkor.* Hong Kong: Odyssey Publications.

Rosenfeld, R. A. 1985. *Farm Women: Work, Farm, and Family Life in the United States.* Chapel Hill, NC: University of North Carolina Press.

Ryder, N. B. 1965. "The Cohort as a Concept in the Study of Social Change." *American Sociological Review* 30: 843–861.

Stark, O. 1991. *The Migration of Labor.* Cambridge, MA: Basil Blackwell.

Tanur, J. M., ed. 1992. *Questions about Questions: Inquires into the Cognitive Bases of Surveys.* New York: Russell Sage Foundation.

Walsh, S. J. 1999. "Deforestation and Agricultural Extensification in Northeast Thailand: A Remote Sensing and GIS Study of Landscape Structure and Scale." In F.A. Schoolmaster, ed., *Proceedings, Applied Geography Conference,* vol. 22: 223–232.

Walsh, S. J., B. Entwisle, and R. R. Rindfuss. 1999a. "Landscape Characterization through Remote Sensing, GIS, and Population Surveys." In Stan Morain, ed., *GIS Solutions in Natural Resource Management: Balancing the Technical-Political Equation* (Santa Fe: OnWard Press), 251–265.

Walsh, S. J., T. P. Evans, W. F. Welsh, B. Entwisle, and R. R. Rindfuss. 1999b. "Scale Dependent Relationships between Population and Environment in Northeast Thailand." *Photogrammetric Engineering and Remote Sensing* 65(1): 97–105.

Wilk, R. R., and R. McC. Netting. 1984. "Households: Changing Forms and Functions." In R. McC. Netting, R. R. Wilk, and E. J. Arnould, eds., *Households: Comparative and Historical Studies of the Domestic Group* (Berkeley: University of California Press), 1–28.

Wofsy, S. C. 2001. "Where Has All the Carbon Gone?" *Science* 292: 2261–2263.

Chapter 6

LINKING PASTORALISTS TO A HETEROGENEOUS LANDSCAPE

The Case of Four Maasai Group Ranches in Kajiado District, Kenya

Shauna B. BurnSilver
Natural Resource Ecology Laboratory, Colorado State University
burnslvr@nrel.colostate.edu
Randall B. Boone
Natural Resource Ecology Laboratory, Colorado State University
Kathleen A. Galvin
Natural Resource Ecology Laboratory and Department of Anthropology, Colorado State University

Abstract Experience gained in looking at land-use change issues over recent decades has shown that human land-use strategies impact and are simultaneously impacted by ecological patterns and processes. In this chapter, we provide an example of a methodology to quantify the linkages between people and environment in a communal resource landscape and detect the impacts of landscape patterns on human land use. Pastoral production strategies in semiarid regions were predicated historically on opportunistic and extensive livestock movements in search of grazing and water across heterogeneous landscapes. However, macroscale political-economic factors that drive land subdivision and economic sedentarization compromise the ability of herders to maintain large-scale and opportunistic grazing patterns by fragmenting the landscape. We used remote sensing, GIS, GPS, and household socioeconomic surveys to: (1) identify a methodology to quantify the ecological heterogeneity of pastoral landscapes in Kajiado District, Kenya, (2) identify the daily spatial scale of pastoral resource use, and (3) illustrate the degree of seasonal variability inherent in this example of a semiarid pastoral system. We defined landscape heterogeneity using NDVI images for wet and dry periods of the year, a 1-km resolution digital elevation model, and a soils layer. We merged heterogeneity layers for wet/dry NDVI, elevation, and soils to form six combinations of heterogeneity indices, then used Monte Carlo assessments to quantify the degree of selection pastoralists made for landscape heterogeneity. Daily pathways did not reveal selection within seasons. Daily path lengths were related to the degree of subdivision and economic sedentarization of households. Integrating annual grazing pathways into these analyses will be a key to better depicting pastoralists' relationships with landscape heterogeneity.

Keywords: landscape heterogeneity, NDVI, heterogeneity indices, randomization tests, pastoral land use, communal land tenure, subdivision, Maasai Pastoralism, Kenya

1. INTRODUCTION

Over the last several decades, the rate and scale of changes in human land use and land cover have increased dramatically (Turner et al. 1990). Experience has shown that local-level environmental and human-driven processes can have local- to global-level ecological, socioeconomic, and cultural consequences. Teasing apart, the directional effects of humans on their environment and environmental effects on human land-use systems require both understanding and quantification of proximate and ultimate system drivers, causal relationships between variables, and the processes that are at play in molding human-environmental interactions over time and space. The coupling of remotely sensed data to social and ecological data has enabled researchers to define some of the specific processes associated with environmental change and human driving mechanisms (Guyer and Lambin 1993; Geoghegan et al. 1998). However, this research has also produced interesting theoretical and methodological challenges. How to link "people to land" or "land to people" in remote sensing analyses is an ongoing question, particularly in systems where the resource base is communally managed and individual households are not directly connected by ownership or leasehold arrangements to particular areas on the landscape over time. As shown by the chapters included in this volume, there is an increasingly strong history of land-use and land-cover change studies that use social science data and remote sensing methodologies to elucidate and model the impacts of human behavior and land use strategies on land cover and processes of landscape transformation (see also Moran et al. 1994; Skole et al. 1994). But the interactions between human land-use systems and the landscapes in which they are situated go strongly in both directions. Human land-use strategies simultaneously shape and are shaped by ecological patterns and processes, always with wider linkages to existing political-economic drivers.

The aim of this chapter is to provide an example of a methodology to quantify the linkages between people and land in a communal resource landscape and to highlight and identify the directional effects of specific attributes of landscape pattern on human land use. We do this by developing a methodology to quantify ecological heterogeneity and its impacts on human and livestock movement across the landscape in the context of a pastoral system in southern Kenya. We first discuss people-land linking issues, the application of landscape ecology theory, and a human ecological conceptual framework as it applies to these questions. We then apply our methodology to a case study of temporal and spatial resource heterogeneity and its effects on pastoral Maasai herding patterns. Finally, we discuss some methodological lessons learned, discuss those methodological questions that remain unresolved, and suggest some future directions for strengthening the methodological connections between remotely sensed landscapes and social science data.

2. ISSUES OF LINKING PEOPLE TO LANDSCAPES

A goal in linking social science data to remotely sensed data is to gain an understanding of human-environment interactions and process-level explanatory mechanisms. One question is: How does one link humans spatially to specific areas of the landscape through time? How are human actors arranged on the landscape, and what are the connections between land units and people? Two general methodological issues arise at this point. First, there is a basic difference between remotely sensed data (i.e., land cover, DEM, NDVI) and social science data (Rindfuss et al. this volume). Land-cover data provides continuous coverage of landscape attributes, whereas social science data is most often point data, and usually sampled according to time and financial constraints of research projects. Second, establishing a 1:1 spatial and temporal relationship between people and the landscape areas or parcels they actually use is crucial, though rarely straightforward. Even in cases where use or ownership rights over land are acknowledged (such as in some agricultural systems), individuals may use multiple parcels in contiguous or noncontiguous areas, and household units divide and change through time (Moran et al. 1994; Fox et. al. 1995). These issues bring added complexity to establishing direct links between landscape spatial attributes and land use.

The situation is compounded further when the goal is to quantify human-environment linkages where the resource base is communal, as is true in the case of many agropastoral and pastoral systems. On a global scale, rangelands and extensive pastoral systems extend over approximately one-third of the Earth's land surface (Galaty 1992), so there is a significant challenge in linking people to land in pastoral ecosystems where some form of communal land tenure defining access to resources is the norm. Similar linking challenges apply for other types of common-pool resources such as forests (Chambers and Leach 1989; Guyer and Lambin 1993; Benjaminsen 1997) and fisheries (McCay 2001).

Access to communal lands for pastoralists is based on a complex set of social, cultural, and historical conditions that has functioned historically to maintain flexible access by pastoral households to resources across space and time (Ostrum 1990; Rutten 1992; Turner 1989). Access is negotiated based on social networks and sets of rules and norms (Bromely and Cernea 1989). A fundamental facet of traditional pastoral systems is that in the face of environmental uncertainty and seasonal fluctuations, households move across the landscape in search of forage and water (Ellis and Swift 1988; Khazanov 1994). The extent and frequency of movement by people and livestock in a pastoral system may vary from nomadic on the high end of the spectrum, through transhumant where households range between customary dry and wet season ranges, to sedentary agropastoralism on the low end (Homewood and Rodgers 1991; Khazanov 1994). However, since the common thread through these land-use patterns is that livestock and/or pastoralists are not stationary on the landscape, it is not straightforward to make the direct temporal and spatial linkages in these systems between people, their animals, and the land units they occupy and use.

There are examples from pastoral systems where researchers have linked people to landscapes without a 1:1 relationship. Lambin et al. (this volume; see also Homewood et al. 2001) synthesized household-level socioeconomic data for similar groups of pastoral land users and used remote sensing to look for general impacts on land cover in another Kenyan pastoral system. A methodological alternative is to use GPS (Global Positioning System) technology to associate the settlement and movement patterns of particular pastoral households with specific grazing areas at different points in time. Rasmussen et al. (1999), in his work with Mongolian pastoralists, and Coppolillo (2000, 2001) with Tanzanian agropastoralists, documented the daily and seasonal movement patterns/settlement locations of sampled pastoral households using GPS technology. Combined with social science data on rules and norms governing movement patterns, pastoralists in a communal resource tenure system are thus linked directly to the landscapes they exploit.

There is, however, a shortfall in this approach. GPS settlement and movement data fails to quantify total impacts on the landscape by all pastoralists in an area across time and space. Measurements of total herd density and frequency of use by pastoralists in an area would supply this type of information. Socioeconomic and GPS data taken at specific time points with sampled pastoral households does not yield this kind of regional landscape-scale data. In contrast, remote sensing data describing a pastoral landscape supplies continuous coverage in all the measured landscape attributes. This is an example of the inherent disconnect between sampled and discrete social science data based on households or community-level data and remotely sensed data coverage that is continuous across landscapes or regions. The challenge remains to generalize sampled data from discrete households to represent land use on a continuous landscape.

3. LANDSCAPE HETEROGENEITY

The conceptual framework used in the Kajiado case study to focus on the directional impact of landscape attributes on pastoral land-use strategies derives from a merging of concepts drawn from ecosystem ecology, landscape ecology, and ecological anthropology. This synthetic framework describes how landscape patterns affect human actions and how human actions, mediated and dictated by wider political-economic drivers, in turn may act to affect ecosystem processes.

Current theoretical work in the ecology of arid lands emphasizes the disequilibrial and heterogeneous nature of these stochastically dominated environments. A new appreciation for the ubiquity of variability in these systems underscores the need for compatible land uses that are flexible and opportunistic enough to deal with such conditions (Ellis and Swift 1988; Behnke and Scoones 1993; Homewood 1995; Niamir-Fuller 1999). This understanding assumes that these ecosystems are complex and dynamic in space and time (Scoones 1999) and that the criteria for success of pastoralists living under the shadow of variability in arid landscapes should be an equally high degree of flexibility and dynamism reflected in their land-use strategies.

Landscape ecology focuses on the effect of landscape patterns on ecological processes (Turner and Gardner 1991). Structure is conceptualized as the spatial relationships between components of the landscape—that is, the spatial distribution of energy, materials, and species in ecosystems (Turner 1989). Ecological structure and spatial patterning of elements in landscapes arise as a product of large-scale ecological and climatic gradients (Quattrochi and Pelletier 1991; Tanser and Palmer 1999). Patterns of ecological heterogeneity arise as combinations of vegetation mosaics and habitat patches of different shapes, size, and degree of isolation from one another (Turner et al. 1990). Temporal and spatial variability in rainfall molds landscapes that are highly variable over both time and space. Therefore, ecological heterogeneity is scale dependent, and this heterogeneity has both temporal and spatial components (Ellis et al. 2001). An important concept linking heterogeneity and pastoral land use is that if spatial scale is increased, this mandates a corresponding increase in heterogeneity, because broadening the scale also increases the number of different elements encountered on the landscape in question (Naeem and Colwell 1991).

Attributes of landscape heterogeneity correspond to a diversity of patches or habitat types, with varying levels of forage quality/quantity over time. Few studies have sought to quantify the ecological variables or landscape patterns characterizing pastoral landscapes and linked this directly as a causal mechanism with human land-use patterns. Coppolillo (2000, 2001) used interviews with agropastoralists, GPS technology, and GIS and Landsat TM imagery to identify landscape characteristics that could help to explain movement patterns. Galvin et al. (2000, 2001) used NDVI measurements and cluster analysis to look at ecological effects of El Niño on pastoral decision making (e.g., marketing behaviors).

Within the framework of ecological anthropology, political-economic variables from global to local scales exert substantial effects on the actions of individuals and groups at the level of local landscapes (Turner and Gardner 1991; Geoghegan et al. 1998). Drivers such as land-tenure policy and increasing interactions with market economy are relevant, extra-local drivers in pastoral systems (Galaty 1992; Little et al. 2001). Policy changes may positively or negatively (or both) affect the range of choices available to land users, ultimately acting to change land-use patterns on the landscape. We used GPS and remote sensing technologies to link pastoral households to a particular landscape and to levels of landscape heterogeneity. Figure 1 illustrates how remote sensing linked human behavior to environmental variables in the case study described here. Human land-use strategies are often molded by climate and the underlying ecological context of landscapes. Humans also modify the environment, causing land-cover change or changes in ecological structure and/or function. Larger-scale political-economic variables in turn drive human impacts on the environment. Remote sensing techniques are an overarching tool to address both ecological affects on human systems and human impacts on land cover and ecological function across time and space.

The focus of the analyses for Kajiado was the impact of the ecological and landscape attributes of heterogeneity and climate on pattern and scale of grazing movement by households. Political-economic factors also act to mediate and change pastoral land-use patterns. Ultimately, the physical expression of these behavior changes will influence the ecological structure and function of the ecosystem in question, although this component of the process is not addressed directly in this chapter.

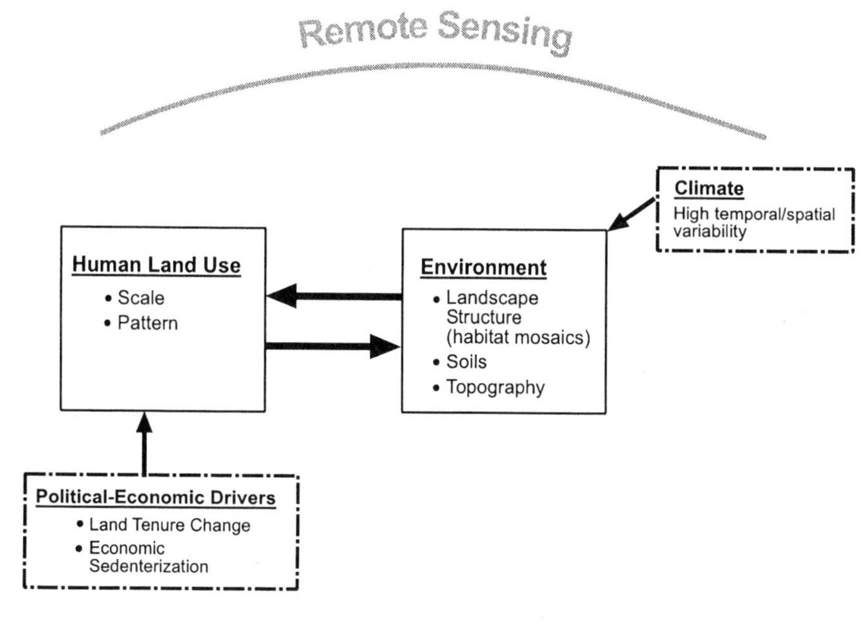

Figure 1. Conceptual model

4. SPATIAL HETEROGENEITY AND THE SCALE OF PASTORAL MOVEMENTS IN FOUR MAASAI GROUP RANCHES

4.1 Background and Research Objectives

Maasai pastoralism is the dominant land use in Kenya in the administrative districts of Kajiado and Narok. Both districts are categorized as arid to semiarid and are characterized by high inter- and intra-annual variability in rainfall (Katampoi et al. 1990). Consequently, the quality and quantity of the forage resource base continuously fluctuates across both time and space. As a result of these fluctuations, pastoralism in these areas was based historically on opportunistic and extensive movements of livestock and people across a heterogeneous landscape to maintain access to sufficient high-quality livestock forage (Behnke and Scoones 1993; Oba and Lusigi 1987; Rutten 1992; Swallow 1994). Maasai in this region fall into the general category of transhumant pastoralists (Khazanov 1994). However, current macroscale socioeconomic and political drivers—namely, land adjudication into group ranches, ongoing subdivision of communal rangelands into privatized parcels, and economic sedentarization of pastoral households—are increasingly fragmenting the pastoral landscape. These political-economic changes in land tenure and economic patterns have compromised the ability of Maasai herders to maintain traditionally extensive grazing patterns. Numerous researchers have pointed out the potential economic and ecological costs of losses in grazing system flexibility (e.g., Grandin et al. 1989; Galaty 1992; Niamir-Fuller 1999; Turner 1989). Few, however, have attempted to quantify the spatial scale of resource use by pastoralists or the ecological characteristics of semiarid and arid environments considered critical to pastoralism in intact versus fragmented landscapes.

4.1.1 Research Objectives

We had several specific objectives in this case study: (1) to use GPS technology and social science techniques to quantify the scale of pastoral grazing movements; (2) then to describe the methodology used to develop a set of preliminary landscape heterogeneity indices and subsequently to overlay spatial grazing movement patterns onto these indices; and finally, (3) to illustrate conceptually the temporal variability inherent in this example of a pastoral system by linking together a series of remote sensing images based on mean biomass levels and their standard deviation. Each of these analyses explored some aspect of the general hypothesis that it is the ecological heterogeneity inherent in intact, large-scale pastoral landscapes that is crucial for the maintenance of pastoralism in highly variable environments. This initial effort to define temporal and spatial landscape heterogeneity is part of a larger study focusing on the spatial and economic components of Maasai pastoral livelihood diversification and land-use intensification strategies (BurnSilver n.d.).

4.2 The Study Region

To accomplish the objectives outlined above, the study needed to include areas with a range of pastoral economic strategies, land-use types, and land-tenure arrangements because the spatial expression of these characteristics impacts the scale of resource use by pastoral households on the landscape. Consequently, the study focused on four Maasai group ranches within the Greater Amboseli Ecosystem: Imbirikani, Olgulului/Lolarashi, Eselengei, and Osilalei. (See Figure 2: Six study areas are shown in italics, and the group ranches containing those areas are shown in a larger regular font. The insets show Kajiado District within Kenya and Kenya within Africa.). Maasai group ranches are extensive tracts of territory with fixed and legal boundaries held by freehold title granted to a group membership of registered individuals (Bekure et al. 1991; Sperling and Galaty 1990). Osilalei is a subdivided group ranch in which pastoralists more or less manage their animals on relatively small *private* parcels of approximately ±100 acres in size. Official subdivision of the other group ranches has not yet occurred. However, informal use rights over small ±2-acre plots for cultivation in high-potential swamplands have been granted to individuals. The study area covers approximately 6,300 km^2 in the southern region of Kajiado District, bounded by Mt. Kilimanjaro to the south, the Chyulu Hills in the east, the Pelewa Hills to the west, and a gradual elevation rise to the Athi-Kibithi Plains in the north. Amboseli National Park is at the southern edge of the study area. This region is well known for its large populations of wildlife that exist side by side with resident Maasai pastoralists and their livestock.

A rainfall gradient generally runs north to south across the study area, extending from 500–600 mm/yr at the northern edge to a low of 350 mm/yr in the southern Amboseli Basin. Rainfall patterns are bimodal. The long and short rains occur March to May and October to December, respectively. Short and long dry seasons extend from January-February and June to mid-October. Vegetation communities in these areas are linked to underlying soil and topographic gradients. Combined with highly variable rainfall patterns, these characteristics translate into an inherently patchy regional landscape, manifested in vegetation mosaics and uneven biomass quality and quantity across space and time. Approximately 92 percent of this landscape has insufficient rainfall for sustained rain-fed agriculture, and as a result land use in these areas is dominated by extensive pastoralism based on a combination of livestock species that includes cattle, goats, and sheep. A series of swamps fed by underground springs extend east-west along the base of Mt. Kilimanjaro. Irrigated horticulture and subsistence agropastoralism are emerging land uses in these areas. Maasai are increasingly sedentarized in their land-use activities around these ecologically “key” resource areas and the infrastructural services that recently have developed around them (e.g., schools, medical facilities, and markets).

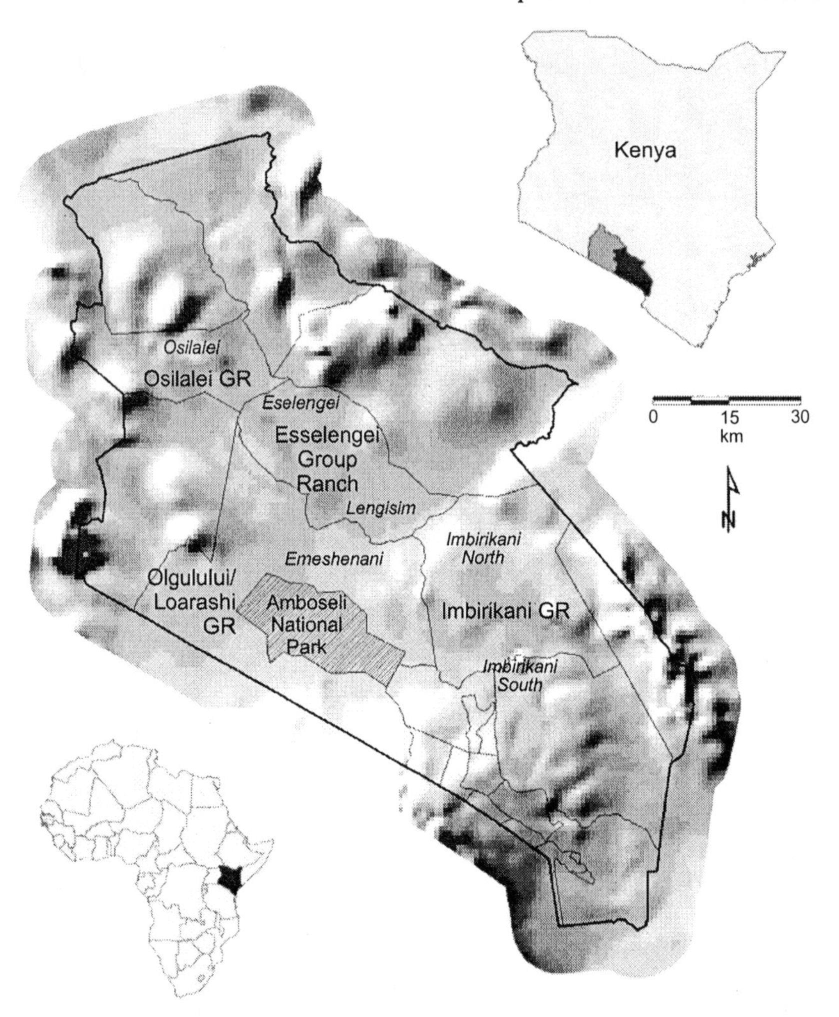

Figure 2. The study region

4.3 Design of Grazing Pathway Sampling

The larger research project focused on six study areas within the four Maasai group ranches (see Figure 2). These study areas were designed to encompass a cross section of Maasai land-use types and land-tenure designations (agropastoral/sedentary, pastoral/extensive, pastoral extensive/ rain-fed agriculture, pastoral/with subdivided grazing parcels). An initial census of all pastoral settlements, or *bomas* (compounds consisting of multiple households), within the six study areas helped to define the "range" of pastoral land-use types present within the area (N = 407). A small sample of elders (n = 5–7) from each area was initially identified, for a total of thirty-nine pastoral household heads. This small sample was stratified by wealth level (rich, medium, poor). We chose household heads as the basic unit of analysis because they function as the locus of economic and movement decisions for a household. A Maasai household *(olmarei)* was defined as a married man and his dependents. Two formal rounds of socioeconomic surveys were conducted with elder male heads of households from these families over the course of the pastoral year; the general goals were to identify the detailed changes in economic strategies, herd size and composition, marketing strategies, and spatial movements that occurred within each pastoral household over the course of successive seasons. These surveys were longitudinal and tracked changes within households and across seasons concerning economic strategies, movements, and livestock production. Additional questions on history of land use and perceptions of subdivision impacts were also added to each survey and asked only once. Additionally, a general land-use and socioeconomic survey (proportionally stratified by neighborhood and wealth level) was carried out with a larger stratified-random sample of household heads across the six study areas (N = 147). Data from the two types of surveys were compatible; the only differences were in degree of detail on economic activities and the fact that the households were only interviewed one time for the general survey.

Our methodology was to first link people to the landscapes they were using. We documented spatial grazing movements of pastoral households in two ways. First, we used a handheld GPS unit to follow the main cattle herds (e.g., not including dairy herds or calf groups) of the small sample of households at two time points: during the long dry season of 2000 and immediately after the cessation of the short rains of 2000–2001 (it continued raining until late January that year). Mean herd sizes for the wealthy (n = 12), medium (n = 15) and poor (n = 12) households followed during the grazing orbits were 157.43 cattle, 30.32 cattle, and 24.20 cattle, respectively. While following the herds, grazing paths were marked at thirty-minute intervals and vegetation data (tree, grass, and shrub percent cover, greenness and height) and herd behavior were also documented at each time point. Secondly, both types of socioeconomic surveys gathered detailed data on the movements of pastoral herds over the extended time period from January 1999 to December 2000. In these surveys, respondents verbally reconstructed their movement patterns on a month-to-month basis, including settlement locations (cattle camps and seasonal—*enkaron*—and permanent—*emparnat—bomas*), grazing endpoint locations, and labor arrangements for herd management. Grazing locations and movements were transformed to spatial data by using community informants from each study area to transfer verbal descriptions and place names onto topographical maps. These maps were then digitized and integrated into a GIS database.

This case study bases analyses of grazing movements and links to ecological heterogeneity only on data from the actual grazing orbits that have GPS readings; it does not include data from the verbally reconstructed pathways. Data for thirty-six pastoral herds followed for a total of sixty-one grazing orbits are included in these analyses (see Figure 3: pathways are overlaid on NDVI wet season plus elevation [left] and dry season plus elevation [right] complexity). The wet season orbits for the Osilalei study area are missing, other households migrated out of the local grazing system, and some herd owners from the same neighborhoods combined herds on a single day's grazing pathway (yielding one pathway for multiple herd owners). Data are from the fall of 2000 and spring of 2001.

Our intention with this small sample of the total spatial movement data was to test the heterogeneity index methodology and refine the techniques as needed. The sixty-one GPS pathways represent only a single-day grazing orbit of pastoralists within two seasons over the course of a full year. There are significant limitations involved with using this limited temporal data set to make conclusions regarding the ultimate access and selection of pastoral households for grazing resources and heterogeneity through time. Financial constraints involved with training and maintaining additional field research assistants limited our ability to gather additional grazing movement data at additional time points while in the field. However, we use these initial analyses to evaluate the methodology we developed.

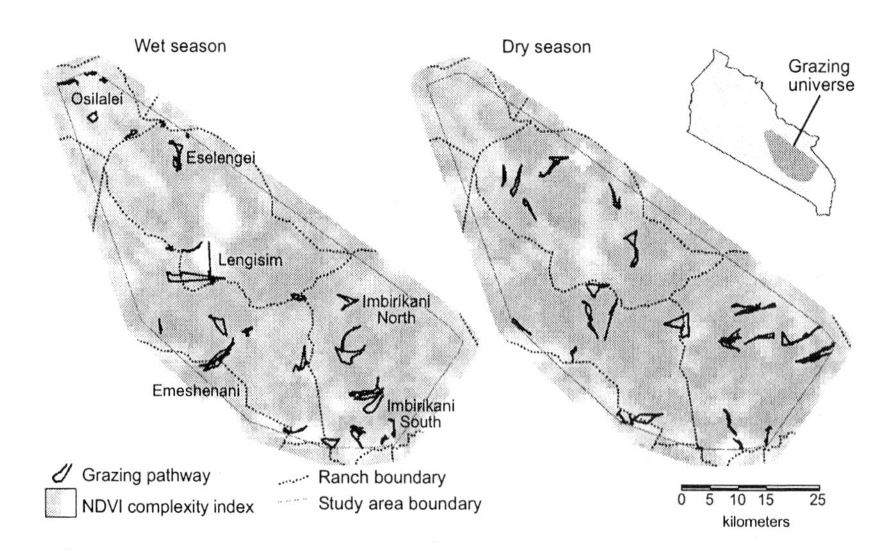

Figure 3. Grazing pathways in southern Kajiado

4.4 Quantifying Spatial Heterogeneity

One-km resolution NDVI images were most important in identifying spatial heterogeneity. Changes in vegetation biomass are reflected in Advanced Very High Resolution Radiometer (AVHRR) Normalized Difference Vegetation Indices (NDVI). The AVHRR satellite captures images with a resolution of 1.1 km^2 cells (resampled to 1 km^2). These images are available globally every ten days (i.e., a dekade) and are termed the Local Area Coverage or LAC data; they are available free of charge from the Global Land 1-km AVHRR program (USGS 1998). However, LAC images are available only for a limited period—from mid-1992 to mid-1996, with some missing periods, including all of 1994.

The LAC mean NDVI images representing thirty-six ten-day periods throughout the year were merged to represent seasonal responses. The logic behind merging two dry seasons and two wet seasons is that herders follow similar grazing patterns on the landscape based on seasonality. Movements and grazing settlement locations between wet and dry seasons differ dramatically, while movements within seasons are similar in pattern but may differ in timing and duration. These two-seasonal responses were the mean NDVI index for each pixel in the image, calculated across the dekades in question (e.g., the mean of seven images—the last dekade in October and three dekades each in November and December—represented mean NDVI for the short wet season). An analogous image was calculated that stored the standard deviation across those images for each pixel. The four seasons (short wet, short dry, long wet, long dry) were collapsed into two using similar calculations, yielding two images for seasonal responses, one for the wet seasons and one for the dry seasons. The wet season image therefore summarizes sixteen dekades (three each for March, April, May, November, and December, and the last dekade of October). The dry season image summarizes twenty dekades (January, February, June, July, August, September, and the first two dekades of October).

In addition to NDVI, we used a 1-km resolution digital elevation model produced by the USGS and distributed by the African Data Dissemination Service (ADDS 2001) and a detailed soils map as layers in the heterogeneity indices. A soils layer for the Amboseli Ecosystem was created by the Kenya Ministry of Agriculture and the Ministry of Tourism and Wildlife. We augmented this for areas not covered using the Soils and Terrain Database compiled by the Republic of Kenya in 1995 and generalized the layer to 1-km resolution.

The layers used (NDVI, elevation, and soils) are helpful as is to interpret decision making by pastoralists, but we sought to compare spatial heterogeneity to those decisions. To represent heterogeneity, we applied focal statistics to the layers. In raster-based GIS analyses, focal statistics summarize the neighborhood of each pixel in an image (Figure 4). We used a circular neighborhood around each pixel—2 km (i.e., two pixels) in radius for NDVI images, 3 km for soils, and 1 km in radius for elevation. These distances were chosen to maximize the complexity of information contained in the resulting images, without oversmoothing the results. Focal statistics used were standard deviation for NDVI and elevation and for soils a count of the number of soil types within the neighborhood (i.e., *variety* is the raster analysis term).

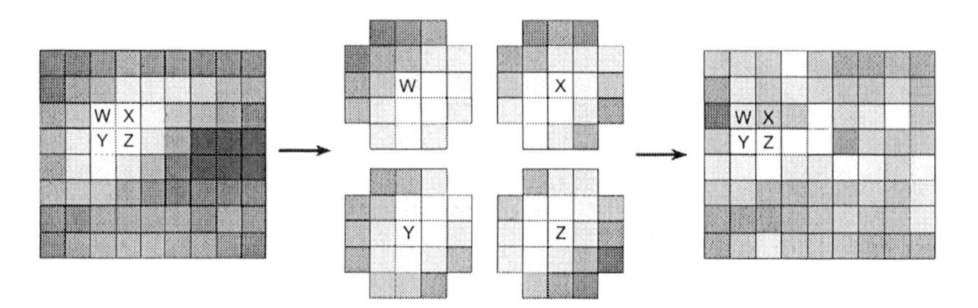

Figure 4. Complexity scores

The complexity layers for NDVI, elevation, and soils may be combined to form heterogeneity indices or may be considered indices as is. We correlated NDVI heterogeneity with pastoral decision making and also combined layers to form indices: (1) wet season NDVI plus elevation; (2) dry season NDVI plus elevation; (3) wet and dry season NDVI plus elevation; and (4–6) each of these (1–3) plus soils. To calculate heterogeneity indices, the complexity layers were each standardized to range from 0 to 1 and the layers were summed.

We used Monte Carlo (or randomization) assessments to quantify the degree of selection pastoralists made for heterogeneity. This entailed comparing the observed complexity associated with a grazing pathway (see Figure 3) with the heterogeneity of areas of the same size and shape but placed randomly. Randomized spatial assessments such as this must define the total area from which random values may be drawn, so that the null models are appropriate—that is, inclusive of areas considered by pastoralists in decision making. This is a critical question in Kajiado, one that is linked to issues of land tenure, negotiated access across Maasai sectional boundaries, and distance to/from water sources. It will require more effort in the future to define the actual grazing universe available to households. But for the analyses reported here, we delineated a general "grazing universe" around the sixty-one observed pathways to define the area considered by pastoralists in making movement decisions (Figure 3). In Monte Carlo assessments, paths were randomly placed so that some portion of their length fell within that grazing universe.

In practice, an ARC/INFO AML was written that selected a path from the group of sixty-one paths, buffered it by 1 km to give it areal dimension, and assessed its complexity for the index of interest (e.g., NDVI deviation or dry season NDVI plus elevation). The pathway was then shifted east-west and north-south by some random amount and rotated by a random amount (0 to 359 degrees). The program then asked if some portion of the randomly placed pathway fell within the grazing universe; if it fell outside this area, the attempt was abandoned, another random placement was calculated, and the process was repeated until a valid placement was found. The heterogeneity for that randomly placed pathway was then calculated and the results stored in a file. This process was repeated 100 times. A small custom program processed the output files to produce ranked scores of the strength of selection (i.e., the rank of the observed heterogeneity versus the randomly placed complexities) and summary statistics.

4.5 Temporal Variability of Resources within the System

Landscape ecology theory and disequilibrial dynamics describe why arid and semiarid systems are heterogeneous in space and time. However, a general analysis of NDVI patterns in southern Kajiado District in the mid-1990s illustrates conceptually what the spatial effects of temporal variability in resource availability might be on pastoral households. To carry out these analyses, we averaged NDVI values across thirty-six available dekades and then calculated standard deviation for the same dekades. These two images were standardized to equal ranges (from 1 to 255, using the mean ±2 standard deviations to define the extremes). We then subtracted the deviation from the mean, yielding a third image that reflects spatial variation relative to average NDVI levels.

4.6 Results and Implications of Analyses

4.6.1 Scale of Pastoral Movements

Analyses of the grazing pathways yielded mean distances traveled by herders within each of the six study areas (see Table 1). Herders in Osilalei and southern Imbirikani traveled the shortest mean distances across both wet and dry seasons. This is unsurprising in that Osilalei households herd their animals on subdivided private parcels. Similarly, households in southern

Imbirikani are sedentarized to a greater degree than other pastoralists and are carrying out irrigated agriculture within the network of swamps running east-west along the base of Mt. Kilimanjaro. The mean value for the Eselengei study area in the late wet season is also low, reflecting the fact that herders during this time are choosing to concentrate their livestock within the Kiboko River floodplain, which is located very close to their wet season settlement areas.

Table 1. Mean distances traveled per pathway by herders in the six study areas (in km)

Study Areas	Osil	Esel	Leng	Emes	SImb	NImb
Dry Season	—	12.49	12.21	10.48	9.84	10.95
Wet Season	5.57	5.98	12.11	10.54	8.63	12.48

Osil: Osilalei (subdivided); Esel: Eselengei; Leng: Lengisim; Emes: Emeshenani; SImb: South Imbirikani (sedentarized); NImb: North Imbirikani

4.6.2 Heterogeneity Indices

We ran analyses on six combinations of the NDVI, elevation, and soils layers. Two of these indices seemed to discriminate landscape heterogeneity within the grazing pathways to a greater degree: These heterogeneity indices are pictured in Figure 5. We present scores for these indices: the Wet season (a) NDVI + Elevation (Het_w_e), and Dry season (b) NDVI + Elevation (Het_d_e) (Table 2). Scores are grouped according to the six study areas. The scores listed represent a number from 1 to 100 and reflect heterogeneity within the observed path versus multiple random paths. High scores indicate that herders chose a path with high heterogeneity, and low numbers reflect that the grazing area for that day was more homogeneous in terms of combined focal deviation NDVI and elevation. Individual and mean heterogeneity scores for each area are presented and indicate that there seem to be wide differences in landscape heterogeneity both across and within the six study areas. Heterogeneity differs widely even with similar starting points within study areas, but changes also according to the grazing direction chosen.

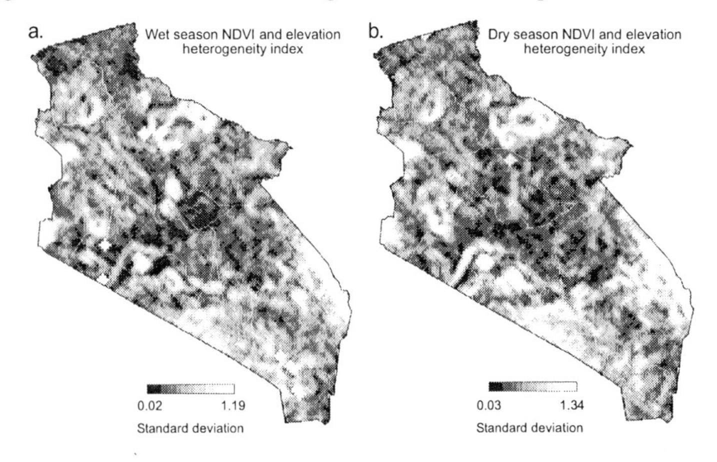

Figure 5. Heterogeneity indices for southern Kajiado District

In analyzing the performance of the heterogeneity indices relative to each other, it was clear also that the addition of the soils layer to the Het_w_e and Het_d_e layers sometimes acted to significantly alter the overall heterogeneity scores for individual pathways (see Table 2). Changes were particularly noticeable in the Emeshenani and N. Imbirikani study areas. The soils layer used for the analyses is made up of polygon categorical data, while both focal NDVI and elevation layers are raster interval data. Therefore, the effects of the soils layers on the overall heterogeneity score for a pathway was strongly linked to whether a herder was traveling through large homogeneous areas of soils or through areas with much change in a small area.

These changes in soil types may or may not be linked to above-ground-level changes in vegetation. For example, landscapes in N. Imbirikani are relatively homogeneous mixes of open grassland and bushland. The soils in that area are highly heterogeneous, yielding a dramatic increase in complexity scores for pathways in that area. Yet the question still stands regarding whether or not this jump in scores based on the presence of a soils data layer reflects a meaningful increase in heterogeneity from the perspective of a pastoralist. This question will be addressed in future analyses.

It is important to recognize that the heterogeneity scores presented here represent grazing movements only at two time points during the course of a year, and therefore they do not fully characterize the degree to which these herders are able to (or choose to) access additional landscape types as they move over the course of the seasons to other settlement and grazing areas. Perhaps the heterogeneity available to herders on any one particular day is not important relative to being able to access heterogeneity over the longer temporal and spatial scales of consecutive, seasonally available grazing areas. Integrating the larger grazing pathway data set into these analyses will better depict pastoralists' relationships with landscape heterogeneity. These future analyses will be more suited to addressing questions regarding the potential negative costs of decreasing spatial scale on the ability of pastoralists to access heterogeneous grazing landscapes. And continued access to heterogeneity again links to larger political-economic changes in land-tenure and sedentarization patterns.

4.6.3 Temporal Variability in Grazing Resources

The results of the analyses pictured in Figures 6a–c illustrate that there are landscape-level spatial differences in average biomass production through time. From the perspective of a pastoralist, this means that the landscapes around some settlement areas will be productive for longer time periods on average than others (for example, the light areas on image 6c), while other locations have low to medium levels of variation through time, relative to average NDVI (for example, the highlands of the Chyulu Hills in N. Imbirikani). Landscape elements of this latter type, in fact, function as drought refuge fallback zones for livestock in times of drought. Other regions such as to the south of Amboseli National Park have low average NDVI levels and high variability through time (dark gray colored areas on Figure 6c). Households settled in these regions would be mandated to move their livestock more frequently in search of better forage.

If we integrate the wider political economic context in Kajiado (e.g., changes in land tenure and subdivision of communal land into private parcels), this analysis suggests that some areas would possibly be better able to sustain sedentary pastoral populations through time than others. For example, Osilalei Group Ranch is on the high end of the rainfall gradient, and the area shows a lower variation in NDVI relation to the average. It is difficult to imagine that pastoral households on other group ranch lands—for example, Emeshanani to the north of Amboseli National Park—would be able to sustain sedentary pastoral activities in the face of medium to high temporal variability relative to average NDVI.

Table 2. Heterogeneity scores for wet and dry season indices

Indices	**Osilalei**		**Eselengei**		**Lengisim**		**Emeshanani**		**S. Imbirikani**		**N. Imbirikani**	
				+soils		+soils		+soils		+soils		+soils
1.1.1.1 Het_d_e			93*		17		4	18	100		2	90
			67		52*		70	92	78→		3	77
			57		57		29	73	86		66	
			53	**45**	41		87	96	91		51	81
			62	**48**	**85**	**76**	8	60	59→		**83**	**54**
									40	**16**	16	48
Mean			**66.4**		**50.4**		**39.6**		**79.3**		**41.6**	
Het_w_e	38	51	89*		90		10	16	92		4	8
	93		12		**75***	**46**	31	55	75→		9	99
	64	89	37	50	**95***	**65**	3	66	82		44	84
	85		40				97	100→	78		7	58
	87	**71**	62	100			32	73	97		46	86
							11	**5**	40		35	66
											23	78
Mean	**73.4**		**48.0**		**86.7**		**17.4**		**77.8**		**30.3**	

Notes:
1. Het_d_e, Het_w_e: Heterogeneity index combining NDVI and Elevation for dry (d) and wet (w) seasons.
2. * Designates that two households were neighbors *(elatia)* and combined their herds for that day's grazing pathway.
3. → Indicates that the herder used household labor or extended family to move their cattle herd to another grazing area; the mean complexity score for each study area includes the complexity scores for those herders who moved their animals to new areas.
4. Numbers in bold indicate values that decrease when the soils layer is added.

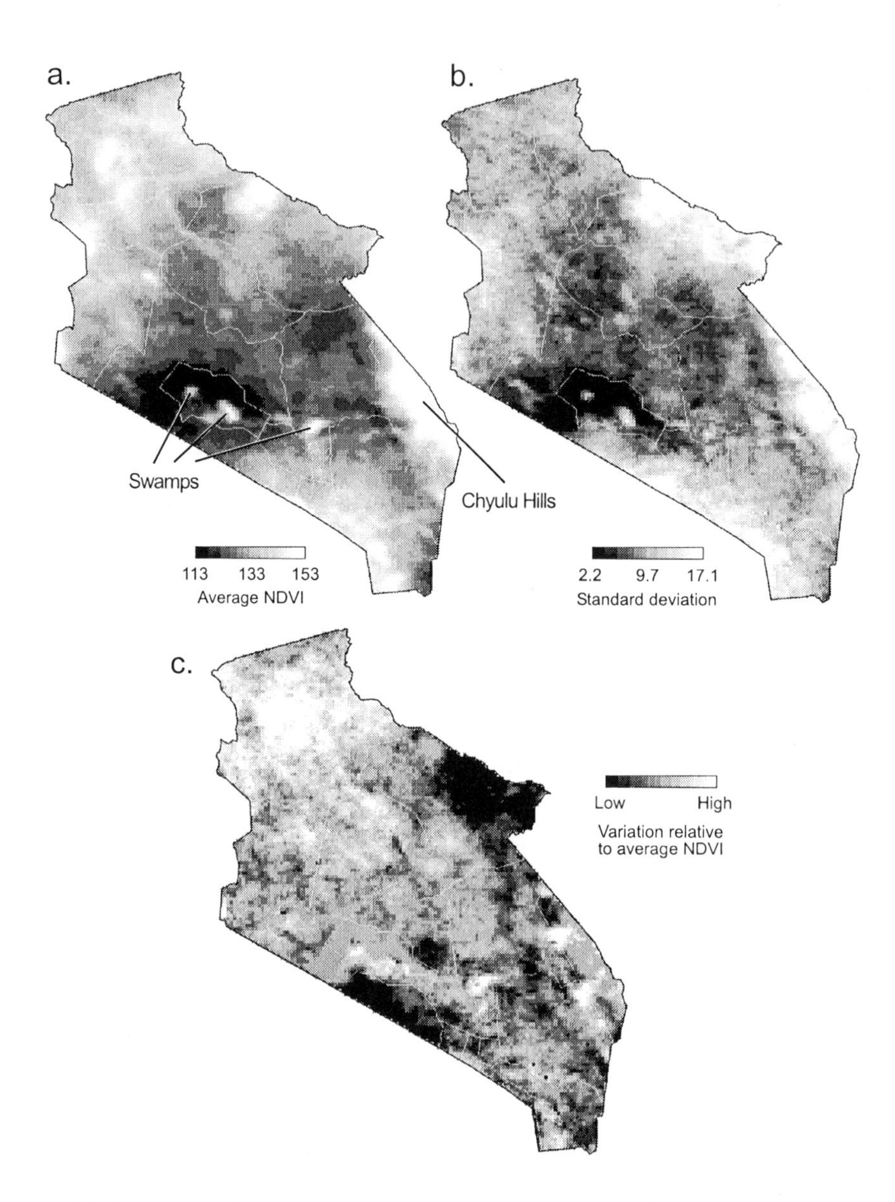

Figure 6. NDVI patterns in southern Kajiado district in the mid-1990s

5. CONCLUSIONS

5.1 Implications of the Kajiado Case Study

Our results confirm that the pastoral grazing landscape in four Maasai group ranches is in fact heterogeneous. It consists of vegetation mosaics and a variety of key resources considered by pastoralists to be of alternatively high and low quality across different seasons. For example, areas of high elevation such as the Chyulu Hills and well-watered areas such as the Amboseli swamps will be considered dry season and drought refuge areas. In contrast, the Kiboko River floodplain south of Lengism is the first area to green-up when it rains, and consequently the pathways of most resident pastoralists in this location coincide and track to this area at the beginning of the rainy season. Maintenance of macroscale pastoral grazing access to these landscape patches through time allows herders to cope with both temporal and spatial variability in resource distribution. Decreases in the spatial scale of grazing access through either land-tenure changes or sedentarization effectively homogenize the pastoral environment and decrease herder flexibility, as well as the ability of pastoralists to cope with variability in local resources. Thus, while ecological theory has moved in the direction of emphasizing the inherent logic of spatially extensive pastoral movements across heterogeneous landscapes, sociopolitical and economic drivers have pushed pastoral systems in the opposite direction—toward fixed forms of land tenure and economic sedentarization, with very little structural flexibility (Galaty 1992; Niamir-Fuller 1999; Swift 1994).

The ultimate goal of further analyses will be to use grazing movements to identify, first, whether grazing pathways reflect the importance of access to temporal and spatial heterogeneity for pastoralists and, second, to identify if decreasing the scale of resource use (e.g., because of land-tenure constraints or economic sedentarization) mandates a loss in access to heterogeneity and actual homogenization of the pastoral resource base.

5.2 Methodological Issues in Retrospect

The methods applied in this Kajiado case study are a work in progress. Methods for representing landscape heterogeneity are still being tested, and more spatial movement and socioeconomic data remains to be integrated into the analyses. These data will increase the spatial and temporal depth of the data set. To date, we have succeeded in linking sampled pastoral households to the communal landscapes that they occupy and use at particular points in time. We also made significant progress in quantifying some of the ecological attributes of pastoral landscapes that are thought to impact human land-use patterns.

GPS technology was effective in linking pastoral households and their livestock movements to the landscape. This technique could be generalized for quantifying other types of human behaviors in landscapes, such as the temporal and spatial distribution of tree cutting/deforestation in forested landscapes. Limitations of the GPS methodology were related only to time and financial considerations. Tracking movements of livestock was time intensive, requiring a trained field assistant to travel with each livestock herd over the course of a full day (~ ten to twelve hours). It was important to measure livestock movements of households during specific time periods within seasons. This mandated putting as many as four people in the field following herds at one time. Procuring multiple GPS units and training actual herd owners in the basic use of GPS technology (e.g., taking waypoint locations) would extend the methodology without incurring exorbitant added financial costs. Another option would be to put GPS collars on livestock within herds, and download waypoints after multiple days of grazing. However, these options would preclude gathering general vegetation data over the course the daily grazing pathway.

Still at issue is the disconnect between remote sensing data that is continuous in space and/or time and household level data that is discrete in time and space. Our initial goal was to quantify the unidirectional effects of spatial heterogeneity on patterns of pastoral movements (Figure 1). In one sense, these particular analyses were not hampered by the mismatch between discrete household data and continuous remote sensing data, since with remote sensing we could represent the relevant attributes of the landscape available to pastoral households in each area. The same would not hold true, however, if the goal becomes to quantify the impacts of pastoral land use on the landscape. This will be a crucial question for Kajiado District in the future, as land privatization gathers momentum in this region. The result could be a fundamental shift from extensive grazing patterns to intensive management of grazing livestock on specific land parcels. Effects of this shift in land-use patterns eventually will loop around to affect landscape structure, function, and processes (Turner and Gardner 1990).

At this point the goal becomes identifying a method of defining *total* impacts (human, livestock, etc.) through linkages between social science data and remote sensing methodologies. It may in fact be impossible to quantify these impacts perfectly. But the challenge is to strike a balance between attributes of both remote sensing and household- or community-level data that would allow researchers to quantify human land-use impacts effectively. For example: a complete census of pastoral *bomas* and livestock movements in regions at more frequent points in time, coupled with an extensive understanding of the rules and norms governing grazing movements in communities, would contribute to transforming discrete social data into a more continuous representation of human impacts on landscapes.

We also plan a straightforward methodological extension to our work, exploring the effects of changing the spatial resolution of underlying images on results. This would allow us to identify the scale at which landscape heterogeneity is important to pastoralists. We used NDVI, DEM, and a soils layer in the current analyses, all with 1-km resolution. However, Landsat TM images are available at reasonable cost and have a resolution of 30 m for spectral bands (15 m for a panchromatic band; 60 m for thermal bands)—much finer than the 1-km resolution data used here. Taking the ratio of two bands of a Landsat satellite image (the near infrared and red bands) and normalizing them yields NDVI values analogous to those used here. We will repeat our analyses using the more spatially resolved Landsat NDVI image (combined with elevation and soils as in the current work) and compare those results to the results reported here. The spatial resolution of the Landsat data will also be degraded, representing coarser satellite images (e.g., from 30 m to 60 m, to 125, 250, 500, and 1000-m resolutions), and again the analyses will be repeated. We hypothesize that the resolution in best agreement with Maasai herders' movements will be at intermediate resolutions. The resolution in best agreement may differ across seasons; prudent selection and planning is more important in the dry season than in the wet. Should selection be at intermediate resolutions (e.g., 250 m), the implications for practical applications may be far-reaching: the MODIS sensor provides daily measures of spectral reflectance at that scale (across the globe), currently at no cost to the user.

In exploratory analyses we will also examine the utility of heterogeneity indices derived from simple addition (e.g., NDVI plus elevation) as used in our current work, with indices created with more complex rules. Perhaps some components of the environment are more important in influencing pastoral decisions, and therefore should be weighted more heavily in heterogeneity indices. Finally, we intend to improve the definition of available lands to pastoralists. The "grazing universe" surrounding observed paths (Figure 3) can likely be made more precise by discarding areas that may be fenced, are inside conservation areas, are too distant from water to be useful to herders, or are outside traditional Maasai sectional boundaries.

The impact of wider political economic phenomena on the land-use strategies of pastoral households was included in the conceptual description of the study, but we did not consider these effects explicitly in the current analyses. Land-tenure change and processes of economic sedentarization are linked to national land-use policies (Bekure et al. 1991; Galaty 1992), yet the wider effects of land-tenure policy on land use are evident at the local scale of Maasai group ranches and individual households. It is probable that household-level characteristics (e.g., degree of economic diversification, wealth level, access to social networks, and access to labor resources) may allow households to mitigate the negative effects on scale of pastoral use represented by sedentarization or changing land tenure (Turner 1989). We plan to integrate these household variables into the spatial analyses to identify if and how fine-scale household characteristics could function to reaggregate herders' spatial access to grazing resources or decrease economic dependence on herding as a production strategy. These additional analyses will again depend on methods that link socioeconomic household data with remote sensing applications.

5.3 Linking Social Science Data and Remote Sensing Methodologies

In the broad context of human-environmental interactions, in which humans both impact and are impacted by their environment, we set out to highlight and quantify the directional importance of landscape patterns on pastoral land-use strategies. Essentially, we hypothesized that access to heterogeneity across space and time *matters* for pastoralists who live in highly variable environments such as the Amboseli ecosystem. The other half of the equation not highlighted in our study is, of course, trying to understand the effects of human land use on land-cover and ecosystem properties. Disentangling the relationships, processes, and causal mechanisms linking social and ecological variables across different temporal and spatial scales is a monumental task. Applying a combination of methodologies drawn from both remote sensing and the social sciences can help us to understand and quantify this iterative process.

ACKNOWLEDGEMENTS

We are indebted to the Il Kisongo and Matapaato Maasai of southern Kenya for sharing knowledge of their lives and their landscapes with us. *Ashe na ling.* We thank the government of Kenya for permission to conduct and publish this research. We thank Jeffrey Worden, who collaborated on gathering settlement census data and GPS data on livestock movements. We would like to thank our colleagues for their conceptual and technical contributions to this work: Jim Ellis, Russ Kruska, Phil Thornton, R. Reid, and M. Coughenour. Research by S. BurnSilver was supported by the U.S. National Science Foundation (BNS-9100132 and SBR-9709762). The research was also supported by the Office of Agriculture and Food Security, Global Bureau, U.S. Agency for International Development, under Grant

No. PCE-G-98-00036-00. The opinions expressed herein are those of the authors and do not necessarily reflect the views of the U.S. Agency for International Development.

REFERENCES

ADDS. 2001. *African Data Dissemination Service, Site Version 2.3.2.* Sioux Falls, SD: U.S. Geological Survey, EROS Data Center.URL: http://edcsnw4.cr.usgs.gov/adds/

Behnke, R. H., and I. Scoones. 1993. "Rethinking Range Ecology: Implications for Rangeland Management in Africa." In R. H. Behnke, I. Scoones, and C. Kerven, eds., *Range Ecology at Disequilibrium: New Models of Natural Variability and Pastoral Adaptation in African Savannas* (London: Overseas Development Institute), 1–30.

Bekure, S., P. N. de Leeuw, B. E. Grandin, and P. J. H. Neate, eds. 1991. "Maasai Herding: An Analysis of the Livestock Production System of Maasai Pastoralists in Eastern Kajiado District, Kenya." ILCA Systems Study 4. Addis Ababa, Ethiopia.

Benjaminsen, T. A. 1997. "Natural Resource Management, Paradigm Shifts, and the Decentralization Reform in Mali." *Human Ecology* 25(1): 121–143.

Boone, R. B., K. A. Galvin, N. M. Smith, and S. J. Lynn. 2000. "Generalizing El Niño Effects upon Maasai Livestock Using Hierarchical Clusters of Vegetation Patterns." *Photogrammetric Engineering and Remote Sensing* 66: 737–744.

Bromley, D. W., and M. M. Cernea. 1989. "The Management of Common Property Natural Resources: Some Conceptual and Operational Fallacies." *World Bank Discussion Papers* 57: 1–66.

BurnSilver, S. B. n.d. (in progress) "Economic Strategies of Diversification and Intensification among Pastoral Maasai: Continuity and Change in Four Group Ranches, Kajiado District, Kenya. Ph.D. dissertation, Colorado State University.

Chambers, R., and M. Leach. 1989. "Trees as Savings and Security for the Rural Poor." *World Development* 17(3): 329–342.

Coppolillo, P. B. 2000. "The Landscape Ecology of Pastoral Herding: Spatial Analysis of Land Use and Livestock Production in East Africa." *Human Ecology* 28(4): 527–560.

———. 2001. "Central-Place Analysis and Modelling of Landscape-Scale Resource Use in an East African Pastoral System." *Landscape Ecology* 16: 205–219.

Ellis, J. E., N. T. Hobbs, R. Behnke, P. Thornton, and R. Boone. 2001. "Biocomplexity, Spatial Scale and Fragmentation: Implications for Arid and Semi-Arid Ecosystems (SCALE)." NSF proposal manuscript.

Ellis, J. E., and D. M. Swift. 1988. "Stability of African Pastoral Ecosystems: Alternate Paradigms and Implications for Development." *Journal of Range Management* 41(6): 450–459.

Fox, J., J. Krummel, S. Yarnasarn, M. Ekasingh, and N. Podger. 1995. "Land Use and Landscape Dynamics in Northern Thailand: Assessing Change in Three Upland Watersheds." *Ambio* 24: 328–334.

Galaty, J. G. 1992. "Social and Economic Factors in the Privatization, Sub-Division and Sale of Maasai Ranches." *Nomadic Peoples* 30: 26–40.

Galvin, K. A., R. B. Boone, N. M. Smith, and S. J. Lynn. 2001. "Impacts of Climate Variability on East African Pastoralists: Linking Social Science and Remote Sensing." *Climate Research* 19: 161–172.

Geoghegan, J., L. Pritchard Jr., Y. Ogneva-Himmelberger, R. R. Chowdhury, S. Sanderson, and B. L. Turner II. 1998. "'Socializing the Pixel' and 'Pixelizing the Social' in Land-Use and Land-Cover Change." In D. Liverman, E. F. Moran, R. R. Rindfuss, and P. C. Stern, eds., *People and Pixels: Linking Remote Sensing and Social Science* (Washington, D.C.: National Academy Press), 51–69.

Grandin, B. E., P. N. de Leeuw, and P. Lembuya. 1989. "Drought, Resource Distribution, and Mobility in Two Maasai Group Ranches, Southeastern Kajiado District." In T. E. Downing, K. W. Gitu, and C. M. Kamau, eds., *Coping with Drought in Kenya: National and Local Strategies* (Boulder: Lenne Rienner Publishers), 245–263.

Guyer, J. I., and E. F. Lambin. 1993. "Land Use in an Urban Hinterland: Ethnography and Remote Sensing in the Study of African Intensification." *American Anthropologist* 95(4): 839–859.

Homewood, K. M. 1995. "Development, Demarcation and Ecological Outcomes in Maasailand." *Africa* 65(3): 331–350.

Homewood, K. M, E. F. Lambin, E. Coast, A. Kariuki, J. Kivelia, M. Said, S. Serneels, and M. Thompson. 2001. "Long-Term Changes in Serengeti-Mara Wildebeest and Land Cover: Pastoralism, Population or Policies?" *PNAS* 98(22): 12,545–12,549.

Homewood, K. M., and W. A. Rodgers. 1991. *Maasailand Ecology: Pastoralist Development and Wildlife Conservation in Ngorongoro, Tanzania*. Cambridge: Cambridge University Press.

Katampoi, K. O., G. O. Genga, M. Mwangi, J. Kipkan, J. Ole Seitah, M. K. van Klinken, and M. S. Mwangi. 1990. *Kajiado District Atlas*. Nairobi: ASAL Programme Kajiado.

Khazanov, A. N. 1994. *Nomads and the Outside World*. Madison: University of Wisconsin Press.

Little, P., K. Smith, B. A. Cellarius, D. L. Coppock, and C. B. Barrett. 2001. "Avoiding Disaster: Diversification and Risk Management among East African Herders." *Development and Change* 32: 387–419.

McCay, B. J. 2001. "Environmental Anthropology at Sea." In C. L. Crumley, ed., *New Directions in Anthropology and Environment: Intersections* (Walnut Creek, CA: Rowman and Littlefield), 254–272.

Moran, E. F., E. Brondizio, P. Mausel, and Y. Wu. 1994. "Integrating Amazonian Vegetation, Land-Use, and Satellite Data." *Bioscience* 44(5): 329–339.

Naeem, S., and R. K. Colwell. 1991. "Ecological Consequences of Heterogeneity of Consumable Resources." In J. Kolasa and S. T. A. Pickett, eds., *Ecological Heterogeneity* (New York: Springer-Verlag), 224–255.

Niamir-Fuller, M. 1999. *Managing Mobility in African Rangelands: The Legitimization of Transhumance*. London: Intermediate Technology Publications.

Oba, G., and W. Lusigi. 1987. *An Overview of Drought Strategies and Land Use in African Pastoral Systems*. London: Overseas Development Institute.

Ostrum, E. 1990. *Governing the Commons: The Evolution of Institutions for Collective Action*. New York: Cambridge University Press.

Quattrochi, D. A., and R. E. Pelletier. 1991. "Remote Sensing for Analysis of Landscapes: An Introduction." In M. G. Turner and R. H. Gardner, eds., *Quantitative Methods in Landscape Ecology* (New York: Springer-Verlag), 51–76.

Rasmussen, M. S., R. James, T. Adiyasuren, P. Khishigsuren, B. Naranchimeg, R. Gankhayag, and B. Baasanjargal. 1999. "Supporting Mongolian Pastoralists by Using GIS to Identify Grazing Limitations and Opportunties from Livestock Census and Remote Sensing Data." *GeoJournal* 47: 563–571.

Rutten, M. M. E. M. 1992. *Selling Wealth to Buy Poverty*. Saarbrucken: Verlag breitenbach.

Scoones, I. 1999. "New Ecology and the Social Sciences: What Prospects for Engagement?" *Annual Review of Anthropology* 28: 479–507.

Skole, D. L., W. H. Chomentowski, W. A. Salas, and A. D. Nobre. 1994. "Physical and Human Dimensions of Deforestation in Amazonia." *Bioscience* 44(May): 314–322.

Sperling, L., and J. G. Galaty. 1990. "Cattle, Culture and Economy: Dynamics in East African Pastoralism." In J. G. Galaty and D. L. Johnson, *The World of Pastoralism: Herding Systems in Comparative Perspective* (New York: Guilford Press), 69–98.

Swallow, B. 1994. *The Role of Mobility within the Risk Management Strategies of Pastoralists and Agropastoralists.* London: International Institute for Environment and Development.

Swift, J. 1994. "Dynamic Ecological Systems and the Administration of Pastoral Development." In I. Scoones, ed., *Living with Uncertainty: New Directions in Pastoral Development in Africa* (Yorkshire: Intermediate Technology Publications), 153–173.

Tanser, F. C., and A. R. Palmer. 1999. "The Application of a Remotely-Sensed Diversity Index to Monitor Degradation Patterns in a Semi-Arid, Heterogeneous, South African Landscape." *Journal of Arid Environments* 43: 477–484.

Turner, B. L., II, R. E. Kasperson, W. B. Meyer, K. M. Dow, D. Golding, J. X. Kasperson, R. C. Mitchell, and S. J. Ratick. 1990. "Two Types of Global Environmental Change." *Global Environmental Change* (December): 14–22.

Turner, M. G. 1989. "Landscape Ecology: The Effect of Pattern on Process." *Annual Review of Ecological Systems* 20: 171–197.

Turner, M. G., and R. H. Gardner. 1991. "Quantitative Methods in Landscape Ecology: An Introduction." In M. G. Turner and R. H. Gardner, eds., *Quantitative Methods in Landscape Ecology* (New York: Springer-Verlag), 3–16.

Turner, S. J., R. V. O'Neill, W. Conley, M. Conley, H. C. and Humphries. 1991. "Pattern and Scale: Statistics for Landscape Ecology." In M. G. Turner and R. H. Gardner, eds., *Quantitative Methods in Landscape Ecology* (New York: Springer-Verlag), 17–50.

USGS. 1998. *Global Land 1-km AVHRR Project.* Sioux Falls, SD: U.S. Geological Survey, EROS Data Center. URL: http://edcwww.cr.usgs.gov/landdaac/1KM/1kmhomepage.htmnl.

Chapter 7

LINKING HOUSEHOLD AND REMOTELY SENSED DATA FOR UNDERSTANDING FOREST FRAGMENTATION IN NORTHERN VIETNAM

Jefferson Fox
Environmental Studies, East-West Center
foxj@eastwestcenter.org
Terry Rambo
Center for Southeast Asia Studies, Kyoto University
Deanna Donovan
Environmental Studies, East-West Center
Le Trong Cuc
Center for Resource and Environmental Studies, National University of Vietnam
Thomas Giambelluca
Geography Department, University of Hawaii
Alan Ziegler
Environmental Engineering and Water Resources Program, Princeton University
Donald Plondke
Environmental Studies, East-West Center
Tran Duc Vien
Center for Agricultural Research and Ecological Studies, Hanoi Agricultural University
Stephen Leisz
Institute of Geography, University of Copenhagen, Denmark and the Center for Agricultural Research and Ecological Studies, Hanoi, Vietnam
Dao Minh Truong
Center for Resource and Environmental Studies, National University of Vietnam

Abstract: This paper describes a research project in northern Vietnam that since 1997 has sought to explore two linked premises.[1] First, socioeconomic policies and technological conditions operating at the household, community, and broader levels both promote and sustain forest fragmentation. Second, fragmentation of forest cover—including changes in the structure, age, and species diversity of the forest fragments—have both positive as well as negative effects on watershed function. To examine these premises, the project collected and linked three types of data. These include spatial data (e.g., topographic maps, aerial photographs, and satellite images), socioeconomic data (e.g., national and regional policies, agricultural practices, household economy, demography, health and nutrition, and individual hopes, fears, and aspirations), and hydrological data (e.g., microclimate, transpiration, and surface runoff data from a 12-ha forest patch). The spatial database served as a framework for analyzing changes in land-cover and forest patterns through time, a tool for integrating the information collected in interviews and policy analysis, and a means of modeling the flow of water and sediment across the landscape. Several papers on this research have been published (Fox et al. 2000; Fox et al. 2001; Rambo and Tran 2001) or are in preparation. In this chapter we briefly summarize the problem statement and theories that guided the research, methods used, and results. Because the primary interest of this workshop is linking social science and remote sensing data, we do not discuss the hydrological research in detail. The hydrological modeling, however, raised further questions for the social economic survey, and consequently we briefly discuss that exercise. In the discussion section we seek to describe and integrate the three lines of inquiry pursued—spatial, socioeconomic, and hydrological modeling—and to answer questions regarding data limitations and quality, practical problems, and methods for improving future research.

Keywords: forest fragmentation, GIS, remote sensing, household surveys, land-use/land-cover change, Vietnam

1. INTRODUCTION

The phenomenon of forest fragmentation has been studied in some detail with regard to the effect of such change on biodiversity and species composition of forest ecosystems (e.g., O'Neill et al. 1988; Lord and Norton 1990; Westman 1990; Nepstad et al. 1991; McClanahan and Wolfe 1993; Belsky and Canham 1994; Enoksson et al. 1995; Rudis 1995; Schelhas and Greenberg 1996). Comparatively little work has yet been done to understand the socioeconomic forces that cause and maintain forest fragments. Even less work has been done on the effects of forest fragmentation, direct and indirect, on the hydrological functions of forests.

In Vietnam, forest covers as little as 17 percent of the landscape (Collins et al. 1991). Shifting cultivation is the most dominant form of agriculture found in most upland areas of Southeast Asia. With a long history of swidden agriculture, the land surface of this region is dominated by secondary vegetation across the full spectrum of the various stages of forest growth. A few researchers have noted the importance of successional vegetation. Potter et al. (1994) suggest that as much as 26 percent of all land in Southeast Asia falls into the "other" category, where *other* is defined as shrub, brush, pasture, waste, and other land-use categories, many of which are actually successional vegetation. Swidden cultivation has not only resulted in a landscape of secondary vegetation, but the forests that remain have been extensively fragmented (Fox et al. 1995).

Forest fragmentation has often been overlooked by social scientists and has rarely been taken into consideration in rural development programs. Clearly, socioeconomic factors play important roles in both the creation and maintenance of forest fragments (Alcorn 1996; Schelhas and Greenberg 1996; Browder 1996; Lebbie and Schoonmaker Freudenberger 1996; and Pinedo-Vasquez and Padoch 1996). Although direct associations between socioeconomic factors and the creation and maintenance of forest fragments are not clearly understood, they have been variously attributed to activities such as shifting cultivation, commercial logging, and infrastructure development, among other factors. Forces appear to exist that promote the maintenance of a patchy landscape even when many rural inhabitants have modified traditional agricultural systems, increased yields, supplemented incomes with earnings from off-farm employment, or migrated to urban areas in search of wage-paying jobs.

The effects of land-cover fragmentation on surface hydrology depend on the relationship between the distribution of land-cover types across the landscape and topography. In a fragmented landscape, overland flow of water (Horton overland flow or HOF) may be reduced whenever a low-infiltration patch (e.g., a compacted surface) lies directly upslope of a high-infiltration patch (e.g., a forest). Some or all of the runoff (sediment) generated on the upslope patch may be infiltrated (deposited) in the lower patch. The basinwide effectiveness of this process depends on the frequency with which patches where water flow is generated lie upslope of high-infiltration patches. This is a function of both the degree of fragmentation and the arrangement of fragments within the watershed. For example, in a landscape with the lowest degree of fragmentation, each land class would be confined to a single contiguous patch. In comparison with a highly fragmented landscape, the contiguous patch would have relatively short upslope or downslope boundaries with contrasting land-cover classes. Two key questions arise from this hypothesis: Are land-cover patches related to topography and, if so, does the resulting arrangement enhance or reduce the buffer effect?

2. STUDY AREA

Researchers at the Center for Research and Environmental Studies at Vietnam National University purposely selected the study site as being representative of villages in the mountainous part of northern Vietnam that practice swidden agriculture. Tat was at that time at the end of the road and had the best forest cover of any of the dozens of villages the researchers were able to reach (Figure 1). Tat, one of ten hamlets in Tan Minh village, is located in Da Bac District, Hoa Binh Province. The study area is within the Da (Black) River Watershed in a region characterized by sinuous, narrow valleys and steep mountains. Elevation is approximately 360 m above sea level, with surrounding mountain peaks rising to 800–950 m. Approximately 20 percent of the Tat landscape has slopes less than 25 degrees, with only the narrow valley floor being flat enough for permanent settlement, roads, and paddy rice. Mountain slopes are typically 40–60 degrees. The climate is subtropical monsoon with hot summer and cold winter seasons. About 90 percent of the annual 1,800 mm of rainfall occurs between May and October. The area was predominantly forested until the 1960s, but today remnant patches exist primarily on steep slopes and inaccessible peaks. Some hilltops and ridgelines are still covered with mature secondary forests, while slopes are occupied by active swidden fields and various stages of young and advanced secondary vegetation (mixtures of grasses, herbs, bamboo, and small trees), mostly former swidden agricultural sites.

Most of the residents of Tat hamlet belong to the Tay ethnic minority group (eighty-six out of ninety-one households). The Tay are the largest of Vietnam's minority populations and speak a language belonging to the T'ai language family that is widely spread across mainland Southeast Asia. The Tay of Tat hamlet, however, are culturally and linguistically quite distinct from the main body of Tay. They belong to a smaller, geographically isolated Tay population of approximately 17,000 individuals found only in Hoa Binh Province, primarily in Da Bac District. During the colonial period, this population was referred to as the Tho Da Bac and now is sometimes called Tay Da Bac.

According to records of the Village Family Planning Clinic, in September 1998, Tat hamlet had a population of 432 persons divided into ninety-one households, while in 1993 there were 389 persons in sixty-nine households. Thus, in five years the population of the hamlet increased by forty-three persons, an annual average increase rate of more than 2 percent. The number of households has increased even more dramatically, with twenty-two new households forming in five years. Formation of new households, even more than an increase in number of individuals, places heavy pressure on natural resources to supply building materials and to provide land for cultivation. In late 1998 the population density of Tat hamlet was fifty-nine persons per km^2. No reliable statistics are available for earlier periods, but one informant recalls that there were only seven households in the valley in the mid-1950s when he was a young boy. Assuming seven members per household, the hamlet's population would have been about fifty persons, representing a mean density of approximately seven persons per km^2.

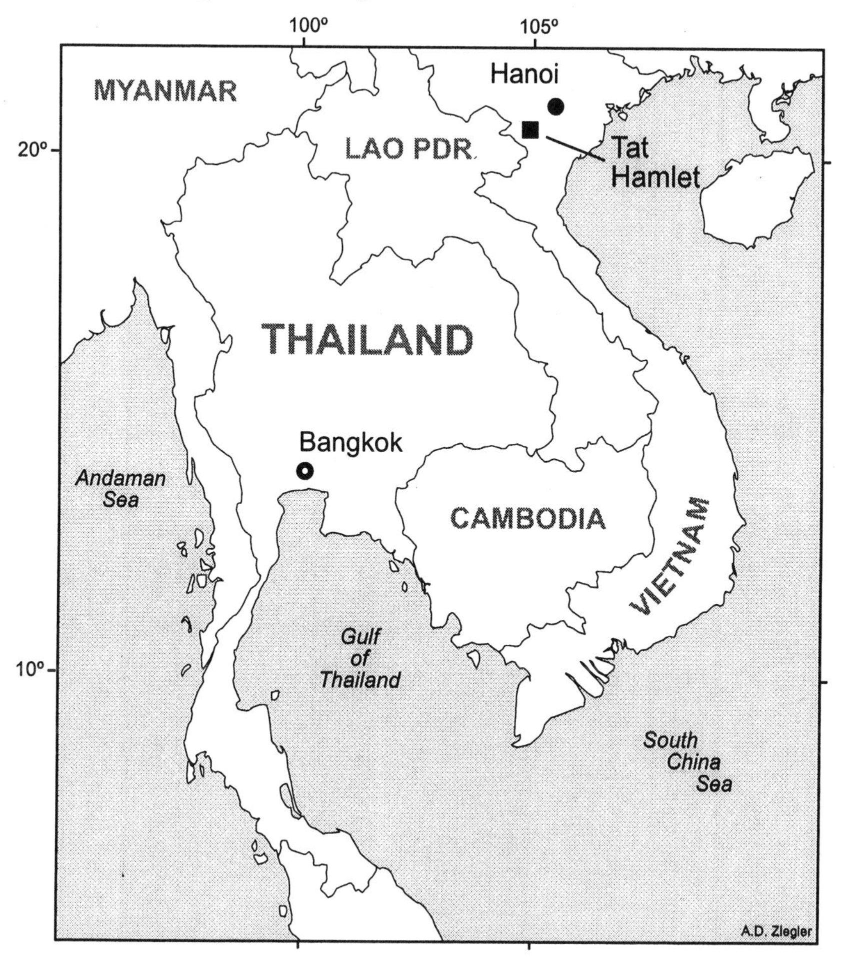

Figure 1. Tat Hamlet: The study area in northern Vietnam

The economy of Tat hamlet is still largely subsistence oriented, although in recent years there has been increasing involvement in the cash economy. There is no market in the hamlet or in Tan Minh village. The nearest market is at Cao Son, about 12 km from Tat. The closest large daily market is in the district capital at Tu Ly. Some products—notably bamboo shoots and other forest products—are collected and sent out to market on buses or logging trucks. Traders from China have even appeared in the hamlet on occasion to buy a variety of wild medicinal plants. Corn, cassava, and canna tubers are sold in bulk either to local middlemen or to traders from outside the village. Cattle and buffalo are sold to Kinh traders who come to the mountains to purchase animals for use in the delta, where draught animals are always in short supply.

Production is almost entirely organized on a household basis. Although most household production is for subsistence purposes and the trade in commodities is very limited, the economy is already strongly market oriented. Farmers report a pressing need to obtain cash, particularly in recent years since the local government began to collect taxes in cash rather than paddy. Informants readily cite prices for different locally produced commodities including livestock, paddy, and fruit. It is lack of physical access to markets and the scarcity of marketable commodities, not any "antimarket mentality," that limits participation in trade.

For as far back as any informants can remember, the Tay of Tat hamlet have been "composite swiddeners" (Rambo 1995). The defining characteristic of composite swiddening is that households simultaneously manage permanent wet rice fields in the valley bottoms and shifting swidden fields on the hill slopes while exploiting wild resources of the forest. Indeed, the household production system of the Tay is notable for its incorporation of a wide range of subsystems. Key subsystems include wet rice fields, home gardens, fishponds, livestock, tree gardens, rice swiddens, and cassava swiddens. Fallow swiddens and secondary forest are also exploited, but administration of these land units is the responsibility of the cooperative. Hence the landscape of Ban Tat is a mosaic of cultivated and fallowed fields, interspersed with protected forest areas. Swidden fields can be found on hill slopes from toe to top.

The household is the main unit of Tay society. It is responsible for organization of labor, management and use of resources, and accumulation and allocation of wealth. Multigenerational extended-family households are the ideal. These are made up of the aged parents, their unmarried sons and daughters, and one or more married sons and their wives and children. In practice, most Tay households are of the nuclear type, although in some cases a widowed parent will live together with their youngest married child. Traditionally, a newly married couple resides for a period ranging from a few months to several years with the bride's parents while accumulating the resources to establish an independent household. Kinship is patrilineal.

Tat hamlet can be characteized as being a loosely structured community. Solidarity is weak and community-level institutions for organizing corporate actions are rudimentary at best. The household is the maximal unit in which trust among members is strong and altruistic behavior expected. Relations among unrelated households are rarely close and spontaneous cooperation among them is infrequent. Consequently, resource management is largely left in the hands of individual households or even individual members of households, and few signs of collective action are evident. To the very limited extent that cooperation among unrelated households occurs, it is usually within the framework of the mass organizations established under state auspices.

During the more than one hundred years since it was first settled, Tat hamlet has passed through four historical periods, each characterized by a different pattern of social organization. These periods are the feudal period, the Resistance War period, the cooperative period, and the current period in which the household is the key unit for management of land and natural resources. In each of these periods the balance of power between the individual households, local community governance institutions, and extralocal governance institutions has varied, with consequent differences in the way in which resources have been managed.

Since 1959, Tat hamlet has been a full-fledged cooperative *(hop tac xa).* As Vietnam downgraded and replaced the collectivized economy with the household economy in the latter part of the 1980s, the management role of the village cooperative was also downgraded. Today the cooperative no longer plays an important role in the management of labor and distribution of agricultural production, but it still retains important functions in regulating land use, taxation, and maintenance of social order. All households in the hamlet must belong to the cooperative in order to live in Tat hamlet.

The cooperative retains control of 55 ha of hill land divided into four tracts. It allows households to clear swiddens on two tracts (approximately 30 ha) and protects the other two tracts (25 ha) that contain old secondary forest. Hamlet residents are allowed to collect bamboo shoots and cut bamboo stems on these lands.

Until recently the two tracts reserved for swiddening were used in an alternating cycle of approximately four years, with one tract being cultivated while the other was in fallow. After the cooperative announced that a tract was open for cultivation, households were free to pick their own plots to clear. Once they selected a plot, the household would erect simple bamboo markers to demarcate its boundaries. After clearing the field, a household could farm it until the cooperative decided that it was time to return the tract to fallow, at which time they would clear a new field in the newly reopened alternate tract. Households retained no residual rights in the abandoned field. Since the beginning of the 1990s, the alternating system has broken down under the pressure of growing population and resulting shortage of land for swiddening. Households now maintain swiddens simultaneously in both tracts. A household may periodically fallow a plot, but it retains control over the plot and may reclear it whenever it is ready to be planted again.

3. THEORY AND HYPOTHESES

A political ecology perspective provides an interdisciplinary approach for systematically and iteratively analyzing relationships among power, culture, and ecosystem resources in a society. Typically, beginning with a set of resource users or a resource conflict, such studies examine a resource's ecological characteristics, the history and contemporary dynamics of its use, and the multiple scales of relations that bear upon and transform its connections to various users (Blaikie and Brookfield 1987; Berry 1989; Fortmann 1990; Peluso 1992; Agarwal 1994; Li 2001; Watts 1997; Neumann 1998). In examining local and extralocal drivers of land use, political ecology is an appropriate framework for our research as it provides the complex approach necessary for this type of endeavor with multiple scales, factors, and actors.

While much has been written about land-use and land-cover change at the case-study level, the larger social, political, economic, and ecological context in which land-use/land-cover change is embedded has been little investigated. Against this theoretical background, we explored four hypotheses: two with respect to national policies, one each with respect to community-level forces and individuals, and one hypothesis with regard to spatial land-cover pattern and overland flow.

At the national level, we hypothesized that (1) *land-tenure policies and* (2) *market access affect forest-use practices.* The first hypothesis emerged from the fact that in 1993 Vietnam revised a Land Law that enabled forestlands to be allocated to households for periods up to fifty years. These long-term leases are formally recognized with the issuance of an official government document known as a "Red Book." The second hypothesis was based on the observation that access to markets is a critical factor for farmers whose agricultural crops are perishable.

At the community level, we hypothesized that (3) *swidden agriculture causes forest degradation and fragmentation but limited deforestation.* We based the hypothesis that swidden causes forest degradation and fragmentation on similar observations from Thailand (Fox et al. 1995).

At the individual level, we hypothesized that (4) *poor villagers exploit forest resources more than richer villagers because they cannot cultivate enough grain or other crops to provide food and other resources to meet household needs.* This hypothesis is based on the observation that, when a household cannot produce enough of the products to meet its needs, it often turns to the forest, perceived as a storehouse of goods that can be traded for cash (Ireson and Ireson 1996).

With regard to spatial land-cover pattern and the topography, we tested the hypothesis that (5) *highly fragmented land cover reduces the efficiency of Horton overland flow and sediment delivery to stream systems.* This hypothesis is based on the observation that relatively narrow buffers (5–20 m) of high-infiltration patches may be sufficient to absorb water-generated upslope on fields or footpaths.

4. METHODS

This project was based on the work of a multidisciplinary team of geographers, anthropologists, foresters, ecologists, and hydrologists assembled to collect and analyze the various types of data gathered. The primary institutions collaborating on the project included the East-West Center (EWC), the Geography Department of the University of Hawaii, and the Center for Natural Resources and Environmental Studies (CRES) of the Vietnam National University, Hanoi. The East-West Center was the lead institution, assuming the primary managerial role in project organization and implementation.

4.1 Spatial Database

The spatial database was developed using 1952 (nominally 1:40,000) aerial photographs as well as the most recent cloud-free Landsat Thematic Mapper image of the study site (June 24, 1995) available from the Royal Thai Remote Sensing Center (see Fox et al. 2001). The 1952 photographs were obtained from the Institut Geographique National (French National Geographic Institute)[2] and were registered and visually interpreted.[3] A supervised classification was conducted for the 1995 Landsat TM image. Results were used to prepare land-cover maps for 1952 and 1995.

Fieldwork was carried out in 1997 and 1998. Prior to the fieldwork, a normalized difference vegetation index (NDVI) analysis and an unsupervised classification of the TM image were done in order to estimate the number of land-cover classes that could be expected to be found in the field and to estimate the optimal number of ground-truth points needed for conducting a supervised classification and accuracy assessment. Land cover could be broken into six to eight classes, running a continuum from agriculture (e.g., rice paddy areas and active swidden) to closed-canopy forest, with the exact makeup of the vegetation along this continuum not known. A minimum of 55 to 60 randomly selected ground-truth points needed to be collected for each land-cover class, or approximately 330 to 480 points in total.

The field site is an isolated area with dense vegetation and steep terrain. It lacked electricity, and researchers could not spend extended time in the field because of local political considerations. In addition, fieldwork was being conducted two to three years after the date of the imagery. Consequently, we developed a field methodology for collecting as many ground-truth points in as random a manner as possible and with as much ancillary information as we could about land cover at the time the image was acquired.

First, an initial survey of the area was done in conjunction with local farmers and resource managers. Farmers were interviewed regarding their farming system and land and resource utilization. Possible paths that could be walked were identified and planning was made to optimize the time spent in the field. Finally, a GPS base station was established.

We conducted a walking survey of the area with local farmers and interviewed them regarding their farming systems, the history of land-cover dynamics in the area, and specific information about individual plots of land.[4] This information was crucial for relating present land cover to land cover found on the historical images. After setting up the GPS base station, we walked transects using GPS receivers to collect ground-reference points. We used a rangefinder, compass, and a clinometer to survey points up to 600 meters from the GPS–surveyed point. Field boundaries, paths, streams, and roads were surveyed and, where possible, community borders. A total of 320 ground-truth points were collected.

We interviewed farmers and resource managers at each ground-truth point about land cover on that piece of land at the time the imagery was collected. If the interviewee could not remember, we noted that information and discussed the general land-cover rotation of the area. Photographs of each ground-truth point and points mapped with the rangefinder were taken and later tied to the reference point in the spatial database. Attribute information such as current land cover, land cover at the time of image acquisition, information about the farming system, and any other information available regarding the history of land cover at each field location was added to the database. All photographs were scanned and included in the database. Other information, such as topographic maps of the area, were digitized and added to the GIS database.

The satellite image was registered, with an RMS error of +/– 1 pixel, to the base map that was created from the collected field survey points. After registering the image to the map base, the field information was laid over the satellite image. Training sets were chosen using between five to ten ground-truth points per class per image to extract spectral training set signatures for each class (a total of sixty-two points were used to derive the training sets). We then conducted a supervised classification of the subset images using Erdas Imagine's maximum likelihood classifier algorithm. Finally, we conducted an accuracy assessment of the classification results. Two types of assessment were done. The first was a strict assessment that classified a checkpoint as correct only if it fell exactly within a correctly classified pixel. The second was a more liberal assessment that accounted for the error propagated through the system. When all possible errors were summed, each reference point could be off by 1.5 pixels—3 to 7 m from the ground survey and up to 30 m (1 pixel) from registration error. The liberal assessment looked at a radius of 1 adjacent pixel around each checkpoint. If any of these pixels were correctly classified vis-à-vis the checkpoint, then the checkpoint was considered correctly classified. For both of the assessments, an error matrix showing the users and consumers accuracy and absolute accuracy was created. Finally, a KHAT statistic for each assessment was calculated and an error analysis of each classification completed. The conservative accuracy assessment was 72 percent with a KHAT of 65 percent. The liberal accuracy assessment was 94 percent with a KHAT of 93 percent.

4.2 Socioeconomic Database

The socioeconomic databases were developed as part of a larger project that looked at social conditions and development trends in five communities in Vietnam's Northern Mountain Region (Le and Rambo 2001). Tat was chosen for this study because it was an intermediate community in terms of a variety of variables (e.g., mixed subsistence and market orientation, accessibility, topography, population density, and so on).

We interviewed national, regional, and district officials, as well as local farmers. Researchers documented changes in national and regional policies influencing land use (e.g., tenure, taxation, credit, import and export regulations) as well as changes in infrastructure (roads and markets). We interviewed residents of the village to learn more about the socioeconomic factors contributing to their decisions regarding resource use and management affecting the creation and maintenance of forest fragments in their area. Key informants were used to assess, among other factors, local peoples' perception of the forest.

We used a wide range of data collection methods—including semistructured interviews with key informants, interviews using structured survey instruments with a random sample of households, and regular observation of household activities—to collect information on village households. Discussion with Tay informants about such observations was the most important source of information.

A team of two experienced researchers with backgrounds in natural and social sciences conducted lengthy informal interviews with key informants chosen for their knowledge of different topics relating to the community as a whole. Although a checklist of topics to be covered guided the questioning, the interviews were deliberately kept informal and additional topics pursued as these emerged in the course of the discussions. Both interviewers took extensive notes during the course of the interviews.

A standardized questionnaire was also designed to elicit information on a wide range of topics relating to the life situation of individual households. Topics included housing conditions and personal possessions, energy sources and water supply, demography and fertility history, health and nutrition, transportation and communications, household economy, social organization, cultural identity, gender relations, and individual hopes, fears, and aspirations. Initially drafted in English and then translated into Vietnamese, the questionnaire was tested with surrogate subjects. Before going to the field, the interviewers, all experienced researchers, were given one day's training in using the household survey instrument, first with one another and subsequently in supervised trials in the hamlet. Researchers conducted interviews in Ban Tat from December 1998 through March 1999.

We used a "Self-Anchoring Striving Scale Questionnaire (SAS), which asks respondents to indicate where their present life situation is located on a ten-rung "ladder of life" (Kilpatrick and Cantril 1960). Respondents were asked to show where their life was five years ago and where they expect it to be five years in the future. Most people see their present life as being better than it was in the past, and they expect their future situation will be even better than it is now.

At the field site, a map of all houses was prepared and an identifying number was assigned to each household. A random sample of forty-two households was selected from the total of ninety-one households in the hamlet.[5] It was virtually impossible to ensure that the interviews were conducted in complete privacy, and usually other members of the family and occasionally a few neighbors were present. More than 350 items of information were solicited from each household in an interview that lasted several hours. In almost a fourth of cases, the individual interviewed was the head of the household; overall, total respondents split almost evenly between men and women. About 97 percent of adults (aged 15–59) reported "farmer" as their main occupation. Interviewers reviewed all completed questionnaires at the end of the day and return visits were made to the household if any information had been omitted from the initial interview (Donovan et al. n.d.).

4.3 Horton Overland Flow Database

A suite of computer simulations using the KINEROS2 runoff model (Smith et al. 1995, 1999) was conducted to assess how the spatial arrangement of the various land-cover types influences hydrologic response during storms. The simulations attempted to quantify the reduction of surface runoff reaching the stream system that results from having a buffer (high-infiltration) patch between an HOF-generating (low-infiltration) patch and a stream. We simulated the situation where a 30 x 30 m field is located immediately above a buffer ranging in length from 2 to 160 m.

We examined this issue by analyzing the land cover and topography of Ban Tat hamlet and the broader Tan Minh Village area. Analysis of the topography was based on a 30-m resolution digital elevation model (DEM) derived for the entire Tan Minh area from a triangulated irregular network (TIN) using Arc/Info version 7.3.1 (ESRI, Inc.). The TIN was constructed previously from a 20-m contour interval topographic coverage that had been digitized from a topographic sheet. Arc/Info functions were used to fill internal sinks in the DEM, determine flow direction from each cell based on the relative elevation of the eight neighboring cells, and compute flow accumulation in each cell. Flow accumulation is the total number of upslope cells that drain into the destination cell and is an indicator of relative position along flow paths in a watershed. A land-cover patch is defined as a zone of contiguous cells having the same land-cover type. Arc/Info was used to identify contiguous land-cover sub areas in each watershed. The count of number of grid cells in each patch was determined using ArcView.

5. DISCUSSION

In northern Vietnam, it is clear that forests have been converted into a mosaic of fields, pastures, and forest patches at various stages of succession, but important questions remain concerning the extent to which changes in land cover affect watershed functions. This project sought to test two linked premises. First, socioeconomic policies and technological conditions operating at the household and broader levels promote (cause) and sustain (maintain) forest fragmentation. Second, fragmentation of forest cover, including changes in the structure, age, and species diversity of the forest fragments, have both positive and negative effects on watershed function.

In terms of linking land use and forest fragmentation, project results suggest that swidden agriculture, the dominant land-use practice, is strongly linked with forest degradation and fragmentation—but not with deforestation. This finding is based on an analysis of land-use/cover change over the forty-three-year period from 1952 to 1995. While the project successively links a land-use practice (swidden) and land-cover change, it cannot establish a one-to-one correlation between individual decision makers and their use of specific plots of land, particularly given the temporary nature of their claims to these plots.

In terms of linking land use and the maintenance of forest fragments, we showed a high degree of dependency of this community on its natural resource base—especially natural forest vegetation. Nearly all households in Ban Tat rely to a greater or lesser degree on uncultivated land—that is, forest fallows and forest land—for income as well as basic foodstuffs and materials. From an analysis of resource products and their sources, we also showed that there is virtually no unutilized land in this landscape. Hence we conclude that if patches of mature forest exist, it is at least partially a result of people managing the landscape to maintain those patches because they perceive them as beneficial individually and as part of the landscape.

In terms of linking land use and the wider political and economic milieu in which it operates, we found that since the late 1980s, when Vietnam began to replace the collectivized economy with a household-based economy, the management of resources has become more complicated. It now involves the village cooperative and local government, the state and its specialized agencies, as well as the individual households. The cooperative has retained control over some forest land, but most land has been allocated to individual households. The state plays an increasingly visible role in forest management by imposing legal restrictions on resource exploitation, supervising allocation of forest lands to households, and collecting taxes on land and resources. The households make their decisions about resource management within this institutional context.

In terms of linking land use with household socioeconomic factors, we found that many factors affect resource management, and the influence of these individual factors will vary over time. Hypothetical factors include access to paddy, access to labor, access to capital, and "connections" stemming from former employment, government service, and ethnic or family ties. Although a diversified portfolio of livelihood strategies is common throughout the community, the technology employed to access resources may vary. Thus while virtually all families take advantage of the wild or uncultivated resources available, richer families may employ cattle to utilize these resources and poorer families may use women and children to collect bamboo shoots or medicinal plants to sell in the market. The type of product exploited is increasingly influenced by market demand now that government policies support economic liberalization and entrepreneurship and rural infrastructure has been improved. The environmental impact of these trends will depend not only on the structure and composition of arrested forest successional vegetation but on the extent and juxtaposition of these patches on the landscape.

For instance, in terms of the impact of forest fragmentation on watershed function, our results suggest that less HOF (and sediment that it carries) reaches streams in Tan Minh than would be the case with identical land cover but less fragmentation.

5.1 Limitations and Analysis of the Data

Despite the relatively high percentage of households in the hamlet that participated, the total number of households in the survey (46 percent sampled) was still small for the analysis that was undertaken. Given the complexity of the decision process regarding resource use, it was difficult to tease out statistical significance in analysis of the many relevant variables. Moreover, we lacked yearlong and multiyear data, which makes it difficult to give definitive conclusions about the exploitation of forest products, many of which are seasonal or periodic depending on weather conditions (e.g., El Niño). The availability of labor, market prices, and transport as well as the family's cash needs may also be important factors influencing a household's decision to collect forest products.

We recognize that the data in this sample were almost wholly recall data and thus can be highly variable with regard to their accuracy. We also acknowledge the possible influence of tax burden or the illegal nature of some forest exploitation activities, including swidden clearance, in a respondent's possible reluctance to full disclosure.[6] Moreover, heads of households—the people most likely to be knowledgeable about income and expenditure—made up only about one-fourth of the total sample. Nevertheless, we feel the provided figures offer a good picture of the relative importance of the various categories of income, and we should not take away from the significance of such relatively rare data.

This project attempted to work within a political ecology framework requiring that we examine the village's ecological characteristics, the history and contemporary dynamics of resource use, and the multiple scales of government and nongovernment programs and economic relations that bear upon and transform connections among various users and their resource base. We collected extensive information on government-sponsored programs in the village since 1945 that were intended to influence resource management (some programs were more effective than others, and none were judged to be resoundingly successful). We did not attempt through our household interviews, however, to link specific resource policies and specific household decisions on land use. In future work we will attempt to see if we can establish a quantifiable relationship between changes in policies and land-use decisions. One problem that will hinder this effort will be establishing a control site, since national policies will presumably affect all villages in similar ways. Efforts to establish a control through time series studies (before and after a change in policy) will be confounded by the fact that land-use decisions are a dynamic process always changing in response to a number of variables, including both social and political policies, as well as economic variables.

Perhaps the major limitation of this project is its small size. The next questions must be: How representative is this hamlet of the others in the village? How representative is this village of other villages in the district or in northern Vietnam as a whole? Some project members have conducted similar research in four other communities in northern Vietnam (Le and Rambo 2001). A first step will be to broaden our analysis to include these communities.

6. PRACTICAL PROBLEMS

6.1 Multicultural and Multidisciplinary Research

The study faced the normal host of problems that arise when researchers from different countries attempt to do collaborative work (i.e., obtaining permits for importing equipment and conducting field research, misunderstandings due to language and culture, difficulties in travel and lodging), but none of these was insurmountable or unique to this study.

All the scientists in this project had previous experience with multicultural and multidisciplinary research efforts combining physical and social science disciplines. Even given this experience, however, we failed to achieve the degree of integration among the different disciplines we sought. The entire research team visited the site for a one-week period when the research was initiated. After that, it became impossible to coordinate the field visits of the various players (i.e., the hydrologists with the ecologist, the economist with the forester, the mapping team with the household survey team). The entire research team met to discuss initial results, to identify overlapping concerns, and to assess where insights from one field of research raised questions that needed to be confirmed or checked by researchers from another field. But it was impossible to develop real-time feedback among the different researchers that would lead to an integrated understanding of the research questions. For example, the hydrologist only recently concluded that low flow accumulation values for more advanced vegetation are found at the tops of hill slopes, on or near the ridgelines. This is an interesting result that the social scientists need to explore further through more semistructured interviews. Given enough time and financial resources to reiterate our research questions, we can perhaps achieve the integration we seek, but it is not easy to achieve—and it is costly.

6.2 Quantitative versus Qualitative Research

This study combined qualitative social science methodologies (e.g., semistructured interviews, constant observation of household activities, sketch maps, and other participatory assessment methodologies) with quantitative methods (e.g., sample household interviews using structured survey instruments). Reiterative participatory assessment techniques provided the researcher with a more nuanced understanding of land-use practices, household economics, resource constraints, and so forth. This understanding then helped the researchers to develop more meaningful hypotheses to test quantitatively through survey methodologies. Both types of methods were useful and ideally should be used together. If forced to choose one or the other type of methodology, the researcher needs to consider whether enough is known about the problem to test meaningful hypotheses, the time and financial resources available to the project, other development objectives that may benefit more from a participatory approach, and a host of other issues. There is no single correct answer as to whether a researcher should use qualitative or quantitative methodologies for understanding land-use/land-cover change. It depends on the situation.

6.3 Individual Households versus the Community

In the introductory chapter of this volume, Rindfuss et al. point out that defining and following households over time is difficult because households are changeable—indeed ephemeral. In our study, we did not attempt to establish a one-to-one relationship between an individual household and their individual pieces of land. We did this for several reasons. First, it was difficult if not impossible to establish such a relationship over a long period of time (more than three years), not only because the households changed but also because the land they used changed with each rotation of the swidden agriculture system.

Second, it was not yet clear to us whether household land-use practices were significantly different from each other so as to make the extra work and expense of tying households to their land parcels useful. If villagers tend to make similar land-use decisions, then we can correlate our understanding of the decision-making process with land-use patterns at the village level. All villagers, whether wealthy or poor, utilize a variety of land-cover types, including forests; all villagers had approximately equal access to markets; and resource tenure did not appear to affect resource use patterns. We hypothesize that these results are due to several facts; namely: (1) Vietnam has only recently opened up to the outside world, and many citizens are still exploring their opportunities for participation in the market; (2) with a long history of communal use of resources, villagers are reluctant to abandon these access rights, especially while newly granted tenure remains contested and untested. As these conditions change and greater economic and political differentiation occurs among villagers, presumably land-use decisions will vary among households as cash crops replace subsistence cultivation, and it will become increasingly important to establish a one-to-one relationship between a household and its land parcels.

Given these difficulties, we chose not to attempt to take household members to their land parcels in order to establish a relationship between their decision-making processes and their land-use practices. We chose instead to conduct an informal survey of the area with local farmers willing to walk with us, interviewing them regarding their farming systems, the history of land-cover dynamics in the area, and specific information about individual plots of land. We used GPS readings to tie all attribute information gained through these interviews—such as current land cover, historical land cover at the time of image acquisition, information about the farming system, and any other information gained in the interview—to the spatial database. This information was crucial for relating present land cover to land cover found on the historical images.

7. OVERVIEW

Looking back, it is easy to say that this project was overly ambitious. It was based on two linked premises that required the input of an international, multidisciplinary group of scientists to examine. It was conducted in an isolated area with dense vegetation and steep terrain that lacked electricity, where researchers could not spend extended time in the field because of local political considerations, and fieldwork was conducted two to three years after the date of the satellite imagery. Better scientific/quantifiable results would have been obtained in a more controlled environment. One of the initial reviewers of the NSF proposal suggested the study should be done in the United States rather than Vietnam. Clearly, in terms of accessing the hydrological implications of forest fragmentation, that reviewer was correct.

Yet the study was useful both as a research and a training exercise. When working in developing countries, it is largely impossible to separate "research" and "development." Research is and should be done in collaboration with development objectives. These include developing local people's awareness of their resource problems and their capacities for resolving these problems themselves, developing the capacity of researchers in national universities and think tanks, and developing institutions that can ask critical questions and conduct scientific research. From these perspectives, this was a fairly successful project. How does one, then, redesign it—and for what ends? Perhaps a major challenge to land-use/land-cover change research is recognizing that it is not solely a scientific/quantitative agenda. Land-use/land-cover change research needs to be done in a participatory manner that helps local communities learn about themselves, learn about their land-use practices and how they are changing, and learn about the political, economic, and social factors that are influencing these changes.

NOTES

---

[1] Financial support for our field research has been provided by grants to the EWC from the Japanese Ministry of Foreign Affairs, the Ford Foundation, the Global Environment Forum, and the U.S. National Science Foundation and by a joint grant to CRES and the EWC from the John D. and Catherine T. MacArthur Foundation. Analysis of remote sensing data on land-cover change in Tat hamlet was supported by a grant to CRES from the Rockefeller Brothers Fund.

[2] The mission was flown in 1952 and 1953 during the dry season (December of each year) and was part of a larger mission to obtain air photos for all of French Indochina. The mission numbers are 19, 20, 57, and 82. The air photos were not in the standard 9 x 9-inch format; they were approximately 5 x 7 inches (12.5 x 17.5 cm). This appears to have been the standard size for the aerial photography missions flown in Indochina. The focal length of the camera used was 125 mm. The flight records indicate that they were flown at an altitude of approximately 5,000 m above terrain (6,100 m above sea level). All of the air photos were taken between 10 A.M. and 3 P.M. The nominal scale of the photographs ranges between 1:40,000 and 1:46,000, varying with the terrain's topography. The photography is panchromatic, vertical aerial photography with 60 percent overlap and between 20 percent and 30 percent side lap (overlap and side lap were determined by measuring the photographs in order to determine the quality of the air photos for stereo viewing).

[3] The air photos were interpreted using a lens stereoscope. In order to help validate the land cover that was identified on the air photos, they were photographically enlarged. The interpreter had already visited the study area and was knowledgeable of the land cover found in the area and the history of the major agricultural fields. It was known, for example, that the core of the main rice paddy area in use today was also cultivated in the 1950s. This knowledge was used in order to identify land cover on the historical photographs. After the air photos were interpreted, locations identifiable on the air photos and on the 1:50,000 maps of the area were marked on the overlay that the interpretation was done on. These marks were used to register the air photo interpretation to the maps. Most of these locations were streams and rivers that were identifiable both on the maps and on the aerial photos. After the registration was done, the interpretation was manually digitized. The digitized interpretation (including interpreted rivers and streams) were overlain on the digitized 1:50,000 maps and a quality check of the rivers and streams was done to verify the quality of the registration.

[4] When the ground truth exercise was done in 1997 and 1998, we asked questions relating to land use/land cover in 1995. If the interviewee could not remember, that information was noted and the land-cover rotation of the area was discussed. This information led to insights as to what the land cover of a particular field would have been in the recent past. Pictures of each reference point and extended point were taken and later tied to the ground reference point in the ground truth database.

[5] The sample was supposed to be forty households, but a mistake was made, so two extra households were included on the list to interview. Forty was the sample size to be used in all of the case-study sites. It was the maximum size that we could do with the time and funding we had (Le and Rambo 2001).

[6] Land tax is assessed on residential land, paddy land, and swidden land—the latter two based on productivity. There are royalty fees due on natural resources exploited, including timber, firewood, bamboo and bamboo shoots, a slaughter fee on cattle and pigs killed for meat, and a trade tax based on sales volume.

REFERENCES

Agarwal, B. 1994. *A Field of One's Own: Gender and Land Rights in South Asia*. Cambridge: Cambridge University Press.

Alcorn, J. 1996. "Forest Use and Ownership: Patterns, Issues, and Recommendations." In J. Schelhas and R. Greenberg, eds., *Forest Patches in Tropical Landscapes* (Washington, D.C.: Island Press), 233–257.

Belsky, J. A., and C. D. Canham. 1994. "Forest Gaps and Isolated Savanna Trees." *BioScience* 44(2): 77–84.

Berry, S. 1989. "Social Institutions and Access to Resources." *Africa* 59(1): 41–55.

Blaikie, P., and H. Brookfield. 1987. *Land Degradation and Society*. London: Methuen.

Browder, J. 1996. "Reading Colonist Landscapes: Social Interpretations of Tropical Forest Patches in an Amazonian Agricultural Frontier." In J. Schelhas and R. Greenberg, eds., *Forest Patches in Tropical Landscapes* (Washington, D.C.: Island Press), 285–299.

Collins, M., J. Sayer, and T. Whitmore, eds. 1991. *The Conservation Atlas of Tropical Forests, Asia and the Pacific*. London: Macmillan for International Union for the Conservation of Nature.

Donovan, D., R. K. Puri, Tran D. V., and A. T. Rambo. n.d. (forthcoming) "Forest and Household Economy in Ban Tat." East-West Center Working Paper. Honolulu.

Enoksson, B., P. Angelstam, and K. Larsson. 1995. "Deciduous Forest and Resident Birds: The Problem of Fragmentation within a Coniferous Forest Landscape." *Landscape Ecology* 10(2): 267.

Fortmann, L. 1990. "Locality and Custom: Non-Aboriginal Claims to Customary Usufructuary Rights as a Source of Rural Protest." *Journal of Rural Studies* 6(2): 195–208.

Fox J., Dao M. T., A T. Rambo, Nghiem P. T., Le T. C., and S. Leisz. 2000. "Shifting Cultivation: A New Old Paradigm for Managing Tropical Forests." *BioScience* 50(6): 521–528.

Fox, J., J. Krummel, S. Yarnasarn, M. Ekasingh, and N. Podger. 1995. "Land Use and Landscape Dynamics in Northern Thailand: Assessing Change in Three Upland Watersheds." *Ambio* 24(6): 328–334.

Fox, J., S. Leisz, Dao M. T., A. T. Rambo, Nghiem P. T., and Le T. C. 2001. "Shifting Cultivation without Deforestation: A Case Study in the Mountains of Northwestern Vietnam." Chapter 17 in Millington, Walsh, and Osborne, eds., *Applications of GIS and Remote Sensing in Biogeography and Ecology* (Boston: Kluwer Academic Publishers).

Ireson, C., and R. Ireson. 1996. "Cultivating the Forest: Gender and the Decline of Wild Resources among the Tay of Northern Vietnam." East-West Center Working Paper No. 6. Honolulu.

Kilpatrick, F., and H. Cantril. 1960. "Self-Anchoring Scaling: A Measure of Individuals' Unique Reality Worlds." *Journal of Individual Psychology* 16: 158–173.

Le T. C. and A. T. Rambo, eds. 2001. *Bright Peaks, Dark Valleys: A Comparative Analysis of Environmental and Social Conditions and Development Trends in Five Communities in Vietnam's Northern Mountains*. Hanoi: National Political Publishing House.

Lebbie, A., and M. Schoonmaker Freudenberger. 1996. "Sacred Groves in Africa: Forest Patches in Transition." In J. Schelhas and R. Greenberg, eds., *Forest Patches in Tropical Landscapes* (Washington, D.C.: Island Press), 300–326.

Li, T. 2001. "Relational Histories and Production of Difference on Sulawesi's Upland Frontier." Journal of Asian Studies 60(1): 41–66.

Lord, J., and D. Norton. 1990. "Scale and Spatial Concept of Fragmentation." *Conservation Biology* 4(2): 197–202.

McClanahan, T. R., and R. W. Wolfe. 1993. "Accelerating Forest Succession in a Fragmented Landscape: The Role of Birds and Perches." *Conservation Biology* 7(2): 279–288.

Nepstad, D., C. Uhl, and E. Serrao. 1991. "Recuperation of a Degraded Amazonian Landscape: Forest Recovery and Agricultural Restoration." *Ambio* 20(6): 248–255.

Neumann, R. 1998. *Imposing Wilderness: Struggles over Livelihood and Nature Preservation in Africa*. Berkeley: University of California Press.

O'Neill, R. V., B. T. Milne, M. G. Turner, and R. H. Garner. 1988. "Resource Utilization Scales and Landscape Pattern." *Landscape Ecology* 2(1): 63–69.

Peluso, N. 1992. *Rich Forests, Poor People: Resource Control and Resistance in Java*. Berkeley: University of California Press.

Pinedo-Vasquez, M., and C. Padoch. 1996. "Managing Forest Remnants and Forest Gardens in Peru and Indonesia." In J. Schelhas and R. Greenberg, eds., *Forest Patches in Tropical Landscapes* (Washington, D.C.: Island Press).

Potter, L., H. Brookfield, and Y. Byron. 1994. "The Eastern Sundaland Region of Southeast Asia." Chapter 10 in J. X. Kasperson, R. E. Kasperson, and B. L. Turner II, eds., *Regions at Risk: Comparisons of Threatened Environments* (Tokyo: United Nations University Press).

Rambo, A. T. 1995. "The Composite Swiddening Agroecosystem of the Tay Ethnic Minority of the Northwestern Mountains of Vietnam." Paper presented at the Regional Symposium on Montane Mainland Southeast Asia in Transition, Chiang Mai University, Chiang Mai, Thailand. November 13–17.

Rambo, A. T., and Tran D. V. 2001. "Social Organization and the Management of Natural Resources : A Case Study of Tat Hamlet, a Da Bac Tay Ethnic Minority Settlement in Vietnam's Northwestern Mountains." *Tonan Ajia Kenkyu* (Southeast Asian Studies) 39(3): 299–324.

Rudis, V. R. 1995. "Regional Forest Fragmentation Effects on Bottomland Hardwood Community Types and Resource Values." *Landscape Ecology* 10(5): 291–307.

Schelhas, J., and R. Greenberg, eds. 1996. *Forest Patches in Tropical Landscapes*. Washington, D.C.: Island Press.

Smith, R. E., D. C. Goodrich, and J. N. Quinton. 1995. "Dynamic, Distributed Simulation of Watershed Erosion: The KINEROS2 and Eurosem Models." *Journal of Soil and Water Conservation* 50(5): 517–520.

Smith, R. E., D. C. Goodrich, and C. L. Unkrich. 1999. "Simulation of Selected Events on the Catsop Catchment by KINEROS2: Report for the GCTE Conference on Catchment Scale Erosion Models." *Catena* 37: 457–475.

Watts, M. 1997. "Black Gold, White Heat: State Violence, Local Resistance, and the National Question in Nigeria." In S. Pile and M. Keith, eds., *Geographies of Resistance* (New York: Routledge), 33–67.

Westman, W. E. 1990. "Structural and Floristic Attributes of Recolonizing Species in Large Rain Forest Gaps, North Queensland." *Biotropica* 22(3): 226–234.

Ziegler, A., T. Giambelluca, D. Plondke, T. Vana, J. Fox, Tran Duc Vien, and M. Nullet. n.d. (in review) "Near-Surface Hydrologic Response in a Fragmented Mountainous Landscape: Tat Hamlet, Da River Watershed, Northern Vietnam." *Journal of Hydrology*.

Chapter 8

LINKING SOCIOECONOMIC AND REMOTE SENSING DATA AT THE COMMUNITY OR AT THE HOUSEHOLD LEVEL

Two Case Studies from Africa

Eric F. Lambin
Department of Geography, University of Louvain
lambin@geog.ucl.ac.be

Abstract One of the major challenges of linking remote sensing and socioeconomic data concerns the definition of the appropriate spatial observation units. This chapter discusses whether it is always beneficial and desirable to link socioeconomic data at the level of households rather than at the level of villages or communities. Some of the pros and cons of linking socioeconomic data at the level of households and at the level of the villages or communities are presented. This is illustrated by two case studies where socioeconomic data were linked with remote sensing data: in southern Cameroon and in the buffer zone of the Masai Mara Natural Reserve in Kenya. In the first case, macroeconomic changes affecting Cameroon have played a fundamental role in the way land-use practices influence the forest cover. In the second case, conversion of rangelands has led to a major decline in nonmigratory wildlife due to the spread of mechanized agriculture on critical spatial locations. This was driven by changes in markets and national land-tenure policies. Political economic theory was used as a framework for both studies. The two studies demonstrated the complementarity of remote sensing and socioeconomic survey data for understanding the causes, processes, and impacts of land-use/land-cover changes. Even though socioeconomic data were always *collected* at the level of households, the two studies have *linked* remote sensing with household survey data at the level of villages or groups of households following a similar land-use strategy rather than at the household level. Reasons for this include logistical constraints in the organization of the field survey and land-tenure system of pastoralist societies. The optimal level of analysis of a joint remote sensing–socioeconomic study depends on the research question and involves a trade-off between the information that can be extracted with a reasonable level of accuracy and the cost of field data collection. Working at the finest possible level is not always the optimal strategy. However, aggregated-level analyses obscure the role of the heterogeneity between actors (e.g., social networks, leadership status, role of education, and differing access to resources, knowledge, and land) in driving land-use changes.

Keywords: remote sensing, land use, deforestation, pastoralism, wildlife, Africa, Cameroon, Masai Mara

1. INTRODUCTION

Several studies have combined socioeconomic household survey data and remote sensing data to better understand processes of land-use change (Guyer and Lambin 1993; Sussman et al. 1994; Behrens et al. 1994; Entwisle et al. 1998; Moran and Brondizio 1998). One of the major challenges of merging these data from heterogeneous sources concerns the definition of the appropriate spatial observation units—that is, the appropriate level of aggregation of information derived from the domains of social phenomena and natural environment (Liverman et al. 1998). While conceptually straightforward, developing the appropriate linkages between household-level and remote sensing data sets can be difficult to implement operationally (Entwisle et al. 1998). In remote sensing, the spatial unit of observation is not directly associated with any unit of observation in the social sciences, such as individuals, households, or communities.

Most of the studies linking remote sensing observations and socioeconomic data have been performed at the scale of administrative units (e.g., Green and Sussman 1990; Skole et al. 1994; Geoghegan et al. 2001). For instance, Wood and Skole (1998) and Pfaff (1996), in their studies of deforestation in the Brazilian Amazon, have aggregated the land-cover data to conform to the administrative units (the *municipio* or county level, in these cases). In this approach, the dependent variable (deforestation derived from remote sensing data) refers precisely to the independent variables (e.g., sociodemographic data obtained through the national census and transport infrastructure density) and vice versa. Entwisle et al. (1998) attempted to link population dynamics derived from field surveys to land-use/land-cover change data in Thailand. They noted the difficulty of relating remotely sensed patterns of land-cover change with field observations of land-use change because people live in nucleated villages away from their fields and because households cultivate multiple noncontiguous plots. For this reason, the integration of the two data sets was performed at the village level.

Aggregated to the village level, household data offer an additional perspective to the remotely sensed land-cover dynamics. The village profiles provide a cross-check on the dynamics observed by remote sensing and can be related to remotely sensed landscape variables (Entwisle et al. 1998; Mertens et al. 2000). But the aggregation of land-cover change data to a coarse resolution leads to a loss of information, as it obscures the variability within the units (Wood and Skole 1998). Moreover, this aggregation can introduce "ecological fallacies" in the interpretation of statistical correlations. When the measures of statistical association are calculated across administrative units, the data do not correspond to the level of the decision units (ibid.).

For these reasons, more recent research efforts have attempted to integrate remote sensing observations and field surveys at finer levels of aggregation—that is, at the scale of households. This property-level linkage between remote sensing and socioeconomic data requires extremely precise fieldwork to georeference every plot used and to ensure a high geometric accuracy when matching pixels with plots. This is achieved at a certain cost. McCracken et al. (1999) successfully analyzed land-cover changes at the property level because each household lived on a well-defined plot and had little impact on adjacent land. Individual household data allow for a better understanding of the land-use practices within each village, as most land-use decisions are made by individuals and households.

Is this expensive field data collection effort worth it? Are there some types of research questions for which village-level analyses of joint remote sensing–household survey data provide the same level of understanding as household-level analyses? The objective of this chapter is to discuss whether it is always beneficial and desirable to link socioeconomic data at the level of households rather than at the level of villages or communities. I first discuss the broad research questions that one attempts to address when linking socioeconomic with remote sensing data. I then highlight some of the pros and cons of linking socioeconomic data at the level of households and at the level of villages or communities. I finally present two case studies where socioeconomic data were linked with remote sensing data at the level of villages or communities. I explain the various factors that have led these studies to be conducted at an aggregated level.

2. BACKGROUND

Most studies linking socioeconomic with remote sensing data are aimed at better understanding human-environment relationships and predicting the impact on land resources of changes in land use. Beyond a simple understanding of drivers of land-use change, such studies also highlight dynamic interactions between land-use changes and processes of landscape transformation (e.g., soil erosion, water availability, modifications in climate patterns induced by land-cover changes, and spread of vector-borne diseases). This can lead to assessments or scenarios of sustainability of land use and vulnerability (or resilience) of communities based on models of human-environment interactions.

Models of land-use change can address a range of questions. For example: Where are land-use changes likely to take place (the *location* of change)? At what rates are changes likely to progress (the *quantity* of change)? What are the feedbacks from land-use change to demographic, social, and economic change? What are the impacts of land-use change on the provision of ecosystem goods and services (Veldkamp and Lambin 2001)? The question on the location of change is often much easier to deal with through models, as it mostly requires identification of the natural and cultural landscape attributes that are the spatial determinants of change—that is, the local proximate causes directly linked to land-use changes. Such models represent the spatial determinants of land-use change, leading to an emphasis on factors such as roads, soil types, or topography. The rate or quantity of change is driven by demands for land-based commodities and is often modeled using an economic framework. The deeper underlying driving forces that control the rates of changes are often remote in space and time and operate at a higher hierarchical level. They often involve macroeconomic transformations and policy changes that influence microdecisions by land managers. Modeling these driving forces often requires the representation of the decision-making processes by actors. The ability to predict the location or quantity of land-use changes and to uncover the proximate or underlying forces driving these changes depends in part on the spatial level of analysis, which is largely determined by the data-collection effort in the field.

3. LINKING REMOTE SENSING AND SOCIOECONOMIC DATA AT THE VILLAGE OR HOUSEHOLD LEVEL

Most studies collect socioeconomic data at the household or individual level. There are exceptions, however, as participatory assessment techniques may be used at the level of groups of land managers. In the latter case, the socioeconomic data can only be linked to remote sensing data at the level at which they have been collected (e.g., village, community, or cooperative of land managers using a given land area) or at a coarser level. When socioeconomic data have been collected at the household or individual level, there is still the option of aggregating the data at the village or community levels before linking them to remote sensing data. The discussion below refers to the level at which socioeconomic data are *linked* to remote sensing data and not to the level at which socioeconomic data are *collected*. It is assumed below that, in all cases, some data have to be collected at the household level.

Decisions to modify land use are taken by land managers—the head of household, the household as a whole, or the head of the enterprise for which a set of households may be working. The household-level linkage of socioeconomic and remote sensing data therefore best captures the actual level of decision making. It also allows a representation of the heterogeneity between land managers operating in a spatial unit. This heterogeneity results from different aspirations, levels of education, management skills, openness to the market, leadership status, and so on. But linking remote sensing and socioeconomic data at the household level comes at a certain cost, as it requires georeferencing every plot of the interviewed households. This operation is labor intensive (i.e., traveling with the household to every plot to collect GPS coordinates or identifying the user of every plot based on detailed maps). Conducting such costly fieldwork may be perfectly justified, but if resources are limited, it is interesting to explore what could be gained or lost from a different resource allocation in alternative study designs.

Household-level linkage is also subject to attribution or georeferencing errors that may contaminate the analysis. While a village-level linkage of socioeconomic and remote sensing data often includes the collection of some data at the household level, only the boundaries of the land used at the village-level—rather than at the level of individual plots—is georeferenced. This is done either by assuming a maximum travel distance to the plots from the houses (which can be estimated via a question in the household survey) or by identifying, for each village, the boundary of land use on a map with a key informant (such as the village chief).

Since it requires aggregation of socioeconomic data, a village-level linkage of socioeconomic and remote sensing data focuses on the dominant responses and behaviors within the village. Therefore, "outliers" do not weigh much on the analysis, as they contribute little to the average value computed from all responses to questionnaires. These outliers may be individuals or households following a singular land-use strategy due to some unique circumstances, and thus they are not necessarily of great interest if one aims at gaining a generalized understanding of land-use dynamics. They may also result from errors and inaccuracies in the interviews at the household level. Note, however, that a household-level linkage of socioeconomic and remote sensing data can also include an analysis and special treatment of outliers.

Still, the village-level linkage may mask key relationships through aggregation (e.g., the role of social networks and institutions within a community). Actually, an analysis based on average values for all households within a village or community does not consider the heterogeneity between actors in a village. Such an analysis is also exposed to the ecological fallacy, which is a logical fallacy inherent in making causal inferences from group data to individual behaviors. There is thus a risk of reaching incomplete—if not misleading—conclusions when village-level links are used.

Another important consideration when designing a study relates to the need to reach a sufficient level of variance between observations, at least for key variables, to allow for an explanatory model to be tested. If all observations are identical with respect to land use, socioeconomic profile, geographic situation, and biophysical environment, little understanding will be gained on what is causing land-use change, as one cannot control for some key variables. In some cases, concentrating an intensive data collection effort—one that would include georeferencing of every plot—in just a few locations may be at the expense of the variance in data. One is more likely to create this variance through a study design that would include many villages in a range of geographic situations. Actually, if resources are limited, there is often a trade-off between the number of villages being analyzed and the depth of field data collection in each of the villages. Maximizing the variance in the data may require allocating the limited resources to a larger geographic spread of the data-collection effort rather than to an intensive, more localized data collection. While this could plead for a study design with a village-level linkage for a large number of villages, two other considerations should be introduced. First, a small sample of households per village—with in-depth data collection and plot-level georeferencing—in a large number of villages could allow for a household-level linkage while still having a large variance. In this case, however, within-village relationships could be missed if the sample of households per village is small. Second, intensive data collection increases the number of variables for which observations are made. Thus, variance may be found at a finer level and for different variables in studies characterized by in-depth household surveys and linkage with remote sensing data at the household level compared to studies with village-level linkage. Note, finally, that every study is different, at least by its specific research question, and it is difficult to propose an optimal study design that would work everywhere.

4. CASE STUDY 1: TROPICAL DEFORESTATION IN CAMEROON

4.1. Objective

Central Africa is one of the deforestation "hot spots" of the planet. In this region as in the other deforestation hot spots, however, the interannual variability in rate of deforestation is high. What is driving these year-to-year variations in the pressure on forest ecosystems? The objective of this study was to better understand the causes and processes of land-use/land-cover changes since the 1970s in the forest zone of the East Province of Cameroon (Mertens and Lambin 2000; Mertens et al. 2000). This was achieved by combining information from an extensive socioeconomic household survey and time series with remote sensing data at a fine spatial resolution. These analyses were carried out against a backdrop of dramatic macroeconomic change that has occurred in Cameroon over the period under study. From the late 1970s until 1985, Cameroon was a booming economy. Beginning in 1985 it has experienced a devastating economic crisis, a structural adjustment program, and, beginning in January 1994, a drastic devaluation of its currency (CFA). How have these transformations influenced population movements, land use, and deforestation in eastern Cameroon?

The study implicitly used a political-economic theory as a background. This influenced the study design, as it was important to relate socioeconomic and land-use change data for "time slices" that corresponded to periods dominated by a certain type of macroeconomic change.

4.2. Data

For the household survey, all villages in the study area were chosen. The study area was defined by the intersection of the areas covered by the different satellite images of the time series. For each of these thirty-three villages, a random sample of a third of all households was selected. This corresponded to 552 households that were present at the time of the survey. All households agreed to respond. This high response rate is probably a characteristic of Africa. The survey focused on: (1) demographic data (household composition, level of education, indigenous/migrant, main, and secondary activities); (2) labor, inputs, and tools; (3) changes in land use (area of crop type over time); (4) production by crop type and land-cover type; (5) marketing strategy (proportion of the production sold and where); (6) clearing and/or increase of size of agricultural plots; (7) land availability, crop preferences, and fallow period; and (8) households' land-use strategies for the near future. These data were related to time series of Landsat and SPOT satellite data, from which maps of land-cover change trajectories were produced using change-detection techniques. A major interest of the field data lies in the retrospective longitudinal framework of the survey. It highlighted the evolution of the household and its land use over three periods related to the key macroeconomic periods and corresponding to the dates of acquisition of remote sensing data.

Data accuracy was assessed using standard techniques. For the remote sensing-based land-cover maps, a random sample of points was selected and analyzed on contemporary aerial photographs. The overall accuracy of the maps ranged from 88 to 95 percent for the different dates. Also, the total area of forest clearing as estimated by remote sensing and household survey (i.e., total area of agricultural plots created or increased during the study period) was compared at the village level ($r = 0.884$). This offered a cross-validation between the two data sets and confirmed the consistency between estimates of land-cover change and land-use change data.

4.3. Method

Statistical analyses between remotely sensed land-cover change data and household information were performed at the scale of the villages. The first step of the analysis was the definition of the village boundaries and the aggregation of the household survey and remotely sensed data at the village level. Two distances were considered in the definition of the boundaries of the agricultural area of each village: (1) the spatial extent of each village into the forest area, and (2) the spatial extent of each village along the road between each pair of villages. Since the villages are close to each other in the study area, the boundary between two adjacent villages along a road was set at a distance from each village such that the ratio of the distance from the village to that boundary and the population of that village was equal for the two villages. In other words, the spatial extent of the agricultural area of a village along the road was proportional to the population of the village. The boundary of the agricultural area of each village from the road into the forest area was defined at a 4-km distance from each village centroid. This distance was assumed to represent the maximum distance traveled from the village to the agricultural plots, based on a field survey (Sunderlin and Pokam 1998). Most farmers walk to their plots. The average size of the villages was 845 ha, most of the villages ranging between 500 and 900 ha.

Different procedures were considered for the aggregation of the household data to the village level, depending on the type of variable (i.e., continuous, categorical, or binary) and the desired information. The sum and the mean functions were applied for continuous variables. Note that the sum value was divided by the sampling coefficient (30 percent) to reflect the total village value. The mode, median, and frequency of occurrence functions were applied for categorical variables. Statistics on land-cover and land-cover changes were extracted for each village for the different periods of study. The frequency of occurrence of deforestation was measured in hectares within each village. These data were related to the socioeconomic variables through multiple regression analysis (Mertens et al. 2000).

4.4. Results

The research results demonstrated that the macroeconomic changes affecting Cameroon have played a fundamental role in the way land-use practices influence the forest cover. The annual rate of deforestation increased after the economic crisis as compared to the previous period. The household survey information enabled identification of the causal relationships and the processes of land-use and land-cover changes. Observations revealed that the beginning of the economic crisis (1986) was associated in time with a strong increase of the deforestation rate related to the population growth, increased marketing of food crops, and modification of farming systems. The beginning of the economic crisis corresponded to a decline in cocoa and coffee prices and in the government provision of input subsidies, motivating food crop expansion in forested areas. Time series of prices for the main crops in the local urban and rural markets were available. Plantain and nonplantain food crops increased in area and became an income source for households, and plantain tended to be established in recently cleared forest areas. The deforestation rate also increased after the devaluation (1994)—but at a much lower rate. The January 1994 devaluation of the CFA made cocoa and coffee prices more attractive again. An increase of the area devoted to coffee was observed after 1993 as compared to the previous period (1985–93). This may have reduced pressure on forest areas by reducing dependence on food crop income.

Deforestation was associated with a modification of the farming systems in the region, as indicated by a decrease in the fallow period on the fields that used to be cultivated in the most extensive way, an increase in the household labor force, and a decrease in the use of agricultural inputs. About 10 percent of the heads of household were allogene (i.e., not being born in the village in which they now reside). Few heads of household migrated to the study villages during the oil boom period of 1980–85, when migrants were heading in large numbers to the big cities rather than to the countryside. The rate of growth of allogene families was twice as fast as the rate of growth of authochtone families (i.e., being born in the village in which they now reside).

The amount of cocoa and coffee production and the proportion sold by the households decreased over time, while an increase in marketing of plantain and nonplantain food crops was observed. The decreasing coffee and cocoa production did not, for the most part, result in the allocation of cocoa and coffee plots to other crops. The cultivators tended to maintain their cocoa and coffee stands in the hope that they could revive their former source of livelihood. This farming approach, which seeks to keep economic options open in the face of an uncertain macroeconomic context, has led to an increase in deforestation.

4.5. Issues Related to Household and Remote Sensing Data Linkages

The spatiotemporal analysis framework developed for the integration of the household survey and remote sensing data allowed for a better understanding of the impact of national-scale factors on decisions by households that are driving land-use/land-cover change processes in the region. Actually, time series of remote sensing data combined with household data aggregated at the level of villages allowed an examination of structural changes related to economic variables that are almost impossible to examine using cross-sectional analysis—whether it be household, village, or spatial. The methodological inability of prior researchers to look at these issues led them to overemphasize variables that are easy to include in cross-sectional analysis (e.g., population and distance to roads and markets) compared with other variables that may be more important (e.g., major price or subsidy changes). This study has highlighted the importance of analyzing processes of change by "time slices" corresponding to periods of macroeconomic change. Research on the causes of deforestation can be greatly enriched by grounding in the macroeconomic context of unfolding development processes and through longitudinal analysis of combined socioeconomic and remote sensing data.

The study was conducted at the village rather than household level mostly due to the logistical organization of the household survey in the field. The household survey was conducted in a short period (three months) during the dry season, when the study site was easily accessible due to the dry season and the farmers were not busy in their fields. It was done by a large number of local interviewers who were not trained in the use of global positioning systems. The small size of the supervising staff located in the field throughout the survey would have made impossible a systematic verification of the quality of georeferencing of field boundaries to ensure the high precision that is required for property-level linkage of remote sensing and household survey data. Also, at the time of fieldwork (1998), there was not enough published scientific evidence of the value of household-level analyses with remote sensing data to convince the funding agency on the additional cost involved. Finally, this survey was part of a broader survey on 500 households all across southern Cameroon as a follow-up of a similar survey in the 1960s. Consistency of the survey with the previous ones and the ones being conducted in the neighboring regions was thus a major constraint.

The advantage of the field survey was that it could be concentrated in a short time period. This guaranteed that the macroeconomic factors influencing land use—which were the target of the study—did not change throughout the course of the household survey. If this would have been the case—for example, with a survey spread over a year or more and conducted by a few surveyors—any macroeconomic change such as a currency devaluation, major policy change, monetary inflation, or period of political instability (of frequent occurrence in Central Africa) would have caused changes in responses by farmers and therefore would have obscured the results of the analysis.

The main disadvantage of the village-level analysis was the inability of the research to uncover the role of heterogeneity between actors within a village in driving some changes. Actually, for some of the surveyed variables, the village-level variance was high. This was, however, taken into account in selecting a method for data aggregation at the village level of household survey variables (mean, mode, median, and frequency of occurrence of a response). Not surprisingly, the results did not identify social networks, power status, education level, or wealth distribution as explanatory variables of deforestation. This is a possible bias related to the study design. Given that methods of central tendency (rather than dispersion) were used, the possible effects of important outliers in each village were missed.

If we had the opportunity to redesign the field data collection of this study, the major difference would be to actually survey the exact boundaries of the area under use for each village rather than to compute it through a slightly arbitrary distance calculation. This would improve the link between people and land at the village level.

5. CASE STUDY 2: LAND-COVER CHANGES AND WILDLIFE DECLINE IN THE SERENGETI-MARA ECOSYSTEM

5.1. Objective

In East African savannas, wildlife tourism is a major source of foreign exchange, but most rural livelihoods depend on farming and herding. Widely perceived habitat loss and wildlife decline are generally attributed to rapid population growth and the spread of subsistence cultivation. However, interannual variability of rainfall on semiarid lands makes directional trends in land-cover change and causal chains hard to establish. In the Serengeti-Mara Ecosystem, a core fortress conservation zone (i.e., an area where any land use except wildlife conservation and viewing is excluded) surrounded by a ring of buffer zones straddles the very different macroeconomic and macropolitical contexts demarcated by the Kenya/Tanzania border

(Homewood et al. 2001; Thompson et al. 2002). The main objective of this research was to identify the proximate causes and driving forces of land-use/land-cover change in the buffer zones around protected areas over the past twenty years. A related objective was to evaluate the impact of land-use/land-cover change on wildlife in the Serengeti-Mara ecosystem. We tested the hypothesis that land-use policies in Kenya have led to land-use/land-cover change around the Masai Mara National Reserve and to declining wildlife numbers (Homewood et al. 2001). What does explain the observed diversification of land-use strategies of Masai pastoralists around the Masai Mara National Reserve in Kenya? How did these new land-use strategies lead to changes in wildlife habitats and decline in wildlife populations? While this study also implicitly used a political-economic theory as a background, it broadened this framework to include social and cultural variables.

5.2. Data

Data were gathered by semistructured interviews over a period of eighteen months (May 1998 to November 1999) in single-round, broad-scale, and detailed multiround surveys at five study sites in Narok District, Kenya (Thompson 2002). The sites were selected to represent the range of land-use situations found in the buffer zone of the Masai Mara National Reserve. The selection took place after an initial survey of the study region by a multidisciplinary team of scientists. The unit of study was the household (*olmarei:* singular), which corresponds to the gate cut in the stockade around the *boma* (homestead) for a married man and his household. *Boma*s are classically made up of groupings of *ilmareta* (plural of *olmarei*), from one in our sample to as many as ten. Traditionally, *boma*s are not clustered in villages and are changing location after several years. Over the study period, however, few movements of homesteads took place. In total, 288 heads of households, randomly selected within each of the five sites, were interviewed. The response rate was very high. The survey covered details of household composition, education level (years of schooling), livestock holdings (cattle and small stock), cultivation of maize and/or wheat, and other types of production/sources of income (shares in tourism facilities, dividends from wildlife associations, wage employment, and remittance from wage earners working away from the household). Data collected by survey was cross-referenced by participant observation, livestock counts, and field measurement during the multiround survey. At a later stage of the survey, information on the sociopolitical leadership status of the head of household (whether the leadership position was none, minor, or senior in the traditional Masai society) was also collected. In practice, decision making related to land use still largely takes place within the traditional sociopolitical structure of the Masai society. On leadership status, information is available for only 132 households. If a household had been involved in agriculture, the area cultivated was calculated as the average of the cultivated area in the last three years. These crop acreages were declared by each household during the survey. The homestead of each household was located with a global positioning system (GPS), which enabled us to represent each household in a geographical information system (GIS). When multiple households in one *boma* were interviewed, these households were represented by one set of GPS coordinates (the *boma*).

A time series of three Landsat images was acquired with a ten-year interval. Change-detection techniques were applied to image pairs, and nine trajectories of change were mapped (Serneels et al. 2001). The land-cover change map was validated through field surveys by visiting each detected area of change. The map was revised after the field visits, as it tended to overestimate change—for example, including short-term vegetation changes related to burning or localized rainfall events that occurred just before the date of one of the Landsat images. The land-cover changes were grouped into changes related to the expansion of mechanized farming, subsistence agriculture, development of permanent settlements (smallholder impact), and vegetation regrowth. The GIS database was created by digitizing the topographic maps and a series of thematic maps, which represent three major categories of factors: those related to the socioeconomic status of the household, accessibility, and natural landscape attributes. The socioeconomic data from households was brought into the GIS database by attributing them to the homestead of the household, which was georeferenced.

5.3. Method

The households were clustered according to land-use strategies, based on the following variables: (1) revenues from shares in tourist facilities and/or from dividends in wildlife associations; (2) acreage under maize cultivation; (3) acreage under wheat cultivation and/or income from leasing land to contractors for wheat cultivation; and (4) wage labor. The k-mean clustering technique was used for this purpose, based on a selection of variables that described the above land-use strategies. We then assessed the determinants of these land-use production choices, based on variables representing socioeconomic status (household size, education level, leadership status, and wealth in livestock equivalents per reference adult), landscape attributes (elevation, slope, agro-ecological zone), and accessibility factors (distance to Masai Mara National Reserve, Narok, mechanized farming, roads, villages, and water). These explanatory variables were tested in a multinomial logit model with the different land-use strategies as dependent variables.

Subsequently, we examined a possible spatial link between households following a given land-use strategy and land-cover changes in the area surrounding their homesteads. We formed buffers of 5 km around each group of households and calculated the percentage of land-cover change, as detected on the satellite images, within each area. While the land-cover change data referred to the period 1988–98, the household survey was conducted in 1998–99 and was only retrospective for the three previous years. Thus, for this last part of the study, the temporal match between remote sensing and household survey data was not perfect. One had to assume that land use measured by the household survey data has been changing more slowly than land cover measured by the remote sensing data.

5.4. Results

In the Masai Mara Ecosystem (Kenya), 60,000 ha of rangelands have been converted to mechanized agriculture, and the total population of nonmigratory wildlife species has declined by 58 percent over the past twenty years. This study showed that, contrary to widely held views, these changes were driven by markets and national land-tenure policies rather than by agropastoral population growth. Spread of mechanized agriculture on critical spatial locations underlined wildlife decline.

Analysis of land-use strategies of pastoralists showed the emergence of clearly distinguishable groups diversifying into small-scale cultivation, large-scale mechanized cultivation, tourism, and a mixture of small-scale cultivation/tourism. In all of these groups, livestock keeping remained a central component of the land-use strategies chosen. The generalized logit model showed that, amongst the group diversifying into mechanized cultivation, distance to the nearest market center followed by agro-ecological zone and slope provided the greatest explanatory power amongst households choosing this land use. This supports economic-based models that explain conversion of rangeland to cultivation as a function of the economic returns achievable from cultivation. However, these accessibility and landscape variables did not fully explain land-use strategies. On the reduced sample for which full socioeconomic variables were available, leadership, education, and wealth (as measured in livestock holdings) were all significantly linked to mechanized cultivation as a choice of land use. Households in the traditional cattle-keeping group were less likely to have a leadership position than households following any of the other land-use strategies. Leadership status and networks could be used to obtain lands at a location that was favorable for development or allowed revenues from distant sources. Likewise, the uptake of agriculture or involvement in tourism was associated with an increasing level of formal education.

Households were often involved in several land-use activities that have an impact on the environment in a wide range around the homestead, so linking households to changes in land cover was difficult. Most land-cover changes were observed in the buffer zones around the households engaged in mechanized farming and, to a lesser extent, in subsistence agriculture. Only mechanized farming as a land-use strategy was strongly linked spatially to participating households. Other land-cover changes could be attributed to the development of permanent settlements.

5.5. Issues Related to Household and Remote Sensing Data Linkages

The introduction of socioeconomic household data provided insights in the decision-making processes on land use by the Masai inhabiting the rangelands surrounding the MMNR. This study increased our understanding of the relative importance of economic versus landscape and social factors that shape land-use/land-cover changes. Land-use strategies followed by the pastoralists were influenced by land-use policies, economic opportunities, social structure of the community, and by the landscape in which people live. Spatial attributes and spatial autocorrelation were important explanatory variables, but they did not fully explain land-use strategies, given the large degree of spatial overlap between the different land-use strategy groups. Other factors, often related with the social status of the households, intervened in the decision-making process about land-use strategies. Even though analyzing the environmental consequences of land-use strategies was difficult, we showed that there certainly was a strong link between the location of the homestead of people involved in mechanized farming and the land-cover changes that this entails. The link with land-cover changes, however, was less obvious for the other land-use strategies. This could be explained by the temporal mismatch of the data sets on land-use and land-cover change.

Few studies had attempted to link socioeconomic data on pastoralist land use to remote sensing data, although Boone et al. (2000) used interviews of Masai head of households to characterize the effect of droughts and El Niño weather patterns on their livestock through changes in vegetation biomass. The Masai living in our study area are primarily pastoralists who use very diverse areas away from their permanent settlements to graze their livestock at certain periods of the year (Homewood and Rodgers 1991). The system of grazing rights is complex. Animals graze on communal land, so a given area is used by different families, often coming from different regions. Linking livestock-related land-cover change data to people was thus difficult. The relevant variable to map is the location of livestock on the landscape rather than the location of fields. Livestock being highly mobile throughout the day and being led to different places every day and every season, the challenge of georeferencing land use at a fine spatial resolution is indeed much greater for such pastoral systems than is the case for permanent cultivation systems. Other patterns of resource use, such as cultivation and income from tourism, may take place closer to the homestead and may thus be more easily related to households.

Given the nature of land use and land tenure of the Masai society, the study was conducted at the level of groups following a similar land-use strategy rather than at the level of households or communities. As most of the land is managed and used communally, attributing the changes in land cover to a single household was impossible. Moreover, there was a large diversity in land-use strategies within a given Masai community living in one location. Actually, income distribution is only in part determined by the community to which Masai belong and by where people have their homesteads, as leadership status in a given community is an important explanatory variable. For this reason, an analysis at the level of communities was not feasible either. Therefore, the analysis was performed at the level of groups of households that follow similar land-use strategies, even if the households belonging to a group were scattered throughout the study area. Analyzing the determinants of land-use decisions for homogenous groups was more insightful than taking location as the unit of analysis. Only for households heavily engaged in mechanized farming was there a strong linkage between location and land-use strategy. In fact, most of these households were clustered in the same region. Still, the rangelands and wood extraction areas of these households were managed in a communal way, and only the cultivated areas could be clearly attributed to a single household.

If we had the opportunity to redesign the field data collection of this study, we would collect more data on the sociopolitical and cultural dimensions of the Masai households, as these variables proved to have a significant influence on land-use decisions. In this study, these variables were added in the course of the survey and were thus not available for all surveyed households. Also, there is a case for a study design that would be more exhaustive spatially—that is, not just five sites but a more representative sample of the entire population of the buffer zone. As location was found to explain only part of the variance in land-use strategy, a research design with samples based on location is likely to cause some gaps in understanding.

6. CONCLUSION

The two studies discussed above have emphasized the complementarity of remote sensing and socioeconomic survey data for improving the understanding of causes, processes, and impacts of land-use/land-cover changes in the region. Even though socioeconomic data were always *collected* at the level of households, the two studies have *linked* remote sensing with household survey data at the level of villages or groups of households following a similar land-use strategy rather than at the household level for different reasons. These included logistical constraints in the organization of the field survey, land management and land tenure of a pastoralist society, and scale of the human-environment system that is investigated.

The optimal level of analysis of a joint remote sensing–socioeconomic study depends on the research question. It involves a trade-off between the information that can be extracted with a reasonable level of accuracy and the cost of field data collection. Working at the finest possible level is not always the optimal strategy. Yet aggregated-level analyses obscure the role of the heterogeneity between actors (e.g., social networks, leadership status, role of education, and differing access to resources, knowledge, and land) in driving land-use changes. It thus tends to oversimplify explanation and modeling of the phenomenon.

REFERENCES

Behrens, C. A., M. G. Baksh, and M. Mothes. 1994. "A Regional Analysis of Bari Land Use Intensification and Its Impact on Landscape Heterogeneity." *Human Ecology* 22(3): 279–316.

Boone, R. B., K. A. Galvin, N. M. Smith, and S. J. Lynn. 2000. "Generalizing El Niño Effects upon Maasai Livestock Using Hierarchical Clusters of Vegetation Patterns." *Photogrammetic Engineering & Remote Sensing* 66: 737–744.

Brondizio, E. S., E. F. Moran, P. Mausel, and Y. Wu. 1994, "Land Use Change in the Amazon Estuary: Patterns of Caboclo Settlement and Landscape Management." *Human Ecology* 22(3): 249–278.

Entwisle, B., S. J. Walsh, R. R. Rindfuss, and A. Chamratrithirong. 1998. "Land-Use/Land-Cover and Population Dynamics, Nang Rong, Thailand." In D. Liverman, E. F. Moran, R. R. Rindfuss, and P. C. Stern, eds., *People and Pixels: Linking Remote Sensing and Social Science* (Washington, D.C.: National Academy Press), 121–144.

Geoghegan J., S. C. Villar, P. Klepeis, P. M. Mendoza, Y. Ogneva-Himmelberger, R. R. Chowdhury, B. L. Turner II, and C. Vance. 2001. "Modelling Tropical Deforestation in the Southern Yucatan Peninsular Region: Comparing Survey ND Satellite Data." *Agriculture, Ecosystems and Environment* 85: 24–46.

Green, G. M., and R. W. Sussman. 1990. "Deforestation History of the Eastern Rain Forests of Madagascar from Satellite Images." *Science* 2: 212–215.

Guyer, J., and E. F. Lambin. 1993. "Land Use in an Urban Hinterland: Ethnography and Remote Sensing in the Study of African Intensification." *American Anthropologist* 95(4): 839–859.

Homewood, K. M., E. F. Lambin, E. Coast, A. Kariuki, I. Kikula, J. Kivelia, M. Said, S. Serneels, and M. Thompson. 2001. "Long-Term Changes in Serengeti-Mara Wildlife and Land Cover: Pastoralists or Policies?" *Proceedings of the National Academy of Science* 98(22): 12,544–12,549.

Homewood, K. M., and W. A. Rodgers. 1991. *Maasailand Ecology: Pastoralist Development and Wildlife Conservation in Ngorongoro, Tanzania.* Cambridge: Cambridge University Press.

Liverman, D., E. F. Moran, R. R. Rindfuss, and P. C, Stern, eds. 1998. *People and Pixels: Linking Remote Sensing and Social Science.* (Washington, D.C.: National Academy Press).

McCracken, S. D., E. S. Brondizio, D. Nelson, E. F. Moran, A. D. Siqueira, and C. Rodriguez-Pedraza. 1999. "Remote Sensing and GIS at Farm Property Level: Demography and Deforestation in the Brazilian Amazon." *Photogrammetric Engineering & Remote Sensing* 65: 1,311–1,320.

Mertens, B., and E. F. Lambin. 2000. "A Spatial Model of Land-Cover Change Trajectories in a Frontier Region in Southern Cameroon." *Annals of the Association of American Geographers* 90(3): 467–494.

Mertens, B., W. D. Sunderlin, O. Ndoye, and E. F. Lambin. 2000. "Impact of Macroeconomic Change on Deforestation in South Cameroon: Integration of Household Survey and Remotely Sensed Data. *World Development* 28: 983–999.

Moran, E. F., and E. S. Brondizio. 1998. "Land-Use Change after Deforestation in Amazonia." In D. Liverman, E. F. Moran, R. R. Rindfuss, and P. C. Stern, eds., *People and Pixels: Linking Remote Sensing and Social Science* (Washington, D.C.: National Academy Press), 94–120.

Pfaff, A. S. P. 1996. "What Drives Deforestation in the Brazilian Amazon: Evidence from Satellite and Socioeconomic Data." World Bank Policy Research Working Paper No. 1772. Washington, D.C.

Serneels, S., M. Y. Said, and E. F. Lambin. 2001. "Land-Cover Changes around a Major East-African Wildlife Reserve: The Mara Ecosystem (Kenya)." *International Journal of Remote Sensing* 22(17): 3,397–3,420.

Skole, D. L., W. H. Chomentowsky, W. A. Salas, and A. D. Nobre. 1994. "Physical and Human Dimensions of Deforestation in Amazonia." *Bioscience* 44(5): 314–322.

Sunderlin, W. D., and J. Pokam. 1998. "Economic Crisis and Forest Cover Change in Cameroon: The Roles of Migration, Crop Diversification, and Gender Division of Labor." Center for International Forestry Research, Bogor. Manuscript.

Sussman, R. W., M. G. Green, and L. K. Sussman. 1994. "Satellite Imagery, Human Ecology, Anthropology and Deforestation in Madagascar." *Human Ecology* 22(3), 333–354.

Thompson, D. M. 2002. "Livestock, Cultivation and Tourism: Livelihood Choices and Conservation in Maasai Mara Buffer Zones, Kenya." Ph.D. thesis, Department of Anthropology, UCL, London.

Thompson D. M., S. Serneels, and E. F. Lambin. 2002. "Land-Use Strategies in the Mara Ecosystem (Kenya): A Spatial Analysis Linking Socioeconomic Data with Landscape Variables." Chapter 3 in S. J. Walsh and K. A. Crews-Meyer, eds., *Linking People, Place and Policy: A GIScience Approach* (Boston: Kluwer Academic Publishers).

Veldkamp T., and E. F. Lambin. 2001. "Predicting Land-Use Change." *Agriculture, Ecosystems & Environment* 85(1–3): 1–6.

Wood, C. H., and D. Skole. 1998. "Linking Satellite, Census, and Survey Data to Study Deforestation in the Brazilian Amazon." In D. Liverman, E. F. Moran, R. R. Rindfuss, and P. C. Stern, eds., *People and Pixels: Linking Remote Sensing and Social Science* (Washington, D.C.: National Academy Press), 70–93.

Chapter 9

HUMAN IMPACTS ON LAND COVER AND PANDA HABITAT IN WOLONG NATURE RESERVE

Linking Ecological, Socioeconomic, Demographic, and Behavioral Data

Jianguo Liu
Department of Fisheries and Wildlife, Michigan State University
jliu@panda.msu.edu
Li An
Department of Fisheries and Wildlife, Michigan State University
Sandra S. Batie
Department of Agricultural Economics, Michigan State University
Richard E. Groop
Department of Geography, Michigan State University
Zai Liang
Department of Sociology, City University of New York – Queens College
Marc A. Linderman
Department of Fisheries and Wildlife, Michigan State University
Angela G. Mertig
Department of Fisheries and Wildlife and Department of Sociology, Michigan State University
Zhiyun Ouyang
Department of Systems Ecology, Research Center for Eco-Environmental Sciences, Chinese Academy of Sciences
Jiaguo Qi
Department of Geography, Michigan State University

Abstract Understanding patterns, processes, and consequences of land-cover change requires close linkages among various ecological, socioeconomic, demographic, and behavioral data. In this chapter, we present an interdisciplinary study of human impacts on land cover and panda habitat in Wolong Nature Reserve (China). Wolong is one of the largest reserves (200,000 ha) designated for giant panda conservation, but it also includes more than 4,000 local residents whose activities have a substantial impact on panda habitat. We have developed a conceptual framework that outlines the rationale for the linkage of various data. Our household and community surveys were linked with remote sensing and geographic information systems at three stages: data collection, data analysis, and systems modeling and simulation. The integration of various sources of data offers useful insight into the underlying mechanisms behind changes in land cover and panda habitat and allows us to project future ecological and demographic changes under different policy scenarios.

Keywords: Wolong Nature Reserve, giant pandas, wildlife habitat, land cover, forest, socioeconomics, demographics, ecology, human behaviors, linkages, households, remote sensing, GIS, modeling, China

1. INTRODUCTION

Land-cover changes are a major aspect of global changes (e.g., see Turner et al. 1994). The extent and intensity of such changes vary across space and time, as do their ecological consequences. Whereas some land-cover changes occur naturally, most of them are caused by human activities. In fact, human activities are the main cause of biodiversity loss and habitat fragmentation as well as rapid changes in ecosystems and landscapes around the world (Ehrlich 1988; Wilson 1988; Vitousek et al. 1997). Even in many of the 30,350 protected areas (such as nature reserves) established to protect natural resources and biodiversity (IUCN 1994), humans are also present and carry out various activities detrimental to biodiversity (Dompka 1996; Liu 2001). For example, in the Wolong Nature Reserve of China, established to protect the world-famous, endangered giant pandas *(Ailuropoda melanoleuca),* land cover, especially forest cover, has undergone significant changes due to activities of local residents (Liu et al. 2001a). Because forest cover is a critical component of panda habitat, loss and fragmentation of forest cover lead to degradation in panda habitat. As a result, the panda range has been dramatically reduced from once including many parts of China and several neighboring countries to including only fragmented areas of southwestern China (Giant Panda Expedition 1974; Schaller et al. 1985; China's Ministry of Forestry and WWF 1989).

The objectives of our project were to detect the spatial and temporal patterns of changes in land cover and panda habitat, to understand the mechanisms behind these patterns, and to develop new policy scenarios and evaluate short- and long-term consequences of different scenarios. To achieve these objectives, a variety of spatial and nonspatial (ecological, socioeconomic, demographic, and behavioral) data need to be closely fused, because patterns and processes of changes in land cover and panda habitat cannot be understood and managed using one type of data alone. The primary goals of this chapter are to present a conceptual framework of linkages and to discuss the methods to fuse various types of data for a holistic understanding of human impacts on land cover and panda habitat.

2. STUDY AREA

Our study area is Wolong Nature Reserve (Figure 1), a reserve established in 1975 for conserving giant pandas. Wolong is located in Sichuan Province in southwestern China (30°45' – 31°25' North, 102°52' – 103°24' East). It is ideal for our study for four main reasons. First, it is one of the largest protected areas (approximately 200,000 ha) designated for conserving the pandas and contains approximately 10 percent of the wild panda population (Zhang et al. 1997). Second, like many other protected areas, there are local residents in Wolong (more than 4,000 local residents in over 900 households in 1998). Third, Wolong is a "flagship" nature reserve and has received exceptional financial and technical support since its creation, both from the Chinese government and many international organizations. Fourth, many biological studies on giant pandas have been conducted (e.g., Hu et al. 1980; Schaller et al. 1985; Johnson et al. 1988; Reid and Hu 1991), and there is a useful record of economic and demographic statistics. These previous studies and data provided a good foundation for our study.

Wolong is situated between the Sichuan Basin and the Qinghai-Tibet Plateau. It has high mountains and deep valleys with elevations ranging from 1,200 to 6,525 m (Figure 1). Wolong encompasses several climatic zones and has high habitat diversity (Schaller et al. 1985). There are more than 2,200 animal and insect species and approximately 4,000 plant species (Tan et al. 1995). In addition to the giant pandas, twelve other animal species and forty-seven plant species in the reserve are also on China's national protection list. Furthermore, the reserve is part of the international Man and Biosphere Reserve Network (He et al. 1996). It is managed by the Wolong Administration Bureau, which reports to both China's State Forestry Administration and Sichuan Province. Under the Administration Bureau, there are two township governments. Each township consists of three villages, and each village includes three to six groups (a group is the lowest administrative unit in China's rural areas, often comprised of approximately a dozen to several dozen households that are geographically close to one another). Farmers comprise the vast majority of local residents. There are various economic activities in the reserve, including agriculture (maize and vegetables are the major crops), fuelwood collection, timber harvesting, house building, collection of Chinese herbal medicines, and road construction and maintenance.

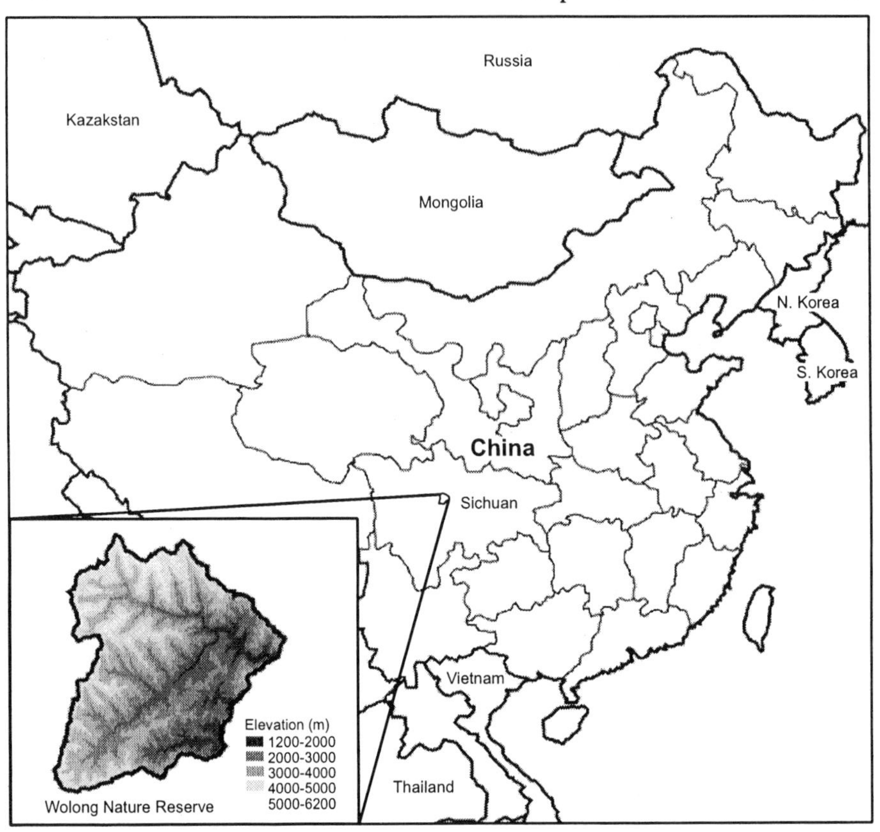

Figure 1. Location map of Wolong Nature Reserve

3. LINKAGES FROM A CONCEPTUAL PERSPECTIVE

The conceptual framework (Figure 2) provides rationales for linking various factors. It consists of three main interrelated components: panda habitat, forest, and humans. Panda habitat is the area that provides food and cover for pandas' reproduction and daily activities. Suitability of panda habitat depends on abiotic and biotic conditions, as well as the degree of human impact (Liu et al. 1999a). Slope and elevation are two major abiotic factors (Schaller et al. 1985; Liu et al. 1999a; Liu et al. 2001a). Pandas prefer flat areas or gentle slopes in order to move around easily. Major biotic factors include bamboo and forest cover. Bamboo is the staple food for the panda and is an understory species in the forests. Like many other wildlife species, the giant panda depends on forest canopy as cover (mainly deciduous broadleaf forests, conifer forests, and mixed conifer and deciduous broadleaf forests) (Schaller et al. 1985; Liu et al. 1999a).

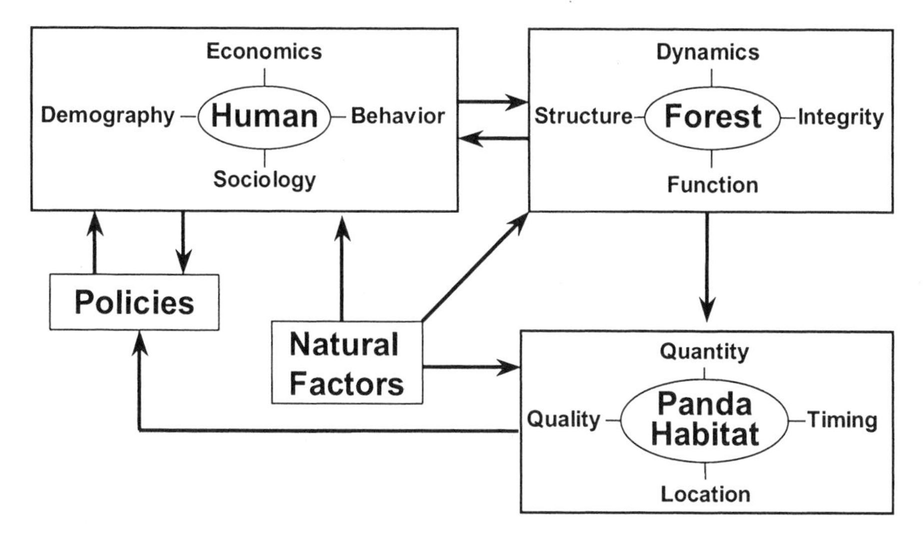

Figure 2. A conceptual framework for studying land cover and panda habitat changes

In the past several decades, human activities have been the primary force that changes the forest ecosystem, thereby altering panda habitat (see Figure 2, Liu et al. 1999a). *Human* in this context refers to local residents but not to those government officials and management staff who develop and implement policies (see below). Human factors include demographic (e.g., population size and distribution, household composition), socioeconomic (e.g., income, expenses, production, consumption, needs and wants, perceptions, and attitudes toward panda conservation), and behavioral variables (i.e., activities) such as timber harvesting and fuelwood collection for cooking and heating. Through human activities, local residents influence forests directly and panda habitat indirectly. Because fuelwood is the major source of energy for cooking and heating (a conservative estimate of fuelwood consumed in the reserve was 10,000 m^3/year, Liu et al. 1999a), fuelwood consumption has led to a significant loss of forests, including changes in forest structure, species composition, and spatial distribution (ibid.). As bamboo is an understory species, changes in forest cover affect the pandas' food supply (Schaller et al. 1985). Because giant pandas also depend on forests for cover, changes in forests can have an important impact on panda habitat quantity, quality, timing (when the habitat is available), and location (e.g., spatial distribution of panda habitat) (Liu et al. 1999a). Although the reserve is about 200,000 ha in size, approximately half of the area is not suitable for the pandas even without human impacts because some regions are above the tree line and other regions have slopes too steep or elevations too high to be suitable for pandas (ibid.).

All three components (human, forest, and habitat systems) in the framework can be directly or indirectly shaped by government policies and other factors. On the one hand, policies are made by government officials rather than by local residents. Such policies can directly affect various aspects of the local human system and indirectly affect the forest and panda habitat. On the other hand, policy-making process and effectiveness can be influenced by local residents and panda habitat conditions. For example, when panda habitat conditions are seriously deficient and human attitudes toward panda conservation are positive, government policies may be more effective and favorable to panda conservation. Loss and fragmentation of panda habitat may prompt the government to design and implement new policies such as out-migration (Liu et al. 1999a), which affect human population and local economy. Furthermore, the human system may be constrained by feedback from the forest system. For instance, after all trees in a forest are harvested, local residents must adopt a different lifestyle without timber and fuelwood. When forests used for fuelwood collection are getting smaller and farther from households, fuelwood collection becomes more difficult. If the difficulties of life in Wolong increase, more people (especially young people) are willing to move out of the reserve through means such as marriage outside Wolong (ibid.). Other factors such as physical environment (e.g., elevation) and landslides also impact humans, forests, and panda habitat. For example, landslides in 1996 killed seven local residents (Yinchun Tan, Wolong Nature Reserve, personal communication, 1999).

The linkages among the three components as well as policies and other factors in the conceptual framework are represented by arrows in Figure 2. For the sake of simplicity, subcomponents within each main component are not connected by arrows, although they are also interrelated. For example, human behaviors are a function of demographic and socioeconomic characteristics. Furthermore, human-environment interactions take place at different scales. For humans, the organizational scale ranges from individuals, households, villages, and townships to the entire reserve community. Concerning land, spatial scales range from plots or pixels (e.g., 1 x 1 m, 80 x 80 m), patches (e.g., forest stand, agricultural land parcel), to the entire reserve landscape. Individuals or households have a direct impact on the environment at small spatial scales. The cumulative and collective effects of individuals, households, and communities (villages and townships) extend to the entire reserve landscape. To understand the linkages among the main components and subcomponents in each main component and their relationships to policies and other factors, it is essential to couple different types of data at various scales.

4. LINKAGES FROM A TECHNICAL PERSPECTIVE

We linked ecological, socioeconomic, demographic, and human behavioral data at three stages: data collection, data analysis, and systems modeling and simulation. For data collection, we combined field studies, interviews, government statistics and documents, information from the literature, and data from remote sensing (satellite imagery and aerial photographs) and global positioning systems (GPS) using GIS and a relational database program (Microsoft ACCESS). In the process of collecting socioeconomic-demographic-behavioral data, we gathered some ecological data (e.g., tree species of fuelwood consumed by households). Similarly, collecting ecological data (e.g., percent of forest canopy cover) also included socioeconomic data (e.g., forest harvesting history such as time and type of harvesting). Socioeconomic-demographic data were often collected simultaneously. For example, in household economic surveys, we also gathered household demographic data. Furthermore, we used ecological data (e.g., land cover from remote sensing imagery) to gather socioeconomic-demographic data (e.g., locations of dwelling units and roads), using the latter (e.g., road intersections as ground control points) to facilitate classification of land-cover types. After the data collection, most of the information was entered into and managed using a relational database program and geographical information systems (including both Arc/Info and ArcView) for analysis.

In the data analysis stage, we used different types of data to explain their interrelationships. For instance, we used ecological data as dependent variables, and socioeconomic-demographic data were treated as independent variables, or vice versa (Liu et al. 1999a; An et al. 2001). We started our analyses with the land because we wanted to find out the patterns of land-cover distribution and temporal changes (Liu et al. 1999a) before attempting to understand the mechanisms underlying the patterns of distribution and changes. We analyzed data using spatial statistics, GIS, and statistical packages such as SAS, SPSS, and LISREL (An et al. n.d.[a]; An et al. n.d.[b]). In the systems modeling and simulation stage, we integrated various types of data and resultant mathematical and statistical models from the data analysis stage into more comprehensive systems simulation models that link demographic-socioeconomic-behavioral-ecological data (Liu et al. 1999a; An et al. 2001). More detailed descriptions about the linkages from a technical perspective are as follows.

4.1 Collection and Analysis of Ecological Data

The use of remotely sensed images has become an indispensable tool in environmental assessment (Lillesand and Kiefer 1994). Remote sensing is particularly valuable for large-scale research, especially in topographically complex regions such as Wolong, where assessment of land cover is both urgent and difficult if carried out by field studies alone (Moran et al. 1994). To measure spatial and temporal patterns of land-cover changes, we have acquired a number of remote sensing imagery: Corona data from 1965 (January 20), Landsat Multi-Spectral Scanner (MSS) data from 1974 (January 3), Landsat Thematic Mapper (TM) data from 1987 and 1997 (September 27), SPOT data from 1998 and 1999, and IKONOS data from 2000. The Corona data were stereo-pair photographs acquired as part of the Corona photo-reconnaissance satellite project (USGS EROS Data Center, Sioux Falls, South Dakota). We scanned the area of interest within each photo into a digital image at 1,200 dpi. We then georeferenced these images to topographic maps of the region. Next we mosaicked these small images to a large single image, examined them visually for inconsistencies, and classified them visually into regions of forest cover and nonforest cover (Liu et al. 2001a). We obtained both Landsat MSS and TM data from China's Satellite Ground Station (Beijing, China). SPOT HRV data (7/18/1998: 20-m multispectral 3 band; 3/17/1999: 10-m panchromatic, 20-m multispectral 4 band) were obtained from SPOT Image Co. (Virginia, USA). We acquired four images of IKONOS data (different times between August 31 and November 16, 2000) with 1-m resolution and 4-m resolution (multispectral image) through the National Aeronautics and Space Administration (NASA). Wolong has a high frequency of cloud cover, making the acquisition of data at regular intervals difficult. The difficulty of obtaining cloud-free imagery necessitated the use of data from different seasons. It is not easy to locate readily recognizable landmarks across the study area. In addition, the complex topography in Wolong further complicated the process of georeferencing the remote sensing data. To facilitate the classification of land cover, we developed a Digital Elevation Model (DEM) from topographic maps and employed a number of GPS–measured ground control points (e.g., major road intersections) to georeference the DEM.

In the summer of 1998, the National Survey Bureau of China established two reference points (with accuracy < 1 cm) in Wolong for us to place two base stations for differential correction of rover data gathered from ground sample sites. In 1998 and 1999, we surveyed approximately 500 ground-truth sites in Wolong (Linderman et al. n.d.). We selected ground-sample sites where access was feasible, with special focus on gathering a statistically representative sample of the various vegetative compositions and ratios, topographic configurations, and bamboo presence. Because of a lack of prior knowledge of observer position relative to the 30-m satellite data grid, each site was a 60-m homogenous square so that at least one complete 30 x 30-m pixel was contained in the site (Linderman et al. n.d.; Liu et al. 2001a). At least two people on a field team estimated and recorded percentage of overstory, mid-story, and understory cover, as well as species composition along with slope, aspect, and elevation. We georeferenced the remote sensing data using highly accurate data (1–5 m) from GPS receivers (Trimble Pathfinder) with real-time differential corrections. We then used the ground-truth data for training and validating a supervised classification of the remote sensing data. Using ERDAS Imagine software or visual classification, we found that forest cover in Wolong had been dramatically reduced from 1965 to 1997 (Liu et al. 2001a; Linderman et al. n.d.).

We employed over 200 independent ground samples to assess the accuracy of the supervised classification using the accuracy assessment tool in ERDAS Imagine software. The supervised classification had an 88 percent overall prediction rate in distinguishing between forested and nonforested areas. Overall accuracy in classifying individual land-cover classes (e.g., deciduous forest, agriculture, grasslands) was 79 percent. Success in predicting specific classes varied primarily due to spectral similarities between certain classes. For example, due to the variability in regrowth and logging activity, deciduous forest and shrub regrowth areas were often spectrally similar, with the majority of the shrub omission errors being misclassified as deciduous forest. Other factors influencing the accuracy of the classification included extreme topography and mixed pixels.

While remotely sensed data can help identify amounts and spatial distributions of forests at large scales, some detailed structure and composition of forests cannot be easily identified. Thus we conducted field surveys (Liu et al. 1999a; Ouyang et al. 2000). We selected random samples of plots (20 x 25 m each) in areas with different forest conditions (e.g., socioeconomic conditions such as places with and without fuelwood collection and timber harvesting). We chose smaller subplots inside the 20 x 25 m plot for sampling shrubs (5 x 5 m, three in each plot) and herbaceous plants (1 x 1 m, 5 in each plot). In each sampling plot or subplot, we recorded canopy closure, species, and size of vegetation (trees, shrubs, bamboo). In our most recent field studies, we also used GPS receivers to record the plot locations and later incorporated the GPS measurements into a GIS for spatial analysis.

4.2. Collection and Analysis of Demographic, Socioeconomic, and Behavioral Data

Demographic, socioeconomic, and behavioral data used in our research were collected either by the Wolong Administration Bureau or by our research team. The Wolong Administration Bureau followed data-collection guidelines set by upper government agencies at the county, province, and national levels, such as the State Statistics Bureau. Although no information about data accuracy is available, the people responsible for data collection received relevant training mandated and provided by upper government agencies. Furthermore, we verified some of the data (e.g., household size) by comparing them with what our team collected and found that the data collected by the Wolong Administration Bureau were good. Using ArcView, Arc/Info, and/or the Microsoft ACCESS relational database, we linked data obtained from the Wolong Administration Bureau with those collected by our research team. For example, for household-level data, names or identification codes of household heads were the linkages among different data sources.

Data collection by our research team began in 1996 (Liu et al. 1999a; Liu et al. 2001a; An et al. 2001), and the most recent survey took place in the summer of 2001. Our data collection followed standard protocols. Our specific methods—though varied depending on the questions and research objectives (e.g., Liu et al. 1999a; An et al. 2001; An et al. n.d.[a]; An et al. n.d.[b])—did not change fundamentally over time so that temporal comparisons would be possible. Although the boundary for a small part of the reserve did change, this portion of the reserve has had no residents and has had no impact on the collection of socioeconomic, demographic, and behavioral data.

In our household surveys, we selected respondents (household heads or other household members, depending on the questions) using simple or stratified random designs. For the pre-interviews (or pilot studies), we used simple random sampling because the goal was to identify all possible major issues and to test the appropriateness of questionnaires (An et al. n.d.[a]). Since simple random sampling could cause omission of some villages in Wolong, we used stratified random sampling (at the village level) to assure that all villages were appropriately represented. Some questions in the formal interviews were modified according to the pilot interviews and preliminary analyses of the data collected. For example, we had expected that money would play an important role in the decision-making process of adolescents leaving parental homes and establishing their own households. However, our pre-interviews indicated that wood and land were the major factors (An et al. n.d.[b]). The pilot studies also allowed us to estimate the amount of time needed for each interview more accurately.

We used face-to-face interviews with local residents because this method is the best and most feasible approach for Wolong (An et al. 2001). Since the majority of the local residents have low educational levels (with an average of 4.2 years/person) (ibid.) and postal service is not readily available to those who live in areas of high elevation, mail surveys would have been extremely problematic. In addition, a telephone survey could not be conducted since the vast majority of local residents have no phones. Face-to-face interviews typically garner higher response (or compliance) rates than other types of surveys (Dillman 1978), and our interviews proved to be highly successful, with nearly 100 percent compliance overall. (Unlike many people in cities of China or developed countries, local residents in Wolong were very cooperative.)

To demonstrate how we conducted cross-sectional interview sessions and then linked them with our remote sensing data, we provide the following example that modeled conditions under which local farmers would be willing to switch from fuelwood (their primary energy source) to electricity for cooking and heating (An et al. n.d.[a]). We hypothesized that this willingness to switch would be explained and predicted by the following four types of variables: (1) demographic features (such as age and gender), (2) economic variables (such as income), (3) electricity price and quality (such as voltage levels), and (4) locational variables (such as whether a specific household is located in Wolong or Gengda Township and distances from household to fuelwood collection sites). The Wolong Administration Bureau has a complete list of all the households in the reserve and household-specific data regarding demographic features and farmland conditions (such as land parcel size). Thus the demographic data needed by our model were directly obtained from the government archives. To cross-check the government data, we also recorded some demographic features in our interview session (see below). The second (economic) and third (electricity price and quality) types of data were elicited from the same interview session. The collection of the fourth (locational) type of data was aided by applying remote sensing techniques. We printed out georeferenced SPOT (1998) maps showing dwelling units and their surrounding features, such as mountain ridges, roads, rivers, and large parcels of land. We then asked the respondents (assisted by our local colleagues) to locate the sites of their fuelwood collection at different times (1970s, 1980s, and 1990s) on the images.

Using the 1996 Wolong Agricultural Census record (containing all the households by villages) as our sampling frame, we conducted our interview session and collected the economic, electricity, and locational data (types 2–4 listed above) that were necessary to construct our quantitative model (An et al. n.d.[a]). A sample of 220 households (about 23 percent of the total number of households) was chosen, reflecting the trade-off between our need for a robust sample and the limitations of time, budget, and manpower. Our stratified sampling process was characterized by proportionally drawing the 220 households from each of the six villages based on its size (N_i, the number of households in village i, i = 1, 2, …, 6). Specifically, for village i, we coded all the households with numbers from 1 to N_i , and then took a random sample of n_i (the sample size in village i) households from a total of N_i households in village i.

Before each interview session started, we gave a brief explanation to the interviewee regarding who we were, the purpose of our research, how they had been selected, the estimated time of the interview, the confidentiality of the interview results, and the voluntary nature of the interview. We collected household socioeconomic and demographic data first—for example, household expense items in 1998, educational levels, ages, genders of all household members, and so on. We then collected the stated preference data by asking the head of each household the price, voltage level, and outage frequency for electricity in the summer of 1999. In the session, to elicit their preferences under different electricity conditions, we designed an approach for realizing random combinations of prices, voltage levels, and outage frequency levels. Seven price levels, three voltage levels, and three outage levels were determined to be appropriate for this study, so we prepared seven price cards, three voltage cards, and three outage frequency cards. The cards in each set were shuffled thoroughly and placed face down. We asked the respondent to pick up one card from the seven price cards, one card from the three electricity outage frequency cards, and finally, one card from the three electricity voltage cards. Then a hypothetical condition regarding electricity price and quality was ready for use. We then asked the respondent (in Chinese): "Under this condition, will you switch from fuelwood to electricity completely?" A 100 percent response rate was achieved in these interview sessions, which is understandable in Chinese rural cultures because rural people are more hospitable to visitors than those in cities. Out of these 220 households, 28 were removed from further analysis because the respondents had problems in answering some of the questions—for example, they could not remember the price of electricity.

Discussion about the model construction and results from these interview data (An et al. n.d.[a]) is beyond the scope of this section. But worthy of mention is that we linked all these data (types 1–4 listed above) and remote sensing data (such as IKONOS) in an ArcInfo database in which household IDs (household heads' names could also be used as IDs, but due to concerns about confidentiality, we did not use them) were used as the database key and all the demographic (such as age, from government archives and cross-checked by our interview data), economic (such as income, from our interviews), and electricity-related data (such as price, from our interviews) were attributes of PAT (Point Attribute Table) in ARC or VAT (Value Attribute Table) in GRID module. The georeferenced remote sensing image provided a good basis for identifying the locations of the households and fuelwood collection sites, creating spatially explicit databases (PAT in ARC and VAT in GRID under Arc/Info), measuring the distances between households and fuelwood collection sites, and displaying spatial patterns of the households and the attributes of these households based on the data contained in the spatially explicit database (PAT or VAT).

4.2.1 Demographic Data

We obtained data regarding human population and households in Wolong from several sources, including Wolong Administration Bureau (annual population reports; 1984–98 records on birth, death, age, sex, in-migration, and out-migration), national census data in 1982 (Wenchuan County 1983), 1996 Agricultural Census (Wolong Nature Reserve 1996), and national census data in 2000 (Wolong Nature Reserve 2000). The 1996 Agricultural Census produced a variety of data (e.g., household size; name, age, and sex of each household member; land area; number of livestock, such as pigs) at the household level. All these micro-level data are now in digital format and are managed in a relational database. We used names of household heads or identification codes to link these data. Because our team members are the only ones who had access to these data in the database, confidentiality was well protected even when the names of household heads were used as the linkage. These data provide us with useful information to study the past dynamics of population size and structure, such as age composition, sex ratio, and educational composition (Liu et al. 1999a, 1999b).

We have been investigating the demographic dynamics, such as household size, household structure, time of household formation and dissolution (e.g., through divorce and marriage), and time of house construction. (Our unpublished data show that from 1975 to 1998, as a net average, twenty-two new households were added to the reserve every year.) For each individual we recorded information such as age, sex, marital status, time of marriage, separation, or divorce if applicable, years of schooling, and relationship to the household head. In household surveys, we interviewed household heads. Sometimes, we also asked knowledgeable family members and neighbors to confirm some of the information provided by the household heads. (Unlike many neighbors in the cities, most neighbors in Wolong talk to each other on a daily or weekly basis; furthermore, neighbors often help each other, especially with activities requiring intensive labor such as fuelwood collection.)

Spatial distributions of households may be very important for understanding the population-environment interrelationships (Bilsborrow and Okoth-Ogendo 1992; Pebley 1998). We used sketch maps (see an example in Figure 3) to identify the relative locations of dwelling units. The sketch maps were obtained from the 1996 Agricultural Census and the 2000 Population Census. These draft drawings helped the censuses avoid missing any households. We have recorded locations of approximately sixty dwelling units using GPS receivers with real-time differential corrections. Besides x- and y-coordinates, we also recorded slope, elevation, aspect, and vegetation conditions around the houses. After differential corrections, the accuracy of location data was within 1–5 m (unpublished data). As dwelling units are also clearly shown on IKONOS imagery (see an example in Figure 4), the location data from GPS measurements are being used as training and validation data for IKONOS images so that we can identify the accurate coordinates of the remaining dwelling units. In addition, we associated the names of household heads with the locations of their dwelling units on the IKONOS imagery. Thus, the spatial data were linked to the socioeconomic and demographic database using the names of the household heads as the linkage key. To help understand temporal dynamics of spatial locations of dwelling units, we also asked about the years when the houses were built and entered this information on the Microsoft ACCESS database.

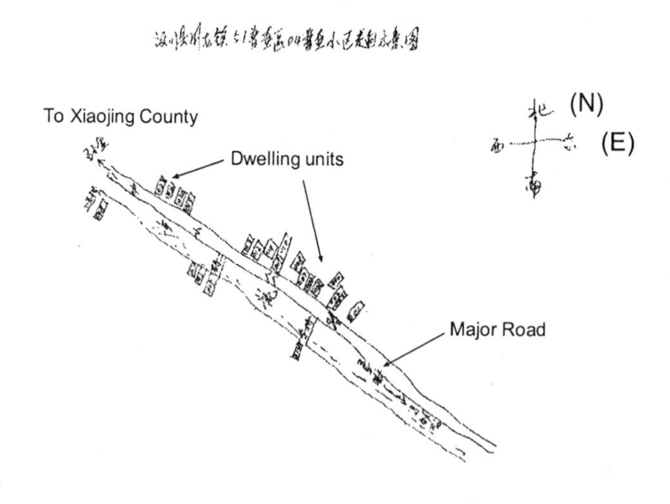

Figure 3. Sketch map showing relative locations of some dwelling units

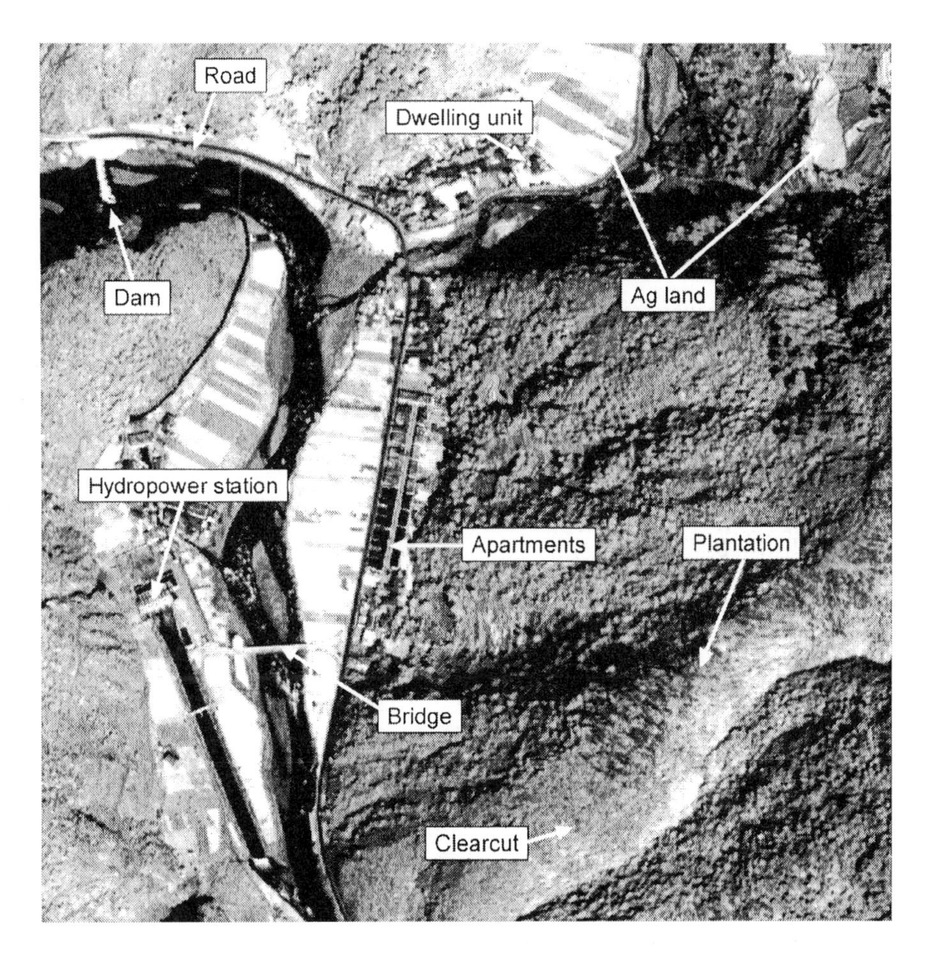

Figure 4. A subset of a panchromatic IKONOS image with a 1-m resolution

Unlike the cases in Brazil (Moran et al. this volume), Ecuador (Walsh et al. this volume), and Thailand (Rindfuss et al. this volume), the exact locations of agricultural land parcels in Wolong have not been recorded using GPS receivers, partially because the vast majority of agricultural land is not far from dwelling units, and the locations of dwelling units can be used as a proxy for locations of agricultural land parcels. Despite the rural economic reform in China implemented in the late 1970s and early 1980s, all land still belongs to the government (Xu et al. 1999). Residents in Wolong were "assigned" parcels of land for farming over a period of time (from several years to a couple of decades). In other words, farmers have the right to use the land for agricultural purposes but do not own the land and do not have the right to sell or transfer any land. The implications of this type of farmer-land relationship for land-cover change are not clear and need to be understood for future research efforts.

4.2.2 Socioeconomic Data

Micro-level factors (socioeconomic conditions at the household level) in each household include amount of annual income, sources of income, amount of expenses, aspects of expenses (e.g., schooling), labor force, land area, area of the house, crop production, number of livestock (e.g., pigs), use patterns of natural resources (e.g., fuelwood), attitudes toward, out-migration, and amount of electricity for cooking, heating, and electronic appliances (such as TVs and radios). Attitudes and perceptions were analyzed on the basis of individuals, whereas other data (e.g., amount of fuelwood consumption) were examined using households as the unit of analysis. We began micro-level socioeconomic surveys in 1996. The sample size depended on specific research objectives. For example, for household fuelwood consumption, we surveyed 220 households (An et al. n.d.[a]). In addition, household socioeconomic conditions (e.g., number of pigs, area of cropland) are available from the 1996 Agricultural Census and the "1999 Wolong Land Contract and Operation Registration" (Wolong Nature Reserve 1999) for each household. These data, including the number of land parcels, cropland area, and types of land (e.g., newly developed land, abandoned land, area returned for reforestation, and area for house building), helped us parameterize or calibrate our models (Liu et al. 1999a; An et al. 2001).

Macro-level socioeconomic factors (factors beyond the household level, such as roads, trade centers, administration buildings, bridges, dams, and schools) also influence population-environment interactions (Shivakoti et al. 1999) because their relationships with households (e.g., households' distances to major roads or trade centers) are different. For example, children in households that are far away from schools may have less chance for education. We began to measure these macro-level socioeconomic factors in 2000. The measurements include these factors' locations and timing of occurrence. We obtained the information regarding the timing of occurrence from the records of the Wolong Administration Bureau, or by interviewing local residents, officials, and other stakeholders. For example, by interviewing schoolteachers we obtained information about the years when the schools (elementary, middle, and high schools) were built. The locations of these factors were measured using GPS receivers, from remote sensing imagery (IKONOS), or from topographic maps. For example, the stream coverage was digitized from the topographic maps and the road coverage was digitized from the IKONOS imagery and verified with GPS samples. Through GIS and spatial statistics, we will be able to link macro-level data with micro-level data. Examples of results from such linkages include distances of dwelling units from roads, schools, forests, and local markets. Similarly, we can measure the distances between residential areas and areas of suitable habitat for pandas as well as the distances between fuelwood collection sites and panda habitat.

4.2.3 Behavioral Data

In 1998, we performed a set of interviews with 329 people (approximately 8 percent of the total population) regarding their activities in the previous year (i.e., 1997) to understand who had a direct impact on the land. Through these interviews we found that the activities of local residents in Wolong were quite diverse, including farming, fuelwood collection, collection of Chinese herbal medicines, road construction and maintenance, small business (e.g., small restaurants), and transportation. With regard to fuelwood collection, approximately 76 percent of the local residents who participated were 25–59 years old, while 21 percent, 2 percent, and 1 percent of the labor for fuelwood collection were 15–24, 60 or older, and 14 or younger, respectively. Furthermore, nearly all the people who collected fuelwood were males. Most women, the elderly, and children do not have a direct impact on forests through fuelwood collection, but their indirect impact must be assessed. For example, households with one or more senior residents consume more fuelwood than those without seniors because the heating season for the elderly is longer (starting early and ending late) (An et al. 2001).

The amount of fuelwood consumption has been measured at the household level (Liu et al. 1999a; An et al. 2001), but we have not yet linked the specific locations of fuelwood collection with individual households due to the enormous difficulty in pinpointing specific locations of harvesting by individual households over time. This difficulty stems from several factors. First, locations of fuelwood collection areas are often several hours (on foot) away from dwelling units owing to the exhaustion of eligible forests near dwelling units, although local residents usually collect fuelwood within administrative boundaries and at locations as close to their dwelling units as possible. Second, the locations of fuelwood collection change over time after the eligible trees in previous locations have been harvested, similar to land-use practices among pastoralists (BurnSilver et al. this volume). Third, fuelwood collection usually takes place in the winter, but our field data collection often occurs in the summer when students in the project are not in class and faculty members are not teaching. Fourth, some residents may not want others to follow them to their sites of fuelwood collection because some of their activities are illegal or against the reserve policy in terms of collection sites, extent of harvesting, species of trees harvested, and sizes of trees harvested. Because of these challenges, we have been testing an alternative approach to identifying locations of fuelwood collection. Thus far, we have interviewed several dozen households and have asked the respondents to indicate approximate locations for fuelwood collection on high-resolution aerial photographs or IKONOS maps. We then made notations on the maps, digitized them, and created location databases and maps using GIS. Currently we are in the process of using field ground-truth data to verify these sites, and we are testing the accuracy of information collected through interviews and high-resolution maps.

4.3. Systems Modeling and Simulation

A major purpose of building systems simulation models is to synthesize information from various sources (Costanza et al. 1991; Liu et al. 1999a; An et al. 2001). Thus far, we have built two major types of models: one at the household level (An et al. 2001) and the other at the reserve level (Liu et al. 1999a). The household-level model (Figure 5) integrates household demography (e.g., household size and age and schooling years of each family member), household economy (land-use activities, income/expense sources, etc.), attitudes toward issues of interest (e.g., childbearing), and fuelwood consumption. An et al. (2001) found that the amount of fuelwood consumption is a function of household size and composition, cropland area, and number of pigs raised. The demographic-socioeconomic-ecological model was verified and validated by an independent set of data. Statistical analyses showed no significant differences between the observed and predicted amounts of fuelwood consumption, indicating that the model mimicked the actual amounts of fuelwood consumption accurately.

The reserve-level model considered the collective impact of all households (Liu et al. 1999a). The model was developed using C++ (Liu 1993) with linkages to a GIS. We used computer simulations to project possible demographic and ecological consequences, given different policy scenarios (see Figure 2). For example, under different birthrates and emigration rates, as well as compositions of emigrants (e.g., younger vs. older residents) and levels of fuelwood consumption, we were able to observe future dynamics of human population size and structure as well as panda habitat over a period of fifty years. Our computer simulations (Liu et al. 1999a) indicated that even if only 22 percent of young people (17–25 years old) relocated (e.g., through going to college, finding jobs elsewhere, and marrying people outside the reserve), the human population size in the reserve would be reduced from 4,300 in 1997 to about 700 in the year 2047, and the giant panda habitats will recover and then increase by 7 percent.

Under the status quo, however, the human population size in the reserve will increase to approximately 6,000 and about 40 percent of the giant panda habitats will be further reduced by the year 2047. The results suggest that relocating young people is socially acceptable, economically efficient, and ecologically sound. Our surveys indicate that most young people are willing to relocate, especially if they can receive higher education (Liu et al. 2001b). Furthermore, even though parents and grandparents generally do not want to relocate themselves, they are highly supportive of the relocation of the youth and feel quite proud if these young people can receive higher education (Liu et al. 1999a; Liu et al. 1999b). Furthermore, the costs for relocating young people would be lower, because relocating one young person is equivalent to moving several seniors; young people can have children, while seniors do not have childbearing capacity. Because young people and adults are the direct force behind land-use and land-cover change, reduction in the number of young people will reduce the number of future adults and thus minimize human impacts on forests and panda habitat in Wolong.

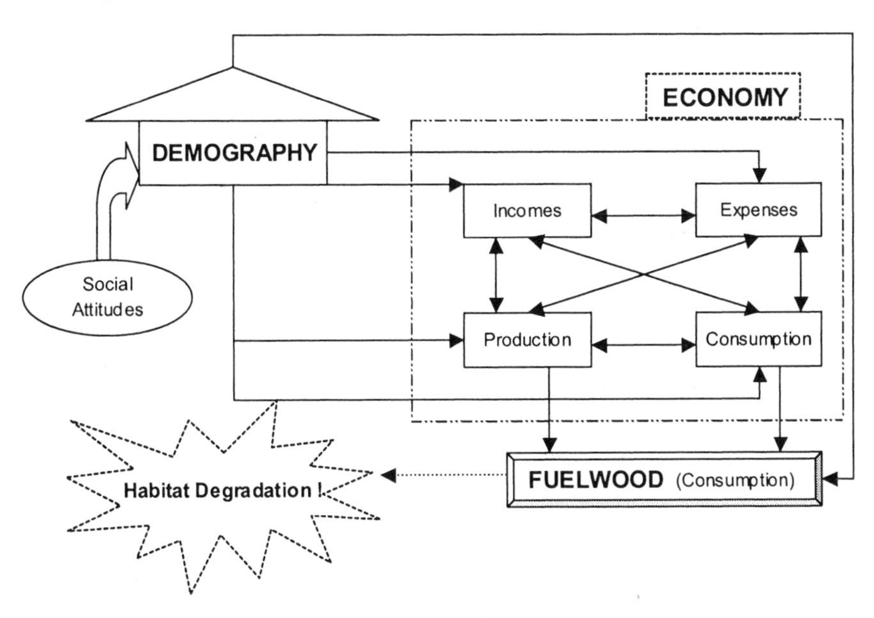

Figure 5. Conceptual structure of a household-level model (An et al. 2001)

5. CONCLUSIONS AND DISCUSSION

Linking ecological, socioeconomic, demographic, and behavioral factors was essential for us to understand the patterns and mechanisms of changes in land cover and panda habitat in Wolong Nature Reserve. The conceptual framework (Figure 2) provided a useful guidance in our research endeavor. The complementary expertise in different disciplines made the linkages technically possible. To a large extent, effective linkages among various factors resulted from close interactions among collaborators in different subject areas. Clear and explicit communication with each other was the key. It took a while for us to learn the essentials (methods, concepts, and theories) of different disciplines. In this learning process, we have also recognized the differences in meaning of the same terms and in standards among disciplines. For example, *population* almost always means *human population* to social scientists, but it means *animal population* to ecologists. Ecologists perceived that an accuracy of 80 percent remote sensing imagery classification was quite low, although this level of accuracy is more than acceptable to the remote sensing community.

Large and topographically complex study areas like Wolong provide both advantages and challenges for technical linkages among various components. On the one hand, our large area offers a heterogeneous environment and allows us to observe a variety of socioeconomic-demographic-ecological conditions. On the other hand, the heterogeneity requires relatively large sample sizes. The scattered distribution of households across high mountains and deep valleys made our surveys very time consuming and physically exhausting. Vehicles could transport researchers to the foothills of the mountains along the only main road. To reach most households and collect most ground-truth data, however, walking or even crawling in the topographically challenging environment—with its various insects, leeches, ticks, and snakes—is the only feasible way.

For interdisciplinary projects like ours, it is probably not remarkable that there are some surprises. Most importantly, the amount of time required to collect some data was more than what we had originally anticipated. In addition, the costs were somewhat underestimated. At present, we are using IKONOS data (1-m resolution) to refine our analyses of land cover for part of the reserve. Ideally we would like to use IKONOS data for the entire reserve analysis if the high-resolution data are accessible without financial constraints. Of course, greater financial support would also enable us to have larger sample sizes for some data. There were, however, some good surprises as well. For example, the data collected and maintained by the reserve administration were more comprehensive than we had hoped, and we were able to obtain and use these data. The sketch maps from the 1996 Agricultural Census and from the 2000 Population Census helped us to identify relative locations of households. These census data also aided us in choosing appropriate strata for our stratified sampling designs. Some of our results were unexpected as well. One major unforeseen outcome was that high-quality panda habitat in the reserve had been lost and had become fragmented more rapidly after the reserve was established (Liu et al. 2001a).

There are many challenges and opportunities in linking ecological, socioeconomic, demographic, and behavioral factors. Advanced technologies such as remote sensing, GIS, and systems modeling are useful tools to make the linkages technically possible, but amounts of time, financial support, and coordination among researchers of different expertise may increase with the extent of linkages. A critical need is to change the culture of academic institutions so that researchers doing interdisciplinary research can be adequately rewarded. Through our study and other studies presented in this book and elsewhere (e.g., ten examples listed in Liu 2001), we are optimistic that these challenges can be met, thanks to several unprecedented opportunities. First, funds are increasingly available from many funding sources. In the United States, these funding sources include the Biocomplexity Initiative at the National Science Foundation, the Population and Environment Program at the National Institute of Child Health and Human Development, and the Land-Use and Land-Cover Change Program at the National Aeronautics and Space Administration. Second, more researchers are interested in this rapidly growing field. Third, institutional culture is changing. More interdisciplinary research efforts are encouraged and supported. Fourth, policy-making processes and natural resource management require output from interdisciplinary research such as ours. It is our hope that the approaches linking various factors presented in this chapter can be applicable (with some modifications, of course) to other places. We believe that the urgent need for and growing interest in such comprehensive and integrated studies will bring to fruition the theories, methods, and applications for linking household and community surveys with remote sensing and GIS.

ACKNOWLEDGEMENTS

We thank the editors (Jefferson Fox, Vinod Mishra, Ronald Rindfuss, and Stephen Walsh) for their helpful comments on an early draft. We also appreciate the logistical support of the Wolong Nature Reserve, especially the assistance from Jinyan Huang, Yingchun Tan, Jian Yang, Hemin Zhang, and Shiqiang Zhou. The remarkable participation and cooperation of local residents were critical to the successful completion of surveys and interviews. This chapter was written while the senior author (Liu) was on sabbatical at the Center for Conservation Biology (CCB) at Stanford University. The hospitality of CCB (especially Carol Boggs, Gretchen Daily, Anne Ehrlich, and Paul Ehrlich) is greatly appreciated. We gratefully acknowledge the financial support from the National Science Foundation (NSF), National Institutes of Health (National Institute of Child Health and Human Development, R01 HD39789), the National Aeronautics and Space

Administration (NASA), the American Association for Advancement of Sciences / John D. and Catherine T. MacArthur Foundation, the St. Louis Zoo, the National Natural Science Foundation of China, the Ministry of Science and Technology of China (G2000046807), and China Bridges International.

REFERENCES

An, L., J. Liu, Z. Ouyang, M. Linderman, S. Zhou, and H. Zhang. 2001. "Simulating Demographic and Socioeconomic Processes on Household Level and Implications for Giant Panda Habitats." *Ecological Modelling* 140: 31–50.

An, L., F. Lupi, J. Liu, M. Linderman, and J. Huang. n.d.[a] (forthcoming) "Modeling the Choice to Switch from Fuelwood to Electricity: Implications for Giant Panda Habitat Conservation." *Ecological Economics* (in press).

An, L, A. Mertig, and J. Liu. n.d.[b] (forthcoming) "Adolescents' Leaving Parental Home in Wolong Nature Reserve (China): Psychosocial Correlates and Implications for Panda Conservation." In review.

Bilsborrow, R., and H. Okoth-Ogendo. 1992. "Population-Driven Changes in Land Use in Developing Countries." *Ambio* 21: 37–45.

China's Ministry of Forestry and WWF. 1989. *Conservation and Management Plan for Giant Pandas and Their Habitat.* Beijing: Ministry of Forestry.

Costanza, R., H. Daly, and J. Bartholomew. 1991. "Goals, Agenda, and Policy Recommendations for Ecological Economics." In R. Costanza, ed., *Ecological Economics: The Science and Management of Sustainability* (New York: Colombia University Press), 1–20.

Dillman, D. 1978. *Mail and Telephone Surveys: The Total Design Method.* New York: John Wiley & Sons.

Dompka, V. 1996. *Human Population, Biodiversity and Protected Areas: Science and Policy Issues.* Washington, D.C.: American Association for the Advancement of Science.

Ehrlich, P. 1988. "The Loss of Diversity: Causes and Consequences." In E. O. Wilson, ed., *Biodiversity* (Washington, D.C.: National Academy Press), 21–27.

Giant Panda Expedition. 1974. "A Survey of the Giant Panda *(Ailuropoda melanoleuca)* in the Wolong Natural Reserve, Pingwu, Northern Szechuan, China." *Acta Zoologica Sinica* 20: 162–173.

He, N., C. Liang, and X. Yin. 1996. "Sustainable Community Development in Wolong Nature Reserve." *Ecological Economy* 1: 15–23.

Hu, J., Q. Deng, Z. Yu, S. Zhou, and Z. Tian. 1980. "Biological Studies of Giant Panda, Golden Monkey, and Some Other Rare and Prized Animals." *Nanchong Teacher's College Journal* 2: 1–39.

IUCN. 1994. *United Nations List of National Parks and Protected Areas.* Cambridge: World Conservation Monitoring Centre and IUCN Commission on National Parks and Protected Areas.

Johnson, K., G. Schaller, and J. Hu. 1988. "Responses of Giant Pandas to a Bamboo Die-Off." *National Geographic Research* 4: 161–77.

Lillesand, T., and R. Kiefer. 1994. *Remote Sensing and Image Interpretation.* New York: John Wiley.

Linderman, M., J. Liu, J. Qi, Z. Ouyang, L. An, and J. Yang. n.d. (forthcoming) "Mapping the Spatial Distribution of Bamboo in a Giant Panda Reserve: Using Remote Sensing Data to Classify Understory Vegetation Cover." In review.

Liu, J. 1993. "ECOLECON: A Spatially Explicit Model for ECOLogical-ECONomics of Species Conservation in Complex Forest Landscapes." *Ecological Modelling* 70: 63–87.

———. 2001. "Integrating Ecology with Human Demography, Behavior, and Socioeconomics: Needs and Approaches." *Ecological Modelling* 140: 1–8.

Liu, J., M. Linderman, Z. Ouyang, L. An, J. Yang, and H. Zhang. 2001a. "Ecological Degradation in Protected Areas: The Case of Wolong Nature Reserve for Giant Pandas." *Science* 292: 98–101.

Liu, J., M. Linderman, Z. Ouyang, and L. An. 2001b. "The Pandas' Habitat at Wolong Nature Reserve: Response." *Science* 293: 603–604.

Liu, J., Z. Ouyang, W. Taylor, R. Groop, Y. Tan, and H. Zhang. 1999a. "A Framework for Evaluating Effects of Human Factors on Wildlife Habitat: The Case of the Giant Pandas." *Conservation Biology* 13(6): 1,360–1,370.

Liu, J., Z. Ouyang, Y. Tan, J. Yang, and S. Zhou. 1999b. "Changes in Human Population Structure and Implications for Biodiversity Conservation." *Population and Environment* 21: 45–58.

Moran, E., E. Brondizio, P. Mausel, and Y. Wu. 1994. "Integrating Amazonian Vegetation, Land-Use, and Satellite Data." *BioScience* 44: 329–338.

Ouyang, Z., J. Liu, and H. Zhang. 2000. "Community Structure of the Giant Panda Habitat in Wolong Nature Reserve." *Acta Ecologia Sinica* 20: 458–462.

Pebley, A. 1998. "Demography and the Environment." *Demography* 35: 377–389.

Reid, D., and J. Hu. 1991. "Giant Panda Selection between Bashania Fangiana Bamboo Habitats in Wolong Reserve, Sichuan, China." *Journal of Applied Ecology* 28: 228–243.

Schaller, G., J. Hu, W. Pan, and J. Zhu. 1985. *The Giant Pandas of Wolong*. Chicago: University of Chicago Press.

Shivakoti, G., W. Axinn, P. Bhandari, and N. Chhetri. 1999. "The Impact of Community Context on Land Use in an Agricultural Society." *Population and Environment* 20: 191–213.

Tan, Y., Z. Ouyang, and H. Zhang. 1995. "Spatial Characteristics of Biodiversity in Wolong Nature Reserve." *China's Biosphere Reserve* 3: 19–24.

Turner, B. L., II, W. Meyer, and D. Skole. 1994. "Global Land-Use/Land-Cover Change: Towards an Integrated Study." *Ambio* 23: 91–95.

Vitousek, P., H. Mooney, J. Lubchenco, and J. Melillo. 1997. "Human Domination of Earth's Ecosystems." *Science* 277: 494–499.

Wenchuan County. 1983. *Statistics of Population Census of 1982*. Sichuan Province, China.

Wolong Nature Reserve. 1996. *Agricultural Census Data*. Wenchuan County, Sichuan Province, China.

———. 1999. *1999 Wolong Land Contract and Operation Registration*. Wenchuan County, Sichuan Province, China.

———. 2000. *Population Census Data*. Wenchuan County, Sichuan Province, China.

Wilson, E. O. 1988. *Biodiversity*. Washington, D.C.: National Academy of Science Press.

Xu, J., J. Fox, X. Lu , N. Podger, S. Leisz and X. Ai. 1999. "Effects of Swidden Cultivation, State Policies, and Customary Institutions on Land Cover in a Hani Village, Yunnan, China." *Mountain Research and Development* 19: 123–132.

Zhang, H., D. Li, R. Wei, C. Tang, and J. Tu. 1997. "Advances in Conservation and Studies on Reproductivity of Giant Pandas in Wolong." *Sichuan Journal of Zoology* 16: 31–33.

Chapter 10

HABITATS, HIERARCHICAL SCALES, AND NONLINEARITIES

An Ecological Perspective on Linking Household and Remotely Sensed Data on Land-Cover/Use Change

George P. Malanson
Department of Geography, University of Iowa
george-malanson@uiowa.edu

Abstract Ecologists are concerned with the extent, timing, and pattern of land-use change. While most of the attention has been focused on remnants as islands, increasing attention is being paid to the surrounding matrix and to connectivity through the matrix. Ecological theory indicates that more attention is needed on the spatial pattern of land-use change, especially the degree to which the change itself depends on spatial pattern. This feedback can lead to nonlinearities in the timing and pattern of change—and so to ecological surprises. Ecological theory also indicates that a hierarchy of building-block processes, the scale of interest, and a broader scale of constraints may be useful—and even necessary—to capture scale-dependent relations between processes and patterns. Ecological and land-use modelers have converged on dynamic spatial simulations as an approach, and both are making advances with agent-based models, which seem particularly appropriate for capturing household-based activity. It is the ability of these models to deal with the feedbacks and nonlinearities seen in ecological systems that gives them their utility.

Keywords habitat destruction, remnants, scale, hierarchy, edge, matrix, simulation

1. INTRODUCTION

An ecological perspective brings two concerns to this volume on linking household and remotely sensed data. Both concerns are broader than the theme of this volume in that they apply to land-cover/land-use change (LCLUC) studies in general, but the linkages reported here might be the best way to address them. The first and most compelling concern is about how studies of LCLUC can contribute to the study of ecology and especially to its application to the conservation and preservation of species, biodiversity, and ecological functions. The second concern, conversely, arises from my perspective as a modeler and is about the ways in which ecology can contribute to studies of LCLUC. What is it about LCLUC that is important from an ecological perspective? While LCLUC has many implications (NRC 1999, 2000), the most important point is the amount of habitat destruction. These broad concerns are emphasized here because the specific concern of linking household and remotely sensed data does not have a good ecological analogue. This limitation also means that the primary research papers cannot be fully critiqued on whether they address ecological concerns, which the projects may do, when they are aimed at specific methodological issues.

2. HABITAT

Habitat destruction is the most important human impact on other species. It is the greatest threat to ecosystem integrity and biodiversity. For example, Wilcove et al. (1998) report that habitat destruction is a contributing factor for 85 percent of imperiled species in the United States, compared to alien species at 49 percent, pollution at 24 percent, overexploitation at 17 percent, and disease at 3 percent. The greater changes are not in the United States (Houghton 1994). Habitat destruction is widespread. We regularly hear messages such as "an area of tropical rainforest the size of Connecticut is lost annually." Some quantitative analyses provide a yardstick for a gross assessment and comparison. Hannah et al. (1994) categorized 27 percent of the world's terrestrial habitat (excluding rock, ice, and barren land) as "undisturbed," but others believe the term has no meaning (e.g., Meyer and Turner 1992). The extremes seen by region and by nation are masked by a global assessment (MacKinnon and MacKinnon 1986). Habitat destruction is also unequal in timing. The most intensive periods of habitat destruction occurred in previous centuries in portions of Europe, East Asia, South Asia, and North America, while South America and parts of Africa and Southeast Asia are now experiencing their most intensive habitat destruction (Skole and Tucker 1993; Laurance et al. 2000). In some regions such as southern Europe, abandoned agricultural land leads to increases in some forms of habitat (Lavorel et al. 1998). Whether one is trying to quantify biodiversity and its loss, minimize destruction, design nature reserves, or restore habitat, the temporal pattern of change is important.

2.1 Remnants

Ecological research attention has focused on the species in the remnants of habitat because the area of habitat destroyed loses most of its plants and animals (e.g., Bierregaard et al. 1992; Laurance et al. 1998a; Laurance et al. 1998b). Remnants are often patches or islands of habitat surrounded by former habitat. Understanding the effects of habitat destruction on remnants is critical to assessing impacts on the diversity of species and community composition. Ambitious, ongoing, long-term experiments have created habitat islands and are tracking their biology. For example, Laurance et al. (1998a, 1998b) have found effects on tree recruitment and mortality, with faster change in general in fragments. They found most effects close to the edges, but did not compare spatial patterns. Remnants are subject to island effects, edge effects, invasions, trophic effects, matrix effects, and their interaction (see Figure 1).

Figure 1. Remnants of tropical forest are surrounded by the matrix of new land use

Island effects are twofold: populations are isolated and their sizes are reduced. The theory of island biogeography (MacArthur and Wilson 1967) provides much of what we want to know about island effects. Isolation leads to lower immigration rates and small population sizes lead to increased extinction rates; together they lead to lower species diversity. Furthermore, a remnant cannot support all species that it held while part of a larger area of habitat, so diversity will decrease through time toward a new equilibrium between local immigration and extinction (MacArthur 1972). A consequence is a time lag between the creation of a remnant by habitat destruction and the local extinction of species (sometimes called an *extinction debt,* indicating that some extant species will inevitably go extinct; Tilman et al. 1994). A time lag can be a problem when attempting to assess the cumulative impact of habitat destruction (see Risser 1988). Another impact is altered disturbance regimes (Malanson 1992); remnants often have different frequencies and intensities of disturbance because of their size and isolation.

Attempts to apply the theory of island biogeography to what would be remnant nature reserves in Amazonia led to the current textbook debate on optimizing the diversity of species in limited remnant areas (Simberloff and Abele 1976). The theory points in favor of single large reserves because it assumes a homogeneous distribution of environment and species.

Remnants are subject to edge effects and the effects of edges on interiors (Murcia 1995). The amount of former habitat lost is thus greater than that simply converted (Chen et al. 1992). The characteristics of edges—for example, their width and difference from the surrounding land cover—change the ecology of the remnant (e.g., Kupfer et al. 1997; Williams-Linera et al. 1998; Stevens and Husband 1998). Still, edge effects are notoriously species-specific (e.g., Schlaepfer and Gavin 2001; Woodward et al. 2001).

Invasions by exotic species may also be enhanced by habitat destruction (Medley 1997). Invasive plant species produce many seeds, they thrive in higher light conditions, and they are widely dispersed; all of these traits make them more likely to establish and thrive on edges than in interiors of forest. Trophic effects can be found when species that require a large territory, such as large predators, cannot persist after their habitat has been fragmented. With their elimination, the species on which they dine (animal or plant) will be affected, and the competitors of the prey are affected in turn. Habitat destruction can make species in remnants more susceptible to the adverse effects of other impacts. Notably, species in remnants face a double bind when rapid global climate change lessens the habitability of a remnant that would force a shift in geographic range—but the shift is inhibited by the reduction in accessible habitat (e.g., Hanson et al. 1989). Scientists and policy makers studying impacts face a double whammy (mental paralysis) when the direct effects of climate change and habitat destruction cannot be differentiated. Turner (1996) concluded that the relative importance of the impacts mentioned above is unknown.

All of the research programs discussed in this volume are aimed directly or strongly at characterizing and explaining the amount and rate of habitat destruction. The potential advances to be seen in combining household survey and satellite remote sensing data are in better projections of the rate of destruction given information on the endogenous and exogenous factors that affect the decisions of the people and the spatially explicit values of land for given uses. Lambin (this volume) indicates that the benefits may be marginal and, in the economic sense, will be increasingly less for more effort. As he notes, information at a coarser scale may capture the dynamics of LCLUC—and knowing how much change and its rate is knowledge at the regional scale (e.g., Kok and Veldkamp 2001). This may be true for the major question on the amount of destruction. For other spatial factors of ecological interest, however, the details of change seen in the household data may be needed. The importance of household decisions probably varies with the importance of communal influence on decisions. Communal influence may be stronger where people live in villages than where they live on their homesteads (tautologically?). The importance of the linkage issue depends on the broader social context.

2.2 Beyond Area: Configurations of Habitat Destruction

The theory of island biogeography has limitations, however. The spatial pattern of habitat change is also important. Metapopulation models focused on the occupancy of patches by single species show that patterns alter the outcomes (e.g., Bascompte and Sole 1996; Holyoak 2000). For example, in a spatially explicit metapopulation model using a gridded landscape, Bevers and Flather (1999) found that threshold effects depended on complicated interactions between spatial pattern and biological parameters, but they were able to conclude that patch shape—elongated vs. blocky—offers colonization and persistence advantages, respectively.

Time lags and extinction rates also depend on the spatial pattern of the habitat—for example, spatial clumping of habitat improves the advantage to better competitors in some models (Tilman et al. 1997) and can alter the order of extinction (Klausmeier 1998; Neuhauser 1998). In my own work (Malanson n.d.[a]), looking for thresholds in species responses to habitat destruction (and finding none), I have found differences with changing patterns of destruction. It has been argued, however, that pattern matters little. Trzcinski et al. (1999) report that spatial pattern had little explanatory power for bird distributions once area was accounted for. Even proponents of this view, however, do not exclude the importance of pattern and connectivity altogether (Fahrig 2001).

The clearest examples of pattern are corridors. Often corridors are considered as separate from some more general pattern, but they may be the essence of a nonhuman's-eye-view of what pattern means. Whether or not corridors are a blind alley—and a sink of resources—for conservation programs is debated (Hobbs 1992; Simberloff et al. 1992; Beier and Noss 1998; Haddad et al. 2000). It would seem, however, that if the definitions of corridors are scaled to the species in question, then they could be important (e.g., Bolger et al. 2001; Mech and Hallett 2001). In some cases linear remnants may have multiple functions: as habitats and as corridors (e.g., Malanson 1993; de Lima and Gascon 1999).

The details of spatial pattern are becoming evident in the studies that link pixels to households. Configuration has been included in earlier work (e.g., Gilruth et al. 1995; Mertens and Lambin 2000; Messina and Walsh 2001), and the potential to expand the data for such models has been discussed (Irwin and Geoghegan 2001). It would seem that it is at this level of spatial detail from the perspective of remnant plant and animal populations that the spatial detail of the processes of LCLUC depend on household decisions. This level is where we may be able to usefully address such questions as: which habitat patches are likely to be perforated, which dissected, which fragmented into pieces, which shrink as edges are converted, which are lost—and whether the remnants will be connected, where corridors will be, and where will the gaps be narrow vs. wide. Among the chapters here, Walsh et al. (this volume) address aspects of configuration in developing a cellular automaton as part of their modeling effort. Fox et al. (this volume) demonstrate that configuration (probably not "fragmentation") is important to watershed function. The dependence of LCLUC on configuration—and thus continuing change in the configuration—needs attention. The scale questions addressed in this volume will need to be taken into account when quantifying configuration.

2.2.1 Spatial Metrics

One approach to including configuration in LCLUC studies is to quantify it using spatial metrics. Landscape ecologists have developed a host of metrics, many of which are either irrelevant or are highly correlated with others (cf. McGarigal and Marks 1995). A careful selection of metrics, however, can provide data that can then be used to describe LCLUC results in ways that communicate to ecologists and can be used as parameters in additional modeling (e.g., Messina and Walsh 2001; Lee et al. 2001). Turner and Geoghegan (this volume) and Walsh et al. (this volume) specifically mention such metrics—which, again, are scale dependent.

2.3 Beyond Remnants

Fox et al. (2000) explain how landscapes with shifting cultivation should be considered as a range of conditions varying from deforestation to forest degradation. Wiens (1997) suggests that the basic species-area relationship of island biogeography will be predictably modified if we take into account the degree to which the area around remnant habitat—the matrix—is habitable or useable (Figure 2). Gascon and Lovejoy (1998) specifically identify edge effects and the characteristics of the matrix as factors that the theory of island biogeography does not address.

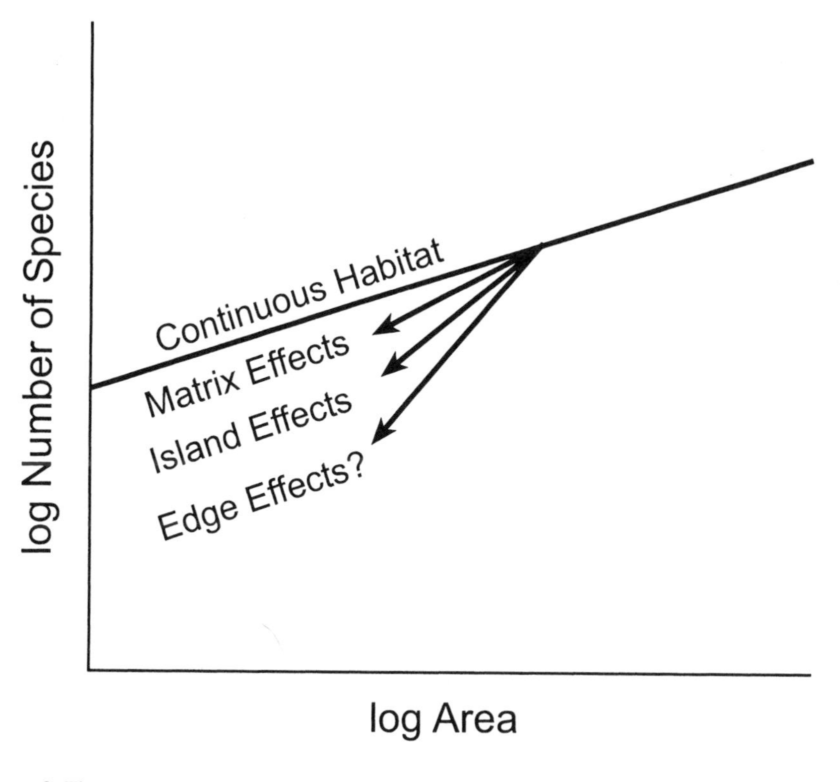

Figure 2. The species-area relationship of island biogeography (after Wiens 1997).

Ecologists have moved beyond two-phase landscapes in some spatially explicit models. A more continuous representation of habitat quality, while retaining discrete cells, may be more realistic for some species. Among other concerns, some species require resources from more than one type of habitat, which may be significantly separated in space (e.g., Pope et al. 2000). Moreover, when considering multiple species, the representation of habitat and habitat quality may differ (Malanson n.d.[b]). This point has been convincingly argued in the past two years (Renjifo 2001; Ricketts 2001). One of the observations that most strongly points toward considering matrix effects is that these surrounding areas are not constant and may, through ecological succession, increasingly resemble the remnant (e.g., Moran and Brondizio 1998).

2.3.1 Connectivity through the Matrix

The most important aspect of the matrix from the viewpoint of ecologists concerned with biodiversity, especially in remnants, is how well species can move among remnants through the matrix. While most attention has been paid to movement through habitat in two-phase landscapes (e.g., Malanson and Cramer 1999), some attempts have been made to use more classes (e.g., Wiegand et al. 1998) or to represent a continuous surface with discrete cells (With 1994; Malanson n.d.[b]). Tischendorf and Fahrig (2000a, b) note that this type of connectivity must be carefully defined because some indices of connectivity increase with fragmentation. For the physical processes of water and sediment transport, Fox et al. (this volume) attempt to characterize landscapes according to their hydrological connectivity. The importance of doing so is immediately obvious in their results, even though their model, by not handling channelization well, may overestimate the effects of small areas (another scale issue). Links through landscapes for people as well as wild and domestic animals, the importance of which is noted by Liu et al. (this volume), need to be quantified and examined.

2.3.2 Lower Quality in Habitat

If we recognize that areas of nonhabitat may be useable—that is, have a value to a species >1—then we must recognize that remnants of habitat may have a value <1, and this difference goes beyond the immediate effects of small size and isolation. It goes directly to the question of land use in addition to land cover. A quiet home in the dark is different from one with a fence, security light, and dog (Theobald et al. 1997). At this level of detail, the common sources of data (i.e., satellite imagery) become only a starting point, and theory and case studies need to be developed and integrated into ecological modeling (e.g., Riebsame et al. 1994).

Theoretical models indicate that the degree to which habitat quality is lessened in remnants, changing 1 to a fraction, will further reduce their value. While cumulative impacts are assessed primarily in the context of increasing the amount of a given impact in an area, such as increasing habitat destruction and resulting fragmentation, multiple impacts are also cumulative. The combined effects of habitat destruction and some types of extractive use of resources may be a double bind for species: habitat is reduced, and the quality of the remnant habitat is lessened. Simulation results indicate that if such combinations lower the reproductive vigor in remnant habitat, the effect will be greatest on those species characterized as superior competitors.

In studies of LCLUC, Lambin (1999) has assessed the remote sensing needs to detect forest degradation, not simple deforestation, and Harcourt et al. (2001) have shown that biota reserves in areas of high human population density suffer higher mortality rates than their size alone would indicate. Most notable in many regions is that the fire frequency in remnants is altered. In the tropics, fire frequency increases (Jones et al. 1995), but it has multiple effects (Eva and Lambin 2000). Clearly, the use of fire is a characteristic that changes the quality of habitat within remnants, its use is related to accessibility and thus to edge, shape, and configuration, and it is a process that may be best studied at the household level.

While most of the work reported in this volume has not assessed the quality of the landscape mosaic from the perspective of plants or animals, focusing instead on changes in land cover and use, it provides a basis for moving in this direction. Turner and Geoghegan (this volume) mention the nature of the matrix and its potential effect on forests. In terms of ecological specificity and assessment of the impact of land use on the quality of habitat within remnants, Liu et al. (this volume) most closely track the use of forest resources in the reserve and its effect on panda habitat (also see An et al. 2001). Their aspects of land-use change following habitat destruction are best tied to household decisions. Finer phenomenological resolution will need to be balanced against classification accuracy.

3. DEALING WITH DYNAMICS

3.1 Scale and Hierarchy

The degree to which the details of spatial pattern discussed above will be ecologically relevant depends on scale (Levin 1992). Rindfuss et al. (this volume) outline the basic principles of an ecological view of scale in a hierarchy. The details of spatial scale that can be seen, for example, in the figures shown by Moran et al. (this volume), Turner and Geoghegan (this volume), and Walsh et al. (this volume) may be relevant to ecological activity if the plants and animals perceive and use the environment at a fine grain; that is, if that level of spatial detail (tautologically) makes a difference (cf. Vos et al. 2001). Whether the scale is relevant must be left to ecologists more familiar with the particular species. We know, for example, that some keystone tropical fig trees are pollinated over great distances (i.e., ca. 10 km; Nason et al. 1998), and Dale et al. (1994a) characterized nine groups of animals in the Brazilian Amazon by their area needs and ability to cross nonhabitat. At first glance, the coarser spatial resolution and scale might seem to miss ecologically relevant detail, but—for the major mammals of their study area, at least—it is probably a good choice.

Some of the work in LCLUC highlights the fact that the nature of the hierarchy of constraint and process may be different than that identified in ecology. For example, Skole et al. (1994) view Amazonian deforestation as driven by processes at a larger scale and constrained by pattern at the local scale. This would seem to be the approach to scale used by Fox et al. (1995). It seems clear that exogenous forces are drivers in LCLUC, and, while these forces could be reconceptualized as changing constraints, as ecologists have done for climatic change (cf. Ellis and Galvin 1994, it may be more useful to restructure the basic approach to a scale hierarchy.

A more formal approach to scale is needed in LCLUC research. Casting the problem in terms of building-block processes and constraints is a starting point that can be borrowed from ecology. How this formalism plays out in LCLUC, however, may be different, and the structure should be adapted and changed as needed.

3.2 Nonlinearity

What was most surprising to me, given the funding opportunities of the NSF biocomplexity initiative, was the lack of explicit consideration of nonlinearity in the chapters included in this volume. Nonlinearity is the core of complexity (Malanson 1999). Ecologists anticipate that ecological responses to the amount of habitat destruction will be highly nonlinear—that is, where incremental change in the landscape will have disproportionately large effects on the biota. This anticipation is based on two areas of theoretical modeling: percolation (Stauffer 1985) and metapopulation (Hanski and Gilpin 1997) research.

Percolation theory is applied to the possibility of organisms and species moving across landscapes (e.g., Malanson and Cramer 1999). Percolation theory models show that habitat destruction has little effect on movement until a threshold or critical point is reached when a gap wide enough to interrupt dispersal is created. In random landscapes—that is, where the spatial patterns of the habitat cells/pathways are random—the threshold is found at ca. 60 percent habitat. Aspects of the geometry of the habitat also change nonlinearly in the vicinity of this threshold; differences are seen in the number of clusters, the size and fractal dimension of the largest cluster, and the mean edge of all clusters (Gardner et al. 1987). Where Mertens and Lambin (2000) found that being near forest edges was a good predictor of deforestation, either the nature of this relation or the process of deforestation would be nonlinear.

Metapopulation models are applied to the number of patches of habitat that will be occupied by a species given possible migration among them. The models show that as the number of patches decreases, extinction rates in them changes only slightly until a threshold is crossed, whereupon extinction rates increase suddenly (Hanski et al. 1996).

Such thresholds or nonlinearities appearing as habitat is decreased depends on the spatial patterns of the landscape and the population dynamics of the species. Modifying the spatial pattern can reduce the abruptness of the percolation threshold and allow dispersal at lower levels. Similarly, spatial pattern lowers the amount of habitat that can support most species (but see the counter argument by Trzcinski et al. 1999). Thus changes in pattern, with reference to those pattern questions raised above under "beyond area," can also produce surprises. Thresholds—for example, a sudden loss of diversity as habitat is decreased—may not become evident until after they are crossed, making recovery through effective restoration difficult.

The population dynamics of these species, most notably fecundity, will also alter these relations. Population dynamics are likely to be affected negatively by land use that degrades habitat quality, decreasing fecundity and increasing mortality. With and King (1999) show that the two forces can interact, where species with the lowest reproductive potential can be more strongly affected negatively by some spatial patterns.

Ecologists concerned with nonlinearities or thresholds are thus interested in the amount of habitat destruction the linkages of process and pattern of which may, as noted above, be captured by a coarser scale and also the spatial pattern of remnants and their habitat quality, the linkages of process and pattern of which again are likely to be captured at the household level.

Some chapters in this volume indicate an interest in and a basis for further consideration of nonlinearity. Moran et al. (this volume) explicitly address historical contingency by examining cohorts of land users. Rindfuss et al. (this volume), while not identifying spatial processes in feedbacks, do recognize that resulting spatial autocorrelation would be important. Walsh et al. (this volume) also discuss spatially autocorrelated behavior and go on to specifically discuss the link between feedback, nonlinearity, and criticality.

Linking household-level survey and remotely sensed data at high resolution can be important to capturing nonlinearity relevant to ecology because these data will be central to defining the critical points in terms of the spatial pattern. Widely spaced observations along a gradient cannot capture criticality.

3.3 Modeling

As an ecological modeler, I had hoped to contribute some ideas on how household and remote sensing data might be used in a modeling framework. Two areas of ecological work are most relevant: individual-based modeling of species on landscapes and inclusion of spatial feedbacks. Individual-based models reveal that dynamics and resulting patterns differ when individual organisms are differentiated from patches or populations of organisms, often because their spatial interaction differs (Huston et al. 1988). Ecosystem function can be largely determined by feedbacks that depend on relative location (Malanson 1999). Individual-based modeling supports the move toward disaggregating data to the household level, and spatial feedbacks will be found in the analysis of high-resolution determination of land use, probably at the scale at which households make decisions.

Lambin (1994) and Lambin et al. (2000) anticipated much of what I would say. Lambin (1994) realized that spatially explicit dynamic landscape models, which include spatial configuration, are able to address our concerns about hierarchy and scale and about spatial and temporal lags. He evaluated models in terms of their ability to analyze the *why, when,* and *where* of deforestation. The *where* should be teased apart to distinguish between *how much* (e.g., Stephenne and Lambin 2001) and *in what pattern.* Veldkamp and Lambin (2001) presented this distinction another way, distinguishing *how much* as a rate question (which must include time) and *where* as location, which could mean *pattern.* They and Lambin et al. (2000) noted the need to address spatial scale and nonlinearity in the future.

The direction that Lambin (1994) pointed for linking models in dynamic spatial simulations (DSS) has been taken up slowly. We can see elements of it in some later work. The key features are incorporating a sequence of LCLUC data in either Markovian or similar form (e.g., Brown et al. 2000) to create the rules for cellular automata (CA). Important in this regard is the analysis determining how spatial configurations are to be part of the rules or outputs (cf. Messina and Walsh 2001).

Lambin (1994) judges that it is necessary to understand the *why* of deforestation a priori before proceeding with DSS. While it is true that the drivers must be specified, it can also be part of the modeling process to eliminate drivers in successive validations. Irwin and Geoghegan (2001) argue that economic drivers can be incorporated more directly in DSS.

Alternatively, Mertens and Lambin (2000) use a logistic multiple regression model and Geoghegan et al. (2001) use logit models to answer the *why* question. The analysis contains spatially explicit detail because the conditions of cells in a neighborhood around each cell are examined to create an independent variable. This answer to the *why* question can be put into the form of a rule in a CA.

DSS can then answer the *when* question. Lambin (1994) notes that this answer depends on assumptions in the parameters. I think this aspect of the problem is one where household data can contribute effectively.

DSS is most appropriate to answer the *how much* and *in what pattern* questions. The term *spatially explicit,* however, covers a wide range of what is and what is not explicit (Malanson 1996). All of the work presented in this volume would fit under this rubric, at least in its broader and less restrictive connotations. By using DSS, the details of *where*—as in the spatial configurations to which plants and animals respond—and which should be indicated by spatial metrics of landscape ecology (e.g., Dale et al. 1994b) can be simulated. Here, household data is important where it links behavior to the spatial pattern—that is, to the pixels. By linking the behavior of households to specific pieces of land, a more accurate understanding and projection of the spatial configuration of LCLUC is possible.

Manson (2000) presented the most tantalizing directions for DSS in which household data would be critical to the model's ability to answer the *when, how much,* and *in what pattern* questions at a fine resolution. He suggested that agent-based modeling be used with cellular automata of the environment. Agent-based models have been used in limited ecological and land-use studies (e.g., Savage et al. 2000; Bousquet et al. 2001). My offering is to indicate that households are most likely to be relevant if one considers them as agents determining the rate and the specific parcels that change, not if one is only trying to improve a probability surface. The crux of the problem is whether a landscape DSS model that uses pattern and feedbacks to create a probability surface tell us most of what we want to know (or accounts for most of the variance), or if more information is needed. An ecological perspective is that incorporating process in a functional model, as opposed to observations in an empirical model, allows projection to new conditions.

4. CONCLUSION

While the chapters of this volume are aimed at explaining why, when, and where changes in land use and land cover will occur, ecologists have attempted to identify where reserves of habitat should be created and maintained (e.g., Rodrigues et al. 2000; Menon et al. 2001). A goal for future work should be to integrate these two approaches so that maps of habitat suitability and future LCLUC can be merged. While one might wish that the areas most likely to change would not coincide with the areas of highest conservation value, this result is not likely. The degree to which the two are related in space may depend on the spatial heterogeneity of the landscape and the response of households to it.

An ecological perspective may also contribute to basic approaches to the study of LCLUC. A more formal approach to scale is needed in LCLUC research. Casting the problem in terms of building-block processes and constraints is a starting point that can be borrowed from ecology. How this formalism plays out in LCLUC, however, may be different, and the structure should be adapted and changed as needed.

One aspect of a hierarchical approach is to consider emergence in the context of complexity theory (Malanson 1999). Emergence is the notion that the pattern and/or behavior of a whole are greater than the sum of its parts. A basic approach to emergent systems is to look for building blocks, at a lower scale, that combine in many possible ways to create patterns/systems at a higher scale. In LCLUC it may be profitable to consider that the many ways in which land is used and changed can arise from a few common building-block processes and a few common constraints rather than from processes and constraints unique to every place. A new view in geography is that place is emergent rather than overspecified.

Complexity theory is also concerned with feedbacks and nonlinearities. Emergence can be the result of feedbacks, and the idea that the whole is greater than the sum of its parts is fundamentally nonlinear. Feedbacks in LCLUC are clear. LCLUC research needs to be aware that nonlinearities—perhaps just beyond the bounds of current observation—may create surprises. Some effort should be devoted to exploring new parameter space, even if only in abstract simulations.

If one looks for emergence, then agent-based models are likely choices as tools. Much of what we think we know about emergence is based on the complex patterns produced by agent-based models where the behavior/processes of the agents are determined by a few simple rules. The complex patterns develop depending on the interactions among the agents and some constraints. In LCLUC, land users as agents will interact with other land users and be constrained by the landscape and larger-scale economics.

I see one primary challenge to directly addressing scale and emergence in LCLUC research. This is the belief that every place is different in ways that do not admit generalization or simplification. It may well be that the emergence of the complexities of places was so deep in the past that it will not be possible to derive satisfying simplification: complexity theory recognizes historical contingency. Any exploration of the possibilities, however, should not be dismissed based on an appreciation of diversity.

REFERENCES

An L., J. G. Liu, Z. Y. Ouyang, M. Linderman, S. Q. Zhou, and H. M. Zhang. 2001. "Simulating Demographic and Socioeconomic Processes on Household Level and Implications for Giant Panda Habitats." *Ecological Modelling* 140: 31–49.

Bascompte, J., and R. V. Sole. 1996. "Habitat Fragmentation and Extinction Thresholds in Spatially Explicit Models." *Journal of Animal Ecology* 65: 465–473.

Beier, P., and R. F. Noss. 1998. "Do Habitat Corridors Provide Connectivity?" *Conservation Biology* 12: 1,241–1,252.

Bevers, M., and C. H. Flather. 1999. "Numerically Exploring Habitat Fragmentation Effects on Populations Using Cell-Based Coupled Map Lattices." *Theoretical Population Biology* 55: 61–76.

Bierregaard, R. O., T. E. Lovejoy, V. Kapios, A. A. DosSantos, and R. W. Hutchings. 1992. "The Biological Dynamics of Tropical Rain Forest Fragments." *BioScience* 42: 859–866.

Bolger, D. T., T. A. Scott, and J. T. Rotenberry. 2001. "Use of Corridor-Like Landscape Structures by Bird and Small Mammal Species." *Biological Conservation* 102: 213–224.

Bousquet, F., C. Le Page, I. Bakam, and A. Takforyan. 2001. "Multiagent Simulations of Hunting Wild Meat in a Village in Eastern Cameroon." *Ecological Modelling* 138: 331–346.

Brown, D. G., B. C. Pijanowski, and J. D. Duh. 2000. "Modeling the Relationships between Land Use and Land Cover on Private Lands in the Upper Midwest, USA." *Journal of Environmental Management* 59: 247–263.

Chen, J. Q., J. F. Franklin, and T. A. Spies. 1992. "Vegetation Responses to Edge Environments in Old-Growth Douglas-Fir Forests." *Ecological Applications* 2: 387–396.

Dale, V. H., S. M. Pearson, H. L. Offerman, and R. V. O'Neill. 1994a. "Relating Patterns of Land Use Change to Faunal Biodiversity in the Central Amazon." *Conservation Biology* 8: 1,027–1,036.

Dale, V. H., R. V. O'Neill, F. Southworth, and M. Pedlowski. 1994b. "Modeling Effects of Land Management in the Brazilian Amazonian Settlement of Rondonia." *Conservation Biology* 8: 196–206.

de Lima, M. G., and C. Gascon. 1999. "The Conservation Value of Linear Forest Remnants in Central Amazonia." *Biological Conservation* 91: 241–247.

Ellis, J., and K. A. Galvin. 1994. "Climate Patterns and Land-Use Practices in the Dry Zones of Africa." *BioScience* 44: 340–349.

Eva, H., and E. F. Lambin. 2000. "Fires and Land-Cover Change in the Tropics: A Remote Sensing Analysis at the Landscape Scale." *Journal of Biogeography* 27: 765–776.

Fahrig, L. 2001. "How Much Habitat Is Enough?" *Biological Conservation* 100: 65–74.

Fox, J., J. Krummel, S. Yarnasarn, M. Ekasingh, and N. Podger, N. 1995. "Land Use and Landscape Dynamics in Northern Thailand: Assessing Change in Three Upland Watersheds." *Ambio* 24: 328–334.

Fox, J., S. Leisz, Dao M. T., A. T. Rambo, Nghiem P. T., and Le T. C. 2000. "Shifting Cultivation: A New Paradigm for Managing Tropical Forests." *BioScience* 50: 521–528.

Gardner, R. H., B. T. Milne, M. G. Turner, and R. V. O'Neill. 1987. "Neutral Models for the Analysis of Broad-Scale Landscape Pattern." *Landscape Ecology* 1: 19–28.

Gascon, C., and T. E. Lovejoy. 1998. "Ecological Impacts of Forest Fragmentation in Central Amazonia." *Zoology: Analysis of Complex Systems* 101: 273–280.

Geoghegan, J., S. C. Villar, P. Klepeis, P. M. Mendoza, Y. Ogneva-Himmelberger, R. R. Chowdhury, B. L. Turner II, and C. Vance. 2001. "Modeling Tropical Deforestation in the Southern Yucatán Peninsular Region: Comparing Survey and Satellite Data." *Agriculture, Ecosystems and Environment* 85: 25–46.

Gilruth, P. T., S. E. Marsh, and R. Itami. 1995. "A Dynamic Spatial Model of Shifting Cultivation in the Highlands of Guinea, West Africa." *Ecological Modelling* 79: 179–197.

Haddad, N. M., D. K. Rosenberg, and B. R. Noon. 2000. "On Experimentation and the Study of Corridors: Response to Beier and Noss." *Conservation Biology* 14: 1,543–1,545.

Hannah, L., D. Lohse, C. Hutchinson, J. Carr, and A. Lankerani. 1994. "A Preliminary Inventory of Human Disturbance of World Ecosystems." *Ambio* 23: 246–250.

Hanski, I., and M. E. Gilpin, eds. 1997. *Metapopulation Biology*. San Diego: Academic Press.

Hanski, I., A. Moilanen, and M. Gyllenberg. 1996. "Minimum Viable Metapopulation Size." *American Naturalist* 147: 527–541.

Hanson, J. S., G. P. Malanson, and M. P. Armstrong. 1989. "Spatial Constraints on the Response of Vegetation to Climate Change." In G. P. Malanson, ed., *Natural Areas Facing Climate Change* (The Hague: SPB Academic), 1–23.

Harcourt, A. H., S. A. Parks, and R. Woodroffe. 2001. "Human Density as an Influence on Species/Area Relationships: Double Jeopardy for Small African Reserves?" *Biodiversity Conservation* 10: 1,011–1,026.

Hobbs, R. J. 1992. "The Role of Corridors in Conservation: Solution or Bandwagon?" *Trends in Ecology and Evolution* 7: 389–392.

Holyoak, M. 2000. "Habitat Patch Arrangement and Metapopulation Persistence of Predators and Prey." *American Naturalist* 156: 378–389.

Houghton, R. A. 1994. "The Worldwide Extent of Land-Use Change." *BioScience* 44: 305–313.

Huston, M., D. DeAngelis, and W. Post. 1988. "New Computer Models Unify Ecological Theory." *BioScience* 38: 682–691.

Irwin, E. G., and J. Geoghegan. 2001. "Theory, Data, Methods: Developing Spatially Explicit Economic Models of Land Use Change." *Agriculture, Ecosystems & Environment* 85: 7–23.

Jones, D. W., V. H. Dale, J. J. Beauchamp, M. A. Pedlowski, and R. V. O'Neill. 1995. "Farming in Rondonia." *Resource and Energy Economics* 17: 155–188.

Klausmeier, C. A. 1998. "Extinction in Multispecies and Spatially Explicit Models of Habitat Destruction." *American Naturalist* 152: 303–310.

Kok, K., and A. Veldkamp. 2001. "Evaluating Impact of Spatial Scales on Land Use Pattern Analysis in Central America." *Agriculture, Ecosystems & Environment* 85: 205–221.

Kupfer, J. A., G. P. Malanson, and J. R. Runkle. 1997. "Factors Influencing Species Composition in Canopy Gaps: The Importance of Edge Proximity in Hueston Woods, Ohio." *Professional Geographer* 49: 165–178.

Lambin, E. F. 1994. "Modelling Deforestation Processes: A Review." Tropical Ecosystem Observations by Satellites, Series B: Research Report 1. Office for Official Publications of the European Community, Luxembourg.

———. 1999. "Monitoring Forest Degradation in Tropical Regions by Remote Sensing: Some Methodological Issues." *Global Ecology and Biogeography* 8: 191–198.

Lambin, E. F., M. D. A. Rounsevell, and H. J. Geist. 2000. "Are Agricultural Land-Use Models Able to Predict Changes in Land-Use Intensity?" *Agriculture, Ecosystems & Environment* 82: 321–331.

Laurance, W. F., L. V. Ferreira, J. M. Rankin-De Merona, and S. G. Laurance. 1998a. "Rain Forest Fragmentation and the Dynamics of Amazonian Tree Communities." *Ecology* 79: 2,032–2,040.

Laurance, W. F., L. V. Ferreira, J. M. Rankin-De Merona, S. G. Laurance, R. W. Hutchings, and T. E. Lovejoy. 1998b. "Effects of Forest Fragmentation on Recruitment Patterns in Amazonian Tree Communities." *Conservation Biology* 12: 460–464.

Laurance, W. F., H. L. Vasconcelos, and T. E. Lovejoy. 2000. "Forest Loss and Fragmentation in the Amazon: Implications for Wildlife Conservation." *Oryx* 34: 39–45.

Lavorel, S., J. Canadell, S. Rambal, and J. Terradas. 1998. "Mediterranean Terrestrial Ecosystems: Research Priorities on Global Change Effects." *Global Ecology and Biogeography Letters* 7: 157–166.

Lee, J. T., S. J. Woddy, and S. Thompson S. 2001. "Targeting Sites for Conservation: Using a Patch-Based Ranking Scheme to Assess Conservation Potential." *Journal of Environmental Management* 61: 367–380.

Levin, S. A. 1992. "The Problem of Pattern and Scale in Ecology." *Ecology* 73: 1,943–1,967.

MacArthur, R. H. 1972. *Geographical Ecology*. Princeton, NJ: Princeton University Press.

MacArthur, R. H., and E. O. Wilson. 1967. *The Theory of Island Biogeography*. Princeton, NJ: Princeton University Press.

MacKinnon, J., and K. MacKinnon. 1986. *Review of the Protected Areas System in the Afrotropical Realm*. Gland, Switzerland: IUCN.

Malanson, G. P. 1992. "Ecology of Fragmented Natural Landscapes: Disturbance Intensity and Spatial Pattern and Scale." *Ekistics* 356: 280–286.

———. 1993. *Riparian Landscapes*. Cambridge: Cambridge University Press.

———. 1996. "Modelling Forest Response to Climatic Change: Issues of Time and Space." In S. K. Majumdar, E. W. Miller, and F. J. Brenner, eds., *Forests: A Global Perspective* (Easton, PA: Pennsylvania Academy of Sciences), 200–211.

———. 1999. "Considering Complexity." *Annals of the Association of American Geographers* 89: 746–753.

———. n.d.[a] (forthcoming) "Extinction Debt Trajectories and Spatial Pattern of Habitat Destruction." *Annals of the Association of American Geographers*.

———. n.d.[b] (forthcoming) "Effects of Spatial Representation of Habitat in Competition-Colonization Models." *Geographical Analysis*.

Malanson, G. P., and B. E. Cramer. 1999. "Ants in Labyrinths: Lessons for Critical Landscapes." *Professional Geographer* 51: 155–170.

Manson, S. M. 2000. "Agent-Based Dynamic Spatial Simulation of Land-Use/Cover Change in the Yucatán Peninsula, Mexico." Paper presented at the 4th International Conference on Integrating Geographic Information Systems and Environmental Modeling: Problems, Prospects, and Needs for Research. Banff, Alberta, Canada, September 2–8.

McGarigal, K., and B. J. Marks. 1995. *FRAGSTATS: Spatial Pattern Analysis Program for Quantifying Landscape Structure*. Corvallis, OR: Oregon State University, Forest Science Department.

Mech, S. G., and J. G. Hallett. 2001. "Evaluating the Effectiveness of Corridors: A Genetic Approach." *Conservation Biology* 15: 467–474.

Medley, K. E. 1997. "Distribution of the Non-Native Shrub *Lonicera maackii* in Kramer Woods, Ohio." *Physical Geography* 18: 18–36.

Menon, S., R. G. Pontius, J. Rose, M. L. Khan, and K. S. Bawa. 2001. "Identifying Conservation-Priority Areas in the Tropics: A Land-Use Change Modeling Approach." *Conservation Biology* 15: 501–512.

Mertens, B., and E. F. Lambin. 2000. "Land-Cover Change Trajectories in Southern Cameroon." *Annals of the Association of American Geographers* 90: 467–494.

Messina, J. P., and S. J. Walsh. 2001. "2.5D Morphogenesis: Modeling Landuse and Landcover Dynamics in the Ecuadorian Amazon." *Plant Ecology* 156: 75–88.

Meyer, W. B., and B. L. Turner II. 1992. "Human Population Growth and Global Land-Use/Cover Change." *Annual Review of Ecology and Systematics* 23: 39–61.

Moran, E. F., and E. S. Brondizio. 1998. "Land-Use Change after Deforestation in Amazonia." In D. Liverman, E. F. Moran, R. R. Rindfuss, and P. C. Stern, eds., *People and Pixels* (Washington, D.C.: National Academy Press), 94–120.

Murcia, C. 1995. "Edge Effects in Fragmented Forests: Implications for Conservation." *Trends in Ecology and Evolution* 10: 58–62.

NRC. 1999. *Human Dimensions of Global Environmental Change*. National Research Committee, Committee on the Human Dimensions of Global Change. Washington, D.C.: National Academy Press.

NRC. 2000. *Global Change Ecosystems Research*. National Research Committee, Ecosystems Panel. Washington, D.C.: National Academy Press.

Nason, J. D., E. A. Herre, and J. L. Hamrick. 1998. "The Breeding Structure of a Tropical Keystone Plant Resource." *Nature* 391: 685–687.

Neuhauser, C. 1998. "Habitat Destruction and Competitive Coexistence in Spatially Explicit Models with Local Interactions." *Journal of Theoretical Biology* 193: 445–463.

Pope, S. E., L. Fahrig, and N. G. Merriam. 2000. "Landscape Complementation and Metapopulation Effects on Leopard Frog Populations." *Ecology* 81: 2,498–2,508.

Renjifo, L. M. 2001. "Effect of Natural and Anthropogenic Landscape Matrices on the Abundance of Subandean Bird Species." *Ecological Applications* 11: 14–31.

Ricketts, T. H. 2001. "The Matrix Matters: Effective Isolation in Fragmented Landscapes." *American Naturalist* 158: 87–99.

Riebsame, W. E., W. B. Meyer, and B. L. Turner II. 1994. "Modeling Land-Use and Cover as Part of Global Environmental Change." *Climatic Change* 28: 45–64.

Risser, P. G. 1988. "General Concepts for Measuring Cumulative Impacts on Wetland Ecosystems." *Environmental Management* 12: 585–589.

Rodrigues, A. S., J. O. Cerdeira, and K. J. Gaston. 2000. "Flexibility, Efficiency, and Accountability: Adapting Reserve Selection Algorithms to More Complex Conservation Problems." *Ecography* 23: 565–574.

Savage, M., B. Sawhill, and M. Askenazi. 2000. "Community Dynamics: What Happens When We Rerun the Tape?" *Journal of Theoretical Biology* 205: 515–526.

Schlaepfer, M. A., and T. A. Gavin. 2001. "Edge Effects on Lizards and Frogs in Tropical Forest Fragments." *Conservation Biology* 15: 1,079–1,090.

Simberloff, D., and L. G. Abele. 1976. "Island Biogeographic Theory and Conservation Practice." *Science* 191: 285–286.

Simberloff, D., J. A. Farr, J. Cox, and D. W. Mehlman. 1992. "Movement Corridors: Conservation Bargains or Poor Investments?" *Conservation Biology* 6: 493–504.

Skole, D. L., W. H. Chomentowski, W. A. Salas, and A. D. Nobre. 1994. "Physical and Human Dimensions of Deforestation in Amazonia." *BioScience* 44: 314–322.

Skole, D. L., and C. Tucker, C. 1993. "Tropical Deforestation and Habitat Fragmentation in the Amazon: Satellite Data from 1978 to 1988." *Science* 260: 1,905–1,910.

Stauffer, D. 1985. *Introduction to Percolation Theory*. London: Taylor and Francis.

Stephenne, N., and E. F. Lambin. 2001. "A Dynamic Simulation Model of Land-Use Changes in Sudano-Sahelian Countries of Africa (SALU)." *Agriculture, Ecosystems & Environment* 85: 145–161.

Stevens, S. M., and T. P. Husband. 1998. "The Influence of Edge on Small Mammals: Evidence from Brazilian Atlantic Forest Fragments." *Biological Conservation* 85: 1–8.

Theobald, D. M., J. R. Miller, and N. T. Hobbs. 1997. "Estimating the Cumulative Effects of Development on Wildlife Habitat." *Landscape and Urban Planning* 39: 25–36.

Tilman, D., C. L. Lehman, and C. Yin. 1997. "Habitat Destruction, Dispersal, and Deterministic Extinction in Competitive Communities." *American Naturalist* 149: 407–435.

Tilman, D., R. M. May, C. L. Lehman, and M. A. Nowak. 1994. "Habitat Destruction and the Extinction Debt." *Nature* 371: 65–66.

Tischendorf, L., and L. Fahrig. 2000a. "On the Usage and Measurement of Landscape Connectivity." *Oikos* 90: 7–19.

———. 2000b. "How Should We Measure Landscape Connectivity?" *Landscape Ecology* 15: 633–641.

Trzcinski, M. K., L. Fahrig, and G. Merriam. 1999. "Independent Effects of Forest Cover and Fragmentation on the Distribution of Forest Breeding Birds." *Ecological Applications* 9: 586–593.

Turner, I. M. 1996. "Species Loss in Fragments of Tropical Rain Forest: A Review of the Evidence." *Journal of Applied Ecology* 33: 200–209.

Veldkamp, A., and E. F. Lambin. 2001. "Predicting Land-Use Change." *Agriculture, Ecosystems & Environment* 85: 1–6.

Vos, C. C., J. Verboom, P. F. M. Opdam, and C. J. F. Ter Braak. 2001. "Toward Ecologically Scaled Landscape Indices." *American Naturalist* 157: 24–41.

Wiegand, T., K. A. Moloney, and S. J. Milton. 1998. "Population Dynamics, Distribution, and Pattern Evolution: Identifying the Fundamental Scales." *American Naturalist* 152: 321–337.

Wiens, J. A. 1997. "Metapopulation Dynamics and Landscape Ecology." In I. Hanski and M. E. Gilpin, eds., *Metapopulation Biology* (San Diego: Academic Press), 43–62.

Wilcove, D., D. Rothstein, J. Dubow, A. Phillips, and E. Losos. 1998. "Quantifying Threats to Imperiled Species in the United States." *BioScience* 48: 607–615.

Williams-Linera, G., V. Dominguez-Gastelu, and M. E. Garcia-Zurita. 1998. "Microenvironment and Floristics of Different Edges in a Fragmented Tropical Rainforest." *Conservation Biology* 12: 1,091–1,102.

With, K. A. 1994. "Using Fractal Analysis to Assess How Species Perceive Landscape Structure." *Landscape Ecology* 9: 25–36.

With, K. A., and A. W. King. 1999. "Extinction Thresholds in Fractal Landscapes." *Conservation Biology* 13: 314–326.

Woodward, A. A., A. D. Fink, and F. R. Thompson. 2001. "Edge Effects and Ecological Traps: Effects on Shrubland Birds in Missouri." *Journal of Wildlife Management* 65: 668–675.

Chapter 11

LINKING PEOPLE AND LAND USE

A Sociological Perspective

William G. Axinn
Department of Sociology and Institute for Social Research, University of Michigan
baxinn@psc.isr.umich.edu
Jennifer S. Barber
Institute for Social Research, University of Michigan

1. INTRODUCTION

Our main aim is to bring a sociological perspective to our consideration of the issues involved in linking social processes to changes in land cover. We focus on the relationships between population change and changes in land use in particular. Needless to say, the discussions of these topics in previous chapters have provided ample demonstration that these issues are complex. The case studies described in the previous chapters also identify several key accomplishments in the field, linking various forms of data to advance the study of people and land-use connections. Our aim in constructing a sociological perspective on the issue is to advance the conceptual tools, analytic approaches, and measurement strategies for addressing the complexities involved in investigating population and land-use relationships.

In order to accomplish this aim, we construct three key streams of theoretical reasoning regarding population and land-use relationships. First, we construct a simplified model of the human determinants of land-use change over time at the local level. Second, we consider the links between social change and changes in environmental consumption behavior that may impact land use. Third, we examine the role of changes in social institutions in the creation of changes in the links between people and land consumption across time and space. We apply these three streams of reasoning to several of the conceptual, analytic, and measurement problems identified in the previous chapters of this book. We review key strengths and weaknesses of the approaches used in the case studies described in previous chapters. Finally, we suggest new approaches designed to build on the strengths of existing research and address some of the weaknesses of current approaches identified in these case studies.

The reasoning we construct is founded on a micro-level view of the population and land-use relationship. This view focuses our attention on the relationships between land use and social behavior of the population at the level of the local community, the household, and the individual person. The perspective we take is explicitly multilevel, acknowledging the importance of connections across levels of aggregation, but we believe the micro-level starting point is particularly important. This is because policies aimed at preventing the degradation of environmental resources by altering population trends or resource use must affect the behavior of individuals and households, and therefore they must influence environmental behavior at the local level. Indeed, many of the case studies described in previous chapters also take a micro-level approach, and this orientation proves particularly fruitful for the investigation of specific mechanisms linking people to land use. Unfortunately, much of the work on the determinants of environmental quality prior to these case studies addresses these issues at a very high level of aggregation—often at regional or even global levels (Bongaarts 1996; Cohen 1995; Davis and Bernstam 1991; Ehrlich et al. 1993; Heilig 1997; Myers 1990; Rees 1996). Much less theory or empirical evidence exists to guide our understanding of local-level mechanisms relating population change to the natural environment.

2. A MICRO-LEVEL MODEL OF LAND USE

2.1 Reciprocal Relationship between Population Change and Land Use

Much of the demographic research community's work on environmental quality has focused on the impact of population parameters on environment quality, including land use (Blaike and Brookfield 1987; Cohen 1995; Davis and Bernstam 1991; Ehrlich et al. 1993; Heilig 1997). The impact of population growth and population redistribution has probably received the greatest attention. Increasingly, however, social scientists are also sensitive to the idea that environmental quality, including land use, may have an important impact on population parameters. Here the impact of environment on out-migration and mortality has probably received the most attention (Hill 1990; Hamilton et al. 1997; Perz 1997). Similarly, Lambin (this volume) finds that leadership, education, and wealth lead to the choice to use land for mechanized cultivation. At the same time, he also finds that mechanized cultivation is associated with increases in education and increased likelihood of a leadership role in the community. Together, these research projects suggest an important reciprocal relationship between population parameters and environmental quality, where each affects the other. But the precise micro-level nature of this reciprocal relationship has proved difficult to identify.

One reason the reciprocal relationship between population change and environment remains difficult to identify is because both causal effects are probably conditioned by the impact of local contextual characteristics. Certainly recent research on important population parameters is consistent with the conclusion that contextual characteristics shape these parameters, at least in part (Axinn and Yabiku 2001; Entwisle and Mason 1985; Entwisle et al. 1989; Casterline 1985; Massey and Espinosa 1997; Sastry 1996). Research on these topics has also shown that local-level contextual characteristics can be particularly important determinants of population processes (Axinn and Fricke 1996; Entwisle et al. 1989). Other recent research also demonstrates that local community characteristics affect environmental outcomes, including both patterns of consumption behavior and land use (Axinn et al. 2001; Axinn and Ghimire 2002; Shivakoti et al. 1999). Together these streams of evidence are consistent with the conclusion that the population-environment relationship at the local community level is conditioned by community social and economic context.

Many of the studies presented and discussed in previous chapters posit a relationship between people and their environment that is conditional on local context. For example, Lambin (this volume) hypothesizes that changes in the Cameroonian economy condition the link between land use/population and deforestation, mainly related to government input subsidies. In another case study, Lambin also hypothesizes that the different macroeconomic and macropolitical contexts in Kenya and Tanzania led to different relationships between land use and land cover—and also between land use and species decline. Similarly, Walsh et al. (this volume), Turner and Geoghegan (this volume), and Rindfuss et al. (this volume) specify the increasing geographic accessibility of the regions in their study—the Ecuadorian Amazon, Mexico's Yucatán Peninsula, and Nang Rong, Thailand, respectively—as key to understanding the land use and deforestation in those regions. Other studies describe other aspects of the social context that condition and/or explain the relationship between people and their environment—from government protection policies for forests and other lands in China and Kenya (Liu et al. this volume; Lambin this volume) to economic crises and currency devaluations in Thailand (Rindfuss et al. this volume). In fact, many of the authors of the studies presented here cite rapid social change as a key reason that they chose their particular study site and as a key reason that the resulting study is interesting and important.

Thus, understanding the impact of community-level population changes on local land use requires careful consideration of local contextual characteristics. Many of the same contextual characteristics that influence population parameters may also influence land use. If this is so, then precise specification of the reciprocal relations between population and environment at the micro level will require a clear understanding of the impact of contextual characteristics on both population processes and land-use changes. In the following section, we consider the impact of contextual change on changes in land use by focusing on the determinants of changes in land use, recognizing that reciprocal relationships between population change and land use are likely to exist.

2.2 Micro-Level Changes in Land Use

Ever since Thomas Malthus (1798) raised concerns regarding population growth and nature's ability to support the growing number of people on the earth, the relationship between population and land use has experienced several levels of theorization. Malthusian population theory and Ricardian classical economic theory, both formulated around 1800, assume constant and unchanging production methods (Bilsborrow 1987). They argue that the growing population puts heavy pressure on land for food production, and the area under cultivation must be increased. As a result, more land will need to be brought under cultivation. Nevertheless, as the population continues to grow in geometric progression, population growth will eventually outstrip the land's capacity to support the growing population.

In contrast, Boserup (1965, further elaborated in 1981) argues that as arable land becomes scarcer relative to population, it is used more intensively. She suggests that as the per unit land area population density increases and the returns to the land per worked hour begin to fall, the necessity for the land to provide for those additional people increases. As a result, the search for higher productivity leads to intensification of land use. Boserup sees a continuum of land extensification at one end to land intensification on the other end.

In a similar vein, Mortimore (1993), who preferred to use the term *land transformation* instead of land-use change, describes land transformation as a three-step process: (1) the transfer of land from one class of management to another; (2) land investment (in soil and water conservation structure, irrigation and drainage structure, management structure as enclosures, productivity management as fertilization, and tree planting and protection);

and (3) land productivity change (in soil chemical, physical, and biological properties and in farm trees, both physiognomy and floristic composition).

Thus, both Boserup and Mortimore suggest that although availability of natural resources can act as a serious constraint on land-use options available to people, land and other available resources have different production potentials. Therefore, depending on population processes (population structure, rate of population growth, density, and household patterns) and local-level community context (markets, transportation and communication, and other institutional arrangements), people may use land in a way that best suits their interests.

Changes in the human population affect land use through changes in behavior. The proximate determinants of land-use patterns are the behaviors that affect use of the land. These include productive behaviors, recreational behaviors, and consumptive behaviors. For example, as human systems of production change—from hunting and gathering, subsistence agriculture, or industrial production—patterns of consuming land change, and these changes alter the use of the land and the nature of the resulting land cover. From this perspective, patterns of consuming land are a fundamental link between human behavior and land use or land cover. Although changing production, recreation, or consumption behavior may drive land-consumption patterns, it is what people do with the land—the way they consume it—that determines the use of land and therefore land cover.

A model designed to predict the land-use consequences of social, economic, or demographic changes must begin by identifying a specific starting state of land-consumption patterns before moving on to identify the likely consequences of specific changes. Differences in the starting state of land-consumption patterns likely create important differences in the factors determining changes in land use. For example, the transition from clearing forested land on the frontier toward sedentary agriculture may involve different determinants than the transition from using land for subsistence-oriented agriculture toward using land for buildings and infrastructure (Axinn and Ghimire 2002). In a settled area characterized by subsistence agricultural production there is an existing pattern of consuming land, usually preserving a good deal of ecological diversity relative to areas characterized by market-oriented production. This is because subsistence-oriented households usually produce fruits and vegetables in addition to cereal crops, and they also maintain common pasture or forest to provide fodder for animals, whereas market-oriented producers generally specialize in a small number of agricultural products. Such a starting state, with subsistence-oriented agricultural uses of land, may by characterized by very different mechanisms of change than a frontier area being cultivated for the first time.

Our objective in this chapter is *not* to identify a specific model of land-use change at the micro level. Many of the case studies reported in previous chapters do that. Instead, our objective is to formulate a general model of land-use change at the micro level in order to evaluate the strengths and weaknesses of various approaches to linking household data to remotely sensed data for the purpose of understanding the impact of the human population on land use. Based on the reasoning outlined above, this general model identifies land-consumption behaviors as key links between other changes in human behavior and land use. These land-consumption behaviors may be driven by productive, recreational, or consumptive behaviors, but it is the nature of the consumption of land that ultimately shapes land use.[1] We think of these land-consumption behaviors as *environmental consumption.* In the following section we consider the ways that social changes—particularly changes in the local community context—produce changes in environmental consumption.

3. SOCIAL CHANGE AND ENVIRONMENTAL CONSUMPTION

Axinn, Barber, and Biddlecom (2001) propose a relatively simple micro-level model of social change and consumption. First we explain that model. Then we build on that model to propose a new model for the study of human interactions with land use change.

3.1 Conceptualizing Environmental Consumption

In a global sense, all consumption is environmental consumption. Ultimately, the natural environment provides all the raw materials that human beings consume; therefore, *all* consumption affects the natural environment. We differentiate, however, between two types of consumption: *direct* and *indirect* consumption. This contrast highlights the proximity between the person consuming the resource and the natural resource itself. So, by *direct* consumption, we mean humans' use of natural resources—including flora, fauna, water, and soil—with their own hands. Cutting down a tree or branch and burning it for heat is an example. Nearly all of the studies presented here include measures of direct consumption—from cutting down forest cover to establish farms (Moran et al. this volume) to fuelwood collection for cooking and heating (Liu et al. this volume). By *indirect* consumption, we mean the consumption of goods that are separated from the natural resource that originally provided the means to create that good. By this definition, almost all consumption in industrialized countries is indirect because natural resources are processed to create the goods that are ultimately consumed. Yet even in a nonindustrialized setting, indirect consumption may be widespread. Purchasing wood from someone else and then burning the wood for heat is an example.

The environmental impact of direct versus indirect consumption depends on the overall magnitude of consumption, but the distinction in types of consumption may also have important consequences. There is some evidence from around the world indicating that when humans consume natural resources directly, they create more effective management systems for those resources that help preserve environmental quality in the long run (Chambers 1997; Douglass 1992; Ostrom 1990). This may indicate that direct consumption leads to less environmental degradation than indirect consumption. This seems reasonable, because direct contact with the environment may make individuals more aware of the potential for environmental degradation and thus may motivate them to manage environmental resources more carefully. This is an idea with which both classical and contemporary environmental sociology appears to concur (Foster 1999).

It is not necessarily the case, however, that direct consumption leads to a lesser impact on the natural environment. Indirect consumption itself does not preclude effective management systems to preserve long-term environmental quality. It may simply take time and investment to develop those systems. Nevertheless, the transition from direct to indirect consumption of environmental resources is a key change in the way that humans interact with the environment at the micro level.

3.2 Conceptualizing Social Change

Many classical sociological treatments of social change focus on changes in the mode of *production* and the implications of those changes for social life (Durkheim 1984; Marx [1867] 1976; Marx [1863–65] 1981). Durkheim argued that improvements in transportation and communication, the spread of monetization, and increased population density all stimulated the division of labor in society. The changing division of labor altered the mode of production, with widespread implications for social organization and social relationships. Marx, on the other hand, focused on the spread of the capitalist mode of production itself, the implications of this mode of production for the relationship between humans and the fruits of their labor, and the consequences of those changes for a broad array of social relationships. Both of these approaches to social change begin with the idea that when technological and institutional contexts change, they alter the character of individuals' daily lives across many different dimensions, not just production. Our conceptualization of social change builds on this foundation by considering the relationship between macro-level social change and a broad array of micro-level social activities, including production.

Historically, most social activities of daily living were organized within the family (Ogburn and Nimkoff 1976; Thornton and Fricke 1987). This included activities such as consumption, residence, recreation, protection, socialization, procreation, and production. Social changes in the technological and institutional context alter the extent to which these social activities are organized within family and kinship units versus outside of those units (Thornton and Fricke 1987; Thornton and Lin 1994). As new nonfamily organizations and services spread at the macro level, the social activities of daily life are reorganized at the micro level, increasingly taking place outside the family (Coleman 1990). The micro-level consequences of changes in the extent to which social activities are organized within families are both broad and dramatic (Coleman 1990; Durkheim 1984; Marx [1867] 1976; Marx [1863–65] 1981; Thornton and Lin 1994). As we argue below, these include dramatic consequences for the nature of environmental consumption.

Thus, the key element in our conceptualization of social change is the idea that the proliferation of nonfamily organizations and services allows multiple domains of individuals' daily lives to become increasingly organized outside the family. For the purpose of linking land use to changes in the human population, the domain of consumption is particularly important. Nonfamily organizations and services—or what Coleman calls *corporate entities*—provide the means to organize consumption outside the family and thus stimulate widespread change in related social activities (Coleman 1990). One example is a shift from making clothes in the home to purchasing clothes in stores. Another is a shift from cooking in the home to eating in restaurants. There are many others (Coleman 1990; Ogburn and Tibbits 1933). We expect the proliferation of nonfamily organizations and services in communities to alter the social context so that more daily activities, including consumption, become organized away from the home and family.

3.3 Linking Social Change and Environmental Consumption

Social changes that promote the organization of daily social life outside of the family are likely to produce a shift from direct to indirect consumption of environmental resources. This idea builds on Marx's concept of *metabolic rift* (Foster 1999; Marx [1867] 1976). Simply put, *metabolic rift* is the idea that the spread of the capitalist mode of production alters both agricultural production and resource consumption so that humans interact less directly with the natural environment from which they derive their sustenance. The greater the intensity of capitalist production, the less direct most humans' interactions are with their natural environment. Both Marx and Engels predicted that these changes would lead to environmental degradation, including degradation of soil, water, and other natural resources (Foster 1999). Environmental degradation is likely because metabolic rift means that individuals' and families' consumption decisions are based on decreasing knowledge of the environmental consequences of those decisions. According to Marx, metabolic rift produces an unsustainable exploitation of the natural environment that degrades soil, water, and other natural resources (Foster 1999; Marx [1867] 1976; Marx [1863–65] 1981).[2]

Our framework builds on this idea and expands our consideration of social change from a focus on modes of production and consumption toward a broader consideration of changes in multiple dimensions of social organization. As discussed above, macro-level social changes that stimulate the spread of nonfamily organizations may alter not only the mode of production, but they may also alter other dimensions of social life, including consumption, socialization, recreation, protection, procreation, and residence. Social life organized through nonfamily organizations promotes intervening social arrangements between natural resources and the consumers of those resources. Family-organized social life, on the other hand, gives individuals more opportunities for direct contact with the resources they consume.

This is clearest in the case of nonfamily consumption-oriented organizations, such as shops and restaurants. These organizations, generally found in marketplaces, process natural resources for use by consumers. The spread of other nonfamily organizations may have similar consequences. Nonfamily productive organizations, such as wage-labor employers, provide access to money, which also promotes indirect environmental consumption. Nonfamily transportation services increase opportunities to consume natural resources that originated a great distance from the home. Nonfamily financial organizations, such as banks, provide opportunities for indirect consumptive transactions. Nonfamily legal organizations, such as courts and police, provide the legal enforcement of contracts, which may also promote indirect environmental consumption. The list could go on, but our general point is that social changes spreading nonfamily organizations and services promote changes in the pattern of consumption from direct to indirect environmental consumption.

Households with access to nonfamily organizations in or near their communities are, therefore, more likely to consume environmental resources indirectly rather than directly. Households with little or no access to nonfamily organizations are likely to consume environmental resources more directly. In a community with no access to these nonfamily organizations, their spread is likely to change the nature of environmental consumption.

4. PEOPLE, LAND USE, AND CONSUMPTION ACROSS TIME AND SPACE

One of the fundamental issues in linking human behavior to changes in land cover is that the decisions and behaviors of people residing in various places around the world may all impact the patterns of land use in any one location. This may happen because the prices of cassava in Europe affect land-use decisions in rural Thailand (Entwisle 2001), because tourists' desires to see the giant panda impact the panda habitat in rural China (An et al. 2001; Liu et al. 2001), or because North American demand for forest and agricultural products stimulates deforestation in rural Brazil (Moran 2001). A key challenge for research linking human behavior and land use is deriving a systematic approach to the spatial dimension of these connections. However, the great variation in distance and nature of the environmental impact of human consumption decisions greatly complicates the effort to find such a systematic approach.

From a sociological perspective, we advocate use of the conceptual tools described above to begin work on this fundamental issue. In a setting with few or no nonfamily organizations and services, as in most case studies presented in this volume, the link between human behavior and the land is likely to be quite close in physical distance. The social organization of environmental consumption behavior is the key link. With few or no nonfamily organizations and services, family members consume what they themselves produce, as in a subsistence agricultural setting. In this situation, individuals' and families' consumption behavior mainly impacts the physical environment and land cover immediately around them. In such a setting it may be sufficient to study the human population residing directly on or adjacent to the land area of interest.

As new nonfamily services and organizations spread through a previously subsistence-oriented setting, they transform the nature of daily social activities and the social organization of consumption behavior. The direction is uniformly toward a greater impact on local resource consumption and land use of those consumption decisions made at great distance. Proximity to markets allows prices in distant places to influence local land use. The expansion of transportation infrastructure allows movement of forest and agricultural products—and tourists—over great distances. New governance and legal institutions guide consumption behavior in places located far away from the decision makers themselves. All of these social changes promote indirect consumption of environmental resources and reduce the relative share of direct consumption of environmental resources. The impact on land use follows, with greater indirect consumption promoting use of land at greater distances.

We simplify this issue by considering land use of two types: nearby land use and distant land use. Direct consumption affects nearby land use, whereas indirect consumption impacts distant land use. The spread of nonfamily organizations and institutions promotes a shift away from direct consumption toward indirect consumption. The result is a parallel shift away from affecting nearby land use toward affecting distant land use. This shift complicates research efforts to link people to land cover and focuses our attention on the social organizations and institutions that link consumption decisions in one location to land-use decisions in another location.

In our efforts to develop approaches for linking people and their behavior—particularly environmental consumption behavior—to land cover and land use, we need approaches for measuring, analyzing, and understanding the social organizations and institutions that link people in one place to land in other places. From the point of view of studying a parcel of land and its uses, we must consider the social organizations and institutions that link the people near the land to people farther away in terms of both time and space. From the point of view of studying a group of people and their impact on land use, we must consider the social organizations and institutions that link those people to parcels of land both nearby and far away. Without consideration of these social organizational links across time and space, we are not likely to achieve any systematic approach to the study of links between human behavior and land-use changes.

Still, identification of the key importance of these social organizations and institutions should not be confused with a solution to the problem of linking people and land. As demonstrated by each of the projects included in this volume, the problem is quite complex. More careful measurement and understanding of these organizations and institutions can be merely a starting point. This starting point, grounded in a sociological perspective on the links between people and land, points us toward specific research strategies for studying the links between people and changes in land use. It does not explain the links between people and changes in land use.

5. KEY LESSONS LEARNED FROM THE CASE STUDIES

The case studies presented in previous chapters of this volume demonstrate a variety of strategies for linking remotely sensed data and household survey data to advance the study of human impacts on land use and land cover across a range of diverse settings. These diverse research endeavors demonstrate some of the key strengths and weaknesses of existing research designs and measurement methods for studying the links between people and changes in land use. Here we briefly review each of these case studies, identifying some of the key strengths and weaknesses of the approaches they use. We then turn to suggestions for new approaches that build on the successes of existing research and begin to address some of the weaknesses we identify.

Chapter 2, by Turner and Geoghegan, focuses on the southern Yucatán peninsular region, the last tropical forest–agriculture frontier in Mexico. The study links TM imagery and aerial photography (data on deforestation) and decision making and farming information from households. The study used a unique stratified sampling design to ensure that *ejidos* from across the region were represented in the sample and also to ensure variability in both ecological conditions and market/road proximity. Thus, a unique feature of this case study is that they paid attention to the relevant macro-level forces in their study design. This is an important strength.

Chapter 3, by Moran, Siqueira, and Brondizio, provides a description of their efforts to link household survey data to remotely sensed satellite imagery for the purpose of studying the impact of household demographic structure on patterns of land clearing. They focus on deforestation in the Brazilian Amazon Basin, a frontier region experiencing a land-use transition from virgin forest to cultivation for agricultural uses. One of the great strengths of their approach is the use of remote sensing–based maps in household interviews so that household members can themselves identify changes in their plots that appear on remote sensing images. This method directly engages household members in making the link between household-based information and remotely sensed information. The implementation of this method is aided by low levels of land fragmentation and joint ownership of land in the region these investigators studied.

The sample design and implementation of household selection procedures raise several fundamental questions of interest to researchers in this arena. First, it would appear that these investigators used a land parcel grid to select their sample of units to be studied. This differs from many other studies in that it is a land-driven sampling strategy rather than a population-based sampling strategy. As a result, the sample is designed to be representative of the land area being settled, but it is not designed to be representative of the population engaged in the settling of that land. None of our currently available study designs accomplish both representation of the population and representation of the land, and this land-oriented strategy is an important complement to population-oriented strategies. Second, the investigators have interviewed a large number of households, but they do not comment on the nonresponse rate for the survey. Though large sample sizes provide many advantages for studying the impact of variations in household structure on land use, survey nonresponse is known to be a key threat to causal inferences based on survey data. Third, the investigators mention that they had difficulty convincing absentee landowners to participate in the study, but they do not mention the number of such refusals or the percentage of the total sample that they represent. These factors are important because nonresponse in a specific subset of the sample is likely to generate biases in the observations. If absentee landowners engage in different land-use practices, as they well might, higher rates of nonresponse in this group will likely bias the results based on completed interviews.

The land-use data collection for this study features linking of multiple remote images over time to create a longitudinal record of land-use patterns. The household survey data collection is a one-time cross-sectional interview including multiple retrospective histories designed to measure changes over time in both household demographic structure and household patterns of using land. These dynamic elements in both the land-use measures and the household measures are an important strength of this study and a key to the investigators' analyses of changes over time. Nonetheless, the one-time cross-sectional design imposes key limitations that the authors identify as important concerns for future research in this area. One example is measurement of property turnovers over time, which they say occur frequently. If the households cultivating the land change over time and key measures are based on retrospective histories of the households occupying the land at the time of the study, it will be impossible to measure factors contributing to land use in the past when another household occupied the parcel in question. This issue is particularly problematic in areas characterized by high levels of migration. Another example is variation in household members' ability to recall land uses in the past. The investigators identify such variations as a problematic feature of their data collection; in fact, variations in measurement error are a well-known threat to conclusions based on household survey data. Both of these types of problems can be addressed through a longitudinal research design that follows the same households over time, repeating measurements periodically. In the population and land-use arena, such longitudinal designs probably require tracking both household and land parcels over time. Nevertheless, as we have argued elsewhere, there are powerful advantages to be gained from employing a longitudinal design in this area of research.

Finally, this study focuses on links between household demographic change and land-use change. It does not address the impact of community change directly. Several remarks made by the authors, however, suggest that variations in community-level factors—particularly proximity to nonfarm employment opportunities—may have important impacts on the relationship between household demographic structure and land use. For example, the authors point out that many young people leave farming to participate in nonfarm employment opportunities. Such moves alter the household demographic structure and may impact land-use practices. So once again, a more detailed understanding of community-level changes and their relationship to households may advance the investigators' analytic aims in important ways.

In an unusually well written and well organized essay, Chapter 4, by Walsh et al., presents a study of deforestation in the Ecuadorian Amazon. This is the first chapter in the volume to combine *longitudinal* data on households with extremely rich and carefully constructed remotely sensed data sets. The remotely sensed data include SIIM products, used to assist householders in making sketch maps, Landsat satellite data, and GPS coordinates to link parcels both near and far from the households. The chapter also clearly articulates the key question addressed in this volume: At what spatial level do social units (e.g., households or communities) or environmental units (e.g., hill slopes or watersheds) influence the composition and spatial structure of the landscape through LCLUC? The authors conclude that in their study area, a mixed type of linking is best given their budgetary constraints. Unfortunately, household surveys were not conducted with households on *solares* (small subdivided areas of larger farms). Thus there may be an important selectivity in the time 2 (1999) sample.

Chapter 5, by Rindfuss et al., describes their procedures for linking households to land parcels in rural Thailand. Like much of the population in poor regions of Asia, Africa, and Latin America, the population in this region of Thailand has been practicing sedentary agriculture for a long time, organized as relatively small, single-household farms. The work the authors describe addresses many common issues in attempting to link people to land in settled agricultural areas, including the complex relationship between ownership and use, the fact that many households use multiple parcels of land, and the fact that some parcels are used by multiple households. Because these issues are both difficult to resolve in empirical research and common to such a large proportion of the population of the world, this chapter will be of interest to many researchers working in this arena.

The chapter has many strong points. One of its greatest strengths is its explicit identification of the fundamental character of the two essential factors to be measured and linked: land and people. Both have a specific character that renders some measurement strategies appropriate and other measurement strategies inappropriate. This chapter offers the clearest and most explicit discussion of these issues in this volume—something many readers will find quite useful. Moreover, this explicit consideration is used by the authors to motivate a measurement strategy that begins from the household level of scale on the human population side of this relationship and links from households to land. This strategy is likely to become a standard practice in the field for research in similar settings.

Another important strength of this chapter is the detailed description of fieldwork procedures used to accomplish these links. Many discrete steps are involved, each laden with key choices impacting the final product; the detailed description of the choices will be extremely useful to most readers of this volume. Like the Moran et al. (this volume) work in Brazil, ultimately the Rindfuss et al. procedures rely on explicit interactions with household members regarding the parcels of land they cultivate. This common feature underscores the point that even when detailed land-cover data are available from remote sensing sources, links to human actors depend on direct interaction with the people on the ground in the area being studied.

Yet another important strength of this chapter is that it describes a research design that is longitudinal in measurement of both people and land. This relatively unique design provides dynamic information on change over time in both land use/land cover and household characteristics and decisions. Because the fundamental processes linking people to land use are themselves dynamic over time, this fully dynamic measurement strategy is of enormous value to researchers attempting to understand the relationships between people and land use. Unfortunately, the strategy for linking households to land parcels described here is specific to one point in time and does not capture the dynamic link between households and parcels that may be created by property turnovers or shifts in parcel use across households. The immense effort required to make the household-to-parcel links described by the authors may render a fully dynamic set of links unfeasible. Nevertheless, measurement of the dynamic nature of the links between people and parcels—matching the dynamic record of change over time among people and among parcels—will ultimately be desirable in the study of human-land interactions.

The focus of this chapter is the links from households to land parcels, and strong theoretical and empirical justifications are provided in support of this focus. The authors also recognize that other forms of social organization outside the household—including communities, governments, markets, and corporations—all impact land use and relationships between households and land use. In fact, the full array of data they collected for this study includes substantial investment in the measurement of community characteristics at multiple points in time. This community-level measurement provides enormous opportunity to investigate the impact of changes in social organization on the relationship between household characteristics and land-use patterns. As we argue below, such investigation, featuring explicit measurement of social organization at the community level, is particularly likely to yield important insights into the relationships between human behavior and land-cover/land-use dynamics.

Chapter 6, by BurnSilver et al., discusses the influence of macroscale socioeconomic and political factors—land adjudication, subdivision of communal rangelands, and economic sedentarization of pastoral households—on the ability of pastoral households to maintain extensive herding patterns among the Maasai in Kajiado and Narok, Kenya. One clear strength of the study is its in-depth measurement of the movement patterns of specific pastoral households over specific grazing areas. These data were collected by following the herds themselves and taking Global Positioning System (GPS) measures with a handheld device at regular intervals. Of course, the challenge, as the authors admit, will be to effectively use such detailed, fine-scaled data.

Another strength of the study is that it considers outside influences on the study area, including government policies (conservation, land tenure, settlement), climate, and environmental characteristics. Thus, the study links the local and the global and is able to incorporate some of the key strengths of both local and global studies. In addition, this study focuses a great deal of attention on the difficulties of linking people to their land when the land is communal, as in agropastoral and pastoral systems.

One weakness of the study is related to study design. This is understandable because, as the authors state, the data were not collected expressly for the purpose of any of these particular case studies. As the authors admit, measures of multiple components of the environment are treated equally in their heterogeneity indices, although some components may be more influential on pastoral decisions. With only six study areas that vary on more than six dimensions, it will be nearly impossible to sort out which components of the environment are more important than others.

Chapter 7, by Fox et al., focuses on forest fragmentation in Vietnam. The case study combines three types of data: spatial, socioeconomic, and hydrological. A key strength of the study is the explicit description of the combination of "qualitative" and "quantitative" methods used. The project took an expressly combinatory approach, which has resulted in a richer data set for analyses. Another aspect of the paper that readers will find useful is the up-front discussion of the difficulty in creating an international, multidisciplinary research team. This project attempted to combine such dramatically different fields as hydrology, ecology, economics, forestry, and survey research and mapping teams. Further, they articulately describe some of the same issues discussed by Lambin in Chapter 8 regarding decisions about whether to link land to people at the household or village level. Although Fox et al. describe this problem in a less general way than Lambin, their discussion of the issues specific to their field site—Vietnam—is likely to provide insight to other researchers facing a similar situation.

An important weakness that Fox et al. explicitly acknowledge in their essay is the timing of fieldwork relative to the available data from satellite imagery; the fieldwork was conducted approximately two to three years after the satellite data were collected. Unfortunately, this means that some conclusions they draw will be limited to exploring the impact of land use on household decision making (at the village level), rather than on how household decisions affect land use.

Chapter 8 is by Lambin. Again, the essay presents relatively unrelated case studies from two study sites in Africa. The first case study focuses on tropical deforestation in Cameroon, seeking to better understand the causes and processes of land-cover/land-use changes since the 1970s and finding that the annual rate of deforestation was greater during a period following economic crisis compared to the period before the economic crisis. The author concludes that the economic crisis contributed to this deforestation. The second case study analyzes land-cover changes and wildlife decline in the Serengeti-Mara ecosystem in Kenya and Tanzania, asserting that a 59 percent decline in wildlife species was due to markets and national land-tenure policies. The vast amount of data collected is a great strength of both case studies. In both, remote sensing data are used to construct a retrospective history of land cover. Another strength is the attention to macro-level forces as potential determinants of land cover.

The introduction to the two case studies focuses on the advantages and disadvantages of using different levels of analysis: household versus village. The discussion mainly focuses on the costs of data collection at the different levels, rather than which analytic unit can best answer the research question. In addition, the case studies suffer from problems commonly associated with the before/after study design. For example, although the authors find that rates of deforestation are higher *after* the economic crisis compared to *before* the economic crisis, it is impossible to know whether the rates of deforestation *would have* increased if there were no economic crisis. In other words, as with any nonexperimental study design, we have no information on the counterfactual situation. In this case study there are more differences between the temporal data points than just the difference on which the authors focus (the economic crisis).

Finally, Chapter 9, by Liu et al., focuses on behavioral influences on the panda habitat in the Wolong Nature Reserve in China. This study highlights a particularly timely and important problem: the destruction of the habitat of the panda, an endangered species. In addition, there are twelve other animal species and forty-seven plant species on China's national protection list. Thus, the study's goals are important.

The study design relies on the same before/after study design found in Chapter 8 and suffers from many of the same limitations. There are, however, special difficulties in studying such a rare animal population (pandas) and in studying such a unique area (a nature preserve for those pandas). Thus the investigators were constrained in their ability to choose a study design. The study could not be conducted experimentally, and the choice of sites could not be randomized. In addition, it would be difficult to find two similar sites that differ only in terms of the key independent variable of interest. As a result, the investigators are forced to rely on a before/after design. Still, a major strength of the project is its attempt to rigorously implement that design, to test its weaknesses, and to account for those weaknesses. For example, simulation models are used to project population size and resulting panda habitats under different policy scenarios. Although the specific assumptions and justifications for the projections are not detailed in this chapter, they are available in other work published by the research team (e.g., Liu et al. 2001).

Overall, the studies contained in this volume are characterized both by important strengths and by important weaknesses. The most common weaknesses among these case studies come from limitations of study design. For example: (1) The number of differences between sites is greater than the number of sites; (2) there is a lack of variance in key independent variables within the study site(s); and (3) there is a lack of information about what *would have happened* in the study site (i.e., the counterfactual) if the key independent variable had not changed. Of course, constraints of budget, time, contacts, and/or research questions limit the study designs employed in many of these case studies. Nonetheless, given these constraints, the case studies in this volume succeed in moving forward the science of linking information on households to information on land use and land cover.

Many of these studies dramatically advance techniques for linking measurements of households and human behavior to measurements of land use and land cover. They use various forms of remote sensing to assemble measures of land use and land cover. Many of the case studies rely on remote sensing to assemble longitudinal or "time-slice" data sets that document changes over time in land-use and land-cover patterns. Others use remote sensing to collect data on a broader geographical area than would be possible using field methods. For example, Liu et al. (this volume) state that "Remote sensing is particularly valuable for large-scale research, especially in topographically complex regions such as Wolong, where assessment of land cover is both urgent and difficult if carried out by field studies alone." They employ a variety of approaches for collecting household data on human behavior. Some of the case studies collect longitudinal data on households; others collect one-time cross-sectional data on households. They then address various techniques for linking the two. Most require some participation of the people being studied in identification of appropriate household-land parcel links. Moran et al. (this volume) and Rindfuss et al. (this volume) provide good examples of techniques for approaching this issue.

A special strength of many of the case studies presented in this volume is the theoretical attention to macro-level influences on the study area coming from outside the study area. Although many of these are described by way of background information only, they are usually described as related to land use. Even the studies that do somehow account for the macro-level characteristics of their study area have not directly measured many of those characteristics. For example, Walsh et al. (this volume) discuss roads, employment opportunities, local marketplace characteristics, banks, and agricultural extension centers as important macro-level characteristics that affect land use in their study area—but they mainly use distance to the nearest road as a proxy for these features. Turner and Geoghegan (this volume) list market conditions, resource institutions, off-farm employment opportunities, attention from local NGOs, and price subsidies as important macro-level characteristics that may influence the relationship between household behaviors and deforestation. Although variance in some of these characteristics is minimized by their study design (i.e., all households experience similar price subsidy opportunities), the study design offers a rich opportunity to *measure* these characteristics and include differences across households as an important component of their study. An important weakness in many of the studies in this volume is the lack of measurement of these macro-level phenomena and the resulting loss of opportunity to include these multiple levels of influence in analyses.

We wish to emphasize the many strong points identified among the case studies reviewed above in approaches for linking household data on people to remotely sensed data on land use. As demonstrated by a number of these case studies, this field has been revolutionized by the application of geographic information analysis (GIA) tools for linking specific households and the data collected on those households to specific parcels of land and to remotely sensed characteristics of that land. These GIA tools allow for an explicit spatial linking of measures of land use and measures of household behaviors. As discussed in the theoretical reasoning presented above, when the social organization of the local context is such that the consumption behavior of local residents primarily affects nearby land use—and the use of land is primarily determined by the population living near that land—this strategy of linking by spatial location works very well. Many of the settings described in these case studies are characterized by such systems of social organization, and the GIA tools for linking household survey measures and remote sensing land-use measures have produced dramatic advances in our understanding of the mechanism through which people alter land use.

Our review of these case studies also identifies some weaknesses in the measurement of institutional and organizational factors linking people to land use across time and space. As the local community context changes so that new forms of social organization allow local residents to impact land use far away and distant residents to impact land use nearby, existing methods for linking people and land use via geographic location are likely to become less satisfying. This is because the link between people and land by location will not capture the full array of people affecting a particular parcel of land or the full range of land parcels affected by a particular group of people. Additional information about organizational structures will be needed in order to form a comprehensive model of human impacts on land. In the following section, we propose some possible approaches for adding measures of change in organizational structures to studies of the links between people and land.

6. STRATEGIES FOR STUDYING SOCIAL CHANGE IN ANALYSES OF LAND USE

Remote sensing offers many important advantages for the study of land-cover change over time. These advantages have been described in the previous chapters. They include, of course, the ability to derive consistent measures across both large land areas and small segments of land and the ability to track changes over both short and long intervals of time. As some of the previous chapters have discussed, remote sensing has important limitations for studying land use, distinct from the advantages for studying land cover. Some of these include the inability to measure multiple levels of land use, particularly when uses close to the ground are blocked by a canopy above, or the inability to measure human governance of the land, including ownership, land-tenure systems, and institutional management.

Remote sensing is particularly unlikely to offer advances in the measurement of social organizations and institutions that shape the nature of daily social activities and the social organization of consumption behavior. Our unprecedented ability to view the physical attributes of land cover from above does not provide any immediate tools for direct measurement of changes over time in the character of organizations and institutions that guide daily social life to affect consumption behavior. Other tools will be needed.

Household surveys and surveys of individuals also offer many advantages for studying human interaction with land. Again, some of these advantages have been highlighted in presentations of ongoing research projects in previous chapters. They include the ability to measure the experiences and activities of individual people and groups, such as households. They also include the ability to measure plans, expectations, and preferences of individual people and collections of people (Fox et al. this volume). Both of these dimensions are extremely valuable for understanding how people use land, how they plan to use land in the future, and factors that may promote changes in land use over time. Some of the previous presentations have also identified important weaknesses of household surveys to study connections between people and land use, including the inability to measure the systems of social organization that link local actors and distant actors, the constantly changing nature of a household (i.e., members joining and leaving the household), and the lack of a scientifically agreed upon definition of a household.

Household and individual surveys may offer some advances in the measurement of social organizations and institutions that shape the nature of daily social activities and the social organization of consumption behavior. In fact, one of the most important contributions the social sciences may be able to make to the interdisciplinary study of links between humans and land cover is the development of new and improved tools for survey measurement. This is because subfields of the social sciences such as survey methodology and cognitive psychology are producing a virtually constant stream of improvements in the tools of survey research. These tools may be particularly helpful in the documentation of changes over time in the daily social activities of individuals and groups of individuals. But household and individual survey methods do not provide any immediate tools for direct measurement of changes over time in the character of institutions, organizations, or services that guide daily social life to affect consumption behavior. Again, other tools will be needed.

Of course, theoretical and empirical attention to the institutions and organizations that link people to the environment is not new in the social sciences, particularly in the arena of studies of common property resources, such as water and forests. In fact, there is a highly developed literature in the area of common property resources that gives particularly careful attention to the structure and dynamics of governance organizations and institutions (Ostrom 1990, 1998; Ostrom et al. 1993; Ostrom and Gardner 1993; Ostrom et al. 1994; Ostrom et al. 1999). Many of the empirical studies in this literature focus on the management of community forest resources and community water resources and are therefore directly relevant to the study of connections between people and land use. Based on the sociological perspective developed above, however, we argue that theoretical and empirical attention to a broad range of institutions and organizations, in addition to governance organizations, will be needed to construct a comprehensive understanding of the links between people and land use. Our empirical focus and measurement strategies will need to move beyond governance to include a broad array of organizations and institutions, including institutions of consumption such as markets; institutions of production such as employers; institutions of socialization such as schools; institutions of recreation and leisure such as movie theaters; institutions of protection such as health care services; and institutions of procreation such as families.

Tools used in the study of common property resource governance and management offer one potential starting point. Researchers studying the governance and management of community irrigation resources, for example, have developed a set of methods sometimes referred to as Rapid Rural Appraisal (Chambers 1988; Chambers and Carruthers 1986). These methods involve a combination of structure similar to survey research and open-ended questioning and observation similar to ethnographic research. They place the researcher directly into the field site to assess local governance institutions, interview local people, and observe local common property resources. The methods are also designed to be quick and provide rapid feedback to program administrators and policy makers. This quickness undoubtedly comes at the expense of rigorous measurement, so that ultimately this approach may not be desirable for long-term scientific inquiry. Likewise, the methods are not designed to capture the full array of characteristics of a broad range of social organization and institutions that may be needed to study the factors linking people to land use. Nonetheless, the multimethod, direct measurement approach is a useful starting point. Many of the studies presented in this volume use a multimethod measurement approach quite successfully, employing combinations of ethnographic, household and individual survey, and remote sensing data.

Consideration of the key dimensions of measurement of changes in social organizations and institutions immediately draws attention to the *dynamics of change over time*. Any examination of the impact of social organizations and institutions on the relationship between people and land use must focus on changes over time in the local social context, as well as changes over time in the social organizations and institutions that make up the context. It is the *change* in this social context that we expect will link people to *changes* in land use. A focus on change over time immediately draws our attention to measurement of the timing and sequencing of changes in the character of organizations and institutions. Methods for measuring these changes over time are a high priority.

Recent advances in the use of calendar-based methods for measuring the timing and sequencing of changes in neighborhoods, local organizations, and institutions hold great promise for this purpose. The Neighborhood History Calendar method is a set of techniques explicitly designed for the measurement of changes over time in the local social context and the character of local social organizations and institutions (Axinn et al. 1997). The method uses an integrated combination of structured survey techniques, unstructured long interviews and observational techniques, and archival techniques to measure the timing and sequencing of changes in the local social context, including changes in the existence and character of local organizations such schools, health care facilities, banks, development programs, and agricultural cooperatives. The attention to timing and sequencing imbedded in this method has many advantages, but both the set of organizations to be measured and the content of the measurement itself must be designed specifically for the study of links between people and land use—something that is not true of existing applications of this method.

Consideration of the key dimensions of changes in social organizations and institutions that drive the selection of the set of organizations to be studied immediately draws attention to the *existence and proliferation of nonfamily organizations*. The sociological perspective we develop above focuses on changes over time in the organization of social activities outside of the home and away from the family through nonfamily organizations and services. This perspective has the advantage of treating many different institutions, organizations, and services with the same analytic status as government institutions. Changes in government structures through the creation and modification of government organizations undoubtedly impact daily social life with dramatic implications for the use of common property resources and land use (Ostrom 1990, 1998). But other nonfamily organizations and institutions, including markets, employers, schools, recreational facilities, health services, and creditors, also have great potential to reshape the nature of daily social life in ways that impact the relationship between people and land use. Comprehensive measurement of the local contextual factors linking people to land will, therefore, be strengthened by an examination of the full array of nonfamily organizations and services that may play a role in linking people and land.

Finally, consideration of what to measure when studying the social organizations and institutions that link people to land use focuses our attention on the *content of organizations and their relationship to environmental consumption*. The sociological perspective we develop here places environmental consumption behaviors as a key proximate determinant of changes in land use. Changes over time in the content of nonfamily social organizations and institutions that alter daily social life in ways that impact consumption are the most likely to affect land use.

Concern for impact on consumption behavior motivates measurement strategies that attend to not only the timing and sequencing of the creation of new nonfamily organizations and institutions but to the timing and sequencing of changes in the content of those organizations and institutions that are related to consumption behavior. So, as nonfamily organizations evolve over time, the specific character of these organizations is likely to change and those changes may alter the relationship between a specific organization and the consumption behavior of those affected by that organization. Thus, documentation of changes over time in the consumption-related character of these organizations will be fundamental to studies of the impact of such organizations on the relationship between people and land use.

Of course, even with new tools for measuring the social context and social organization linking people to land, issues of linking these contextual measures to people and land will still remain. The same GIA tools that have proved so successful for linking household data to land-use data in many of the studies described here may also be useful for linking new types of data on organizations and institutions to both households and land. New community measurement designed to understand which parcels of land people affect will need to be linked to households. New community measurement designed to understand which people affect specific parcels of land will need to be linked to land. Thus the same GIA tools described in many of the chapters of this book, which are so useful for linking people and land when people affect land near their households, may also be useful in linking measures of social context to either people or land when people affect land farther away through various forms of social organization. In fact, there are already examples of success using GIA tools to link social context to human behavior in order to incorporate contextual measures into explanations of behavior (Entwisle et al. 1996). As new tools for measuring the social context of human and land-cover interaction become available, it is quite likely these same GIA techniques will be a fundamental tool for linking together these various forms of measurement.

7. CONCLUSION

The methods for integration of data from remote sensing techniques for the measurement of land cover and household survey techniques for the measurement of the human population presented in this volume have revolutionized the study of human impacts on land use and land cover. In particular, the use of geographic information analysis (GIA) techniques has allowed researchers to link information on people to land use near those people. As demonstrated in several of the case studies presented in previous chapters, this approach can yield dramatic advances in our understanding of the mechanisms linking the behavior of people to the use of land. This is especially true when the local community context limits the impact of the local population to land use nearby.

As local community contexts change so that local people can affect land use far away and people far away can affect local land use, the simple geographic link between data on individuals and households and data on land use will become less likely to yield satisfying answers to questions regarding the mechanisms linking changes among people to changes in land use and land cover. We argue that direct measurement of the social organizations and institutions that link people in one location to environmental consumption and land use in another location will be needed to understand the links between people and land.

These arguments are based on a sociological perspective that begins by examining the determinants of land use at the local level. This examination reveals the fundamental importance of environmental consumption behavior as a proximate determinant of land use, linking other social, economic, and demographic changes to land use. Next we propose a simple model of changes over time in consumption patterns at the local level. This model emphasizes the crucial role of social change in promoting changes in the social organization of consumption patterns, even as economic and demographic factors determine the total volume of consumption. Finally, we argue that the same changes in social organization that promote a shift from direct environmental consumption to indirect environmental consumption also promote a shift from the use of land nearby to the use of distant land. This shift in physical proximity of land use has dramatic implications for the links between people and their consumption of land-based resources over time and space. Growing physical distance between people and the land-cover patterns they affect also undermines our efforts to link people to land use and land cover using a simple combination of remote sensing measurement and household survey measurement linked through GIA tools.

Instead, we argue that new methods for measuring changes over time in the existence and consumption-related content of local nonfamily social organizations and institutions need to be developed and integrated into a comprehensive strategy for studying the links between people and land cover. We advocate multimethod approaches for measuring these organizations and institutions. Such approaches will combine the high structure of survey methods and the flexibility of long interviews and direct observation to document both the existence and character of nonfamily organizations and institutions. Such approaches will also build on calendar-based methods for the measurement of the timing and sequencing of changes over time in the character of nonfamily organizations and services. Although existing methods provide useful starting points, methods that focus specifically on the consumption-related characteristics of nonfamily organizations and institutions are not currently available. Thus, from a sociological perspective, a key issue on the research agenda for studying the links between people and land cover is the development of new methods for measuring the social organizations that link people to distant land uses—and integration of that measurement into existing frameworks for linking household surveys and remote sensing of land cover.

NOTES

[1] We present the simplest model here as a starting point. We encourage construction of more complex models, building on this simple model. Our model is intended to emphasize the idea that more distant causes may work through proximate determinants, such as land consumption patterns, and may affect land use differently, depending on their effects on more proximate determinants.

[2] Note that Marx was neither the only scholar to espouse these ideas nor the first. As much as a century earlier, for example, Adam Smith proposed similar ideas (Smith 1776)

ACKNOWLEDGEMENTS

The research reported here was supported by a generous grant from the National Institute of Child Health and Human Development (NICHD grant # R01-HD33551). We wish to thank the staff of the Population and Ecology Research Laboratory for their many contributions to the ideas discussed here. We also wish to thank Dirgha Ghimire, Ann Biddlecom, Lisa Pearce, and Scott Yabiku for their contributions to the ideas discussed here and the research on which those ideas are based.

REFERENCES

An, L., J. Liu, Z. Ouyang, M. Linderman, S. Zhou, and H. Zhang. 2001. "Simulating Demographic and Socioeconomic Processes on Household Level and Implications for Giant Panda Habitats." *Ecological Modelling* 140: 31–49.

Axinn, W. G., J. S. Barber, and A. E. Biddlecom. 2001. "Social Change, Household Size, and Environmental Consumption." Institute for Social Research Working Paper. University of Michigan.

Axinn, W. G., J. S. Barber, and D. J. Ghimire. 1997. "The Neighborhood History Calendar: A Data Collection Method Designed for Dynamic Multilevel Modeling." *Sociological Methodology* 27: 355–392.

Axinn, W. G., and T. E. Fricke. 1996. "Community Context, Women's Natal Kin Ties, and Demand for Children: Macro-Micro Linkages in Social Demography." *Rural Sociology* 61: 249–271.

Axinn, W. G., and D. J. Ghimire. 2002. "Population and Environment: The Impact of Fertility on Land Use in an Agricultural Society." Institute for Social Research Working Paper. University of Michigan.

Axinn, W. G., and S. T. Yabiku. 2001. "Social Change, the Social Organization of Families, and Fertility Limitation." *American Journal of Sociology*. 106(5): 1,219–1,261.

Bilsborrow, R. 1987. "Population Pressure and Agricultural Development in Developing Countries: A Conceptual Framework and Recent Evidence." *World Development* 15(2): 183–203.

Blaike, P., and H. Brookfield, eds. 1987. *Land Degradation and Society*. New York: Routledge Kegan & Paul.

Bongaarts, J. 1996. "Population Pressure and the Food Supply System in the Developing World." *Population and Development Review* 22(3): 483–503.

Boserup, E. 1965. *The Conditions of Agricultural Growth: The Economics of Agrarian Change under Population Pressure*. Chicago: Aldine Press.

———. 1981. *Population and Technological Change: A Study of Long-Term Trends*. Chicago: University of Chicago Press.

Casterline, J. B. 1985. *The Collection and Analysis of Community Data*. Voorburg, Netherlands: International Statistical Institute.

Chambers, R. 1988. *Managing Canal Irrigation: Practical Analysis from South Asia*. New York: Cambridge University Press.

———. 1997. *Whose Reality Counts? Putting the First Last*. London: Intermediate Technology.

Chambers, R., and I. D. Carruthers. 1986. "Rapid Appraisal to Improve Canal Irrigation Performance: Experience and Options." Research Paper No. 3. Digana, Sri Lanka: International Irrigation Management Institute.

Cohen, J. 1995. *How Many People Can the Earth Support?* New York: Norton.

Coleman, J. S. 1990. *Foundations of Social Theory*. Cambridge: Harvard University Press.

Davis, K., and M. Bernstam, eds. 1991. *Resources, Environment, and Population*. Oxford: Oxford University Press.

Douglass, M. 1992. "The Political Economy of Urban Poverty and Environmental Management in Asia: Access, Empowerment and Community Based Alternatives." *Environment and Urbanization* 4(2): 9–32.

Durkheim, E. 1984. *The Division of Labor in Society*. New York: Free Press.

Ehrlich, P., A. Ehrlich, and G. Daily. 1993. "Food Security, Population, and Environment." *Population and Development Review* 19(1) :1–32.

Entwisle, B. 2001. "Population and Land Use in Nang Rong, Thailand." Paper presented at the Population Association of America Annual Meetings, Washington, D.C. March 29–31.

Entwisle, B., J. B. Casterline, and H. Sayed. 1989. "Villages as Contexts for Contraceptive Behavior in Rural Egypt." *American Sociological Review* 54: 1,019-1,034.

Entwisle, B., and W. Mason. 1985. "Multilevel Effects of Socioeconomic Development and Family Planning Program on Children Ever Born." *American Journal of Sociology* 91: 616–649.

Entwisle, B., R. R. Rindfuss, D. K. Guilkey, A. Chamratrithirong, S. R. Curran, and Y. Sawangdee. 1996. "Community and Contraceptive Choice in Rural Thailand: A Case Study of Nang Rong." *Demography* 33(1): 1–11.

Foster, J. B. 1999. "Marx's Theory of Metabolic Rift: Classical Foundations for Environmental Sociology." *American Journal of Sociology* 105: 366–405.

Hamilton, L., C. Seyfrit, and C. Bellinger. 1997. "Environment and Sex Ratios among Alaskan Natives: An Historical Perspective." *Population and Environment* 18(3): 283–299.

Heilig, G. K. 1997. "Anthropogenic Factors in Land-Use Change in China." *Population and Development Review* 23(1): 139–168.

Hill, A. 1990. "Demographic Responses to Food Shortages in the Sahel." In G. McNicoll and M. Cain, eds., *Rural Development and Population: Institutions and Policy* (New York: Oxford University Press), 168–192.

Liu, J., M. Linderman, Z. Ouyang, and L. An. 2001. "The Panda's Habitat at Wolong Nature Reserve: Response." *Science* 293: 603–604.

Liu, J., M. Linderman, Z. Ouyang, L. An, J. Yang, and H. Zhang. 2001. "Ecological Degradation in Protected Areas: The Case of Wolong Nature Reserve for Giant Pandas." *Science* 292: 98–101.

Malthus, T. [1798] 1966. *First Essay on population, 1798*. New York: St. Martin's Press.

Marx, K. [1867] 1976. *Capital: A Critique of Political Economy*. Vol. 1. New York: Vintage.

———. [1863–65] 1981. *Capital: A Critique of Political Economy*. Vol. 3. New York: Vintage.

Massey, D. S., and K. E. Espinosa. 1997. "What's Driving Mexico–U.S. Migration? A Theoretical, Empirical, and Policy Analysis." *American Journal of Sociology* 102: 939–999.

Moran, E. F. 2001. "The Development Cycle of Domestic Groups and Deforestation in the Amazon." Paper presented at the Population Association of America Annual Meetings, Washington, D.C. March 29–31.

Mortimore, M. 1993. "Northern Nigeria: Land Transformation under Agricultural Intensification." In C. L. Jolly and B. B. Torry, eds., *Population and Land Use in Developing Countries* (Washington, D.C.: National Academy Press), 42–69.

Myers, N. 1990. "The World's Forests and Human Populations: The Environmental Interconnections." *Population and Development Review* 16 (Supplement: "Resources, Environment, and Population: Present Knowledge, Future Options"): 237–251.

Ogburn, W. F., and M. F. Nimkoff. 1976. *Technology and the Changing Family*. Westport, CT: Greenwood Press.

Ogburn, W. F., and C. Tibbitts. 1933. "The Family and Its Function." In E. A. Ross, *The Principles of Sociology* (New York: Henry Holt), 421–432.

Ostrom, E. 1990. *Governing the Commons*. Cambridge: Cambridge University Press.

Ostrom, E. 1998. "A Behavioral Approach to the Rational Choice Theory of Collective Action." Presidential Address, American Political Science Association 1997. *American Political Science Review* 92(1): 1–22.

Ostrom, E., J. Burger, C. B. Field, R. B. Norgaard, and D. Policansky. 1999. "Revisiting the Commons: Local Lessons, Global Challenges." *Science* 284: 278–282.

Ostrom, E., and R. Gardner. 1993. "Coping with Asymmetries in the Commons: Self-Governing Irrigation Systems Can Work." *Journal of Economic Perspectives* 7(4): 93–112.

Ostrom, E., R. Gardner, and J. Walker. 1994. *Rules, Games, and Common-Pool Resources*. Ann Arbor: University of Michigan Press.

Ostrom, E., L. Schroeder, and S. Wynne. 1993. *Institutional Incentives and Sustainable Development*. San Francisco: Westview Press.

Perz, S. 1997. "The Environment as a Determinant of Child Mortality among Migrants in Frontier Areas of Pará and Rondonia, Brazil, 1980." *Population and Environment* 18(3): 301–324.

Rees, W. 1996. "Revisiting Carrying Capacity: Area-Based Indicators of Sustainability." *Population and Environment* 17(3): 195–215.

Sastry, N. 1996. "Community Characteristics, Individual and Household Attributes, and Child Survival in Brazil." *Demography* 33: 211–229.

Shivakoti, G. P., W. G. Axinn, P. Bhandari, and N. B. Chhetri. 1999. "The Impact of Community Context on Land Use in an Agricultural Society." *Population and Environment* 20: 191–213.

Smith, A. 1776. *An Inquiry into the Nature and Causes of the Wealth of Nations*. London: Strahan and Cadell.

Thornton, A., and T. E. Fricke. 1987. "Social Change and the Family: Comparative Perspectives from the West, China, and South Asia." *Sociological Forum* 2: 746–779.

Thornton, A., and H.-S. Lin. 1994. *Social Change and the Family in Taiwan*. Chicago: University of Chicago Press.

INDEX

E

F

G

H

I

K

L

Q

R

S

T

U

V

W